RESEARCH IN HUMAN CAPITAL AND DEVELOPMENT

Volume 3 • 1983

HEALTH AND DEVELOPMENT

Editorial correspondence pertaining to articles to be published should be sent to:

>   Professor Ismail Sirageldin, Series Editor
>   Department of Population Dynamics
>   The Johns Hopkins University
>   School of Hygiene and Public Health
>   615 N. Wolfe Street
>   Baltimore, Maryland 21205

All other correspondence should be sent to:

>   JAI Press Inc.
>   36 Sherwood Place
>   P.O. Box 1678
>   Greenwich, Connecticut 06830

RESEARCH IN HUMAN CAPITAL AND DEVELOPMENT

A Research Annual

HEALTH AND DEVELOPMENT

Editors: DAVID SALKEVER
ISMAIL SIRAGELDIN
The Johns Hopkins University

ALAN SORKIN
University of Maryland

VOLUME 3 • 1983

Greenwich, Connecticut   London, England

Copyright © 1983 by JAI PRESS INC.
36 Sherwood Place
Greenwich, Connecticut 06830

JAI PRESS INC.
3 Henrietta Street
London WC2E 8LU
England

All rights reserved. No part of this publication may be reproduced, stored on a retrieval system, or transmitted in any form or by any means, electronic, photocopying, filming, recording or otherwise without prior permission in writing from the publisher.

ISBN: 0–89232–166–0

Manufactured in the United States of America

CONTENTS

LIST OF CONTRIBUTORS   vii

INTRODUCTION
*David Salkever, Ismail Sirageldin and
Alan Sorkin*   ix

PART I. CONCEPTS AND MEASURES

A CONCEPTUAL MODEL OF HEALTH
Hector Correa   3

CONCEPTUAL FRAMEWORK FOR THE PLANNING
OF MEDICINE IN DEVELOPING COUNTRIES
Peter Newman   29

HEALTH DEVELOPMENT: A DISCUSSION
OF SOME ISSUES
Oscar Gish   57

PART II. HEALTH IN HUMAN CAPITAL FORMATION

ADOLESCENT HEALTH, FAMILY BACKGROUND,
AND PREVENTIVE MEDICAL CARE
Linda N. Edwards and Michael Grossman   77

AN ECONOMIC ANALYSIS OF THE DIET, GROWTH
AND HEALTH OF YOUNG CHILDREN IN THE
UNITED STATES
Dov Chernichovsky and Douglas Coate   111

THE DEMAND FOR PRENATAL CARE AND THE
PRODUCTION OF HEALTHY INFANTS
Eugene Lewit   127

EARLY LIFE ENVIRONMENTS AND ADULT
HEALTH: A POLICY PERSPECTIVE
Anthony E. Boardman and Robert P. Inman   183

SUMMARY AND DISCUSSION OF PART II
David S. Salkever 209

PART III. HEALTH IN DEVELOPMENT

CORRELATES OF LIFE EXPECTANCY IN
LESS DEVELOPED COUNTRIES
Robert N. Grosse and Barbara H. Perry 217

THE POWER OF HEALTH
Wilfred Malenbaum 255

HEALTH PLANNING IN SUDAN
Ronald J. Vogel and Nancy T. Greenspan 271

HEALTH EXPENDITURE IN A RACIALLY
SEGREGATED SOCIETY: A CASE STUDY
OF SOUTH AFRICA
M. D. McGrath 329

HEALTH MANPOWER IN RELATION TO
SOCIOECONOMIC DEVELOPMENT AND
HEALTH STATUS
T. Fülöp and W. A. Reinke 329

SUMMARY AND DISCUSSION OF PART III
Alan L. Sorkin 353

AUTHOR INDEX 359

LIST OF CONTRIBUTORS

Anthony E. Boardman　　　Faculty of Business and Commerce
　　　　　　　　　　　　　University of British Columbia

Dov Chernichosvky　　　　Health and Welfare, Economics and
　　　　　　　　　　　　　Administration, Ben-Gurion
　　　　　　　　　　　　　University of the Negev, Israel

Douglas Coate　　　　　　Department of Economics
　　　　　　　　　　　　　Rutgers University

Hector Correa　　　　　　Graduate School of Public and
　　　　　　　　　　　　　International Affairs
　　　　　　　　　　　　　University of Pittsburgh

Linda Edwards　　　　　　Queens College and Graduate
　　　　　　　　　　　　　School of the City University of
　　　　　　　　　　　　　New York, and The National Bureau
　　　　　　　　　　　　　of Economic Research, New York

Tamas Fülöp, Director　　Division of Health Manpower
　　　　　　　　　　　　　Development, World Health
　　　　　　　　　　　　　Organization, Switzerland

Oscar Gish　　　　　　　 Department of Community Health
　　　　　　　　　　　　　Addis Ababa University, Ethiopia

Nancy T. Greenspan　　　 Health Care Financing
　　　　　　　　　　　　　Administration
　　　　　　　　　　　　　Washington, D.C.

Robert N. Grosse　　　　 Department of Health Planning and
　　　　　　　　　　　　　Administration, School of Public
　　　　　　　　　　　　　Health, University of Michigan

LIST OF CONTRIBUTORS

Michael Grossman — Graduate School of the City University of New York, and The National Bureau of Economic Research, New York

Robert P. Inman — Departments of Finance and Economics, The Wharton School, University of Pennsylvania

Eugene Lewit — Department of Medicine and Office of Primary Health Care Education, UMDNJ–New Jersey Medical School

Wilfred Malenbaum — Department of Economics, The Wharton School, University of Pennsylvania

M. D. McGrath — Department of Economics, University of Natal, South Africa

Peter Newman — Department of Political Economy, Johns Hopkins University

Barbara H. Perry — Department of Health Planning and Administration, School of Public Health, University of Michigan

William A. Reinke — Department of International Health, Johns Hopkins University

David S. Salkever — Department of Health Services Administration, Johns Hopkins University

Alan Sorkin — Department of Economics, University of Maryland

Ronald J. Vogel — Department of Public Policy, Planning and Administration, College of Business and Public Administration, University of Arizona, Tucson

INTRODUCTION

What is the role of health in development? This general question has been gaining urgency in recent years as governments increasingly are required to justify social expenditures in terms of expected returns. Public expenditures on health vary greatly among countries. For example, in 1977 among the developing countries Tanzania spent 38 percent more while Pakistan spent 77 percent less than might be expected based on norm expenditures as developed in (1). Large differences also exist among industrial countries, Sweden spending 27 percent more while the United States and Greece spent 30 percent less than the norm. Given this wide variation in public expenditures on health, it is natural to ask to what extent is human capital created and the quality of human input raised by these expenditures? There are no straightforward answers to this more policy-specific question. It is not a difficult task to demonstrate that, although improved, the health of the great majority of the people in developing countries has remained poor despite relatively high levels of expenditure in that sector. Indeed, there is a considerable body of literature which attempts to identify the economic returns from investments in human capital. For example, studies in the "growth accounting" tradition, although attributing an important role to human capital in explaining growth, do not necessarily explain the causal role of human capital (2). More specifically, one needs to illustrate conceptually as well as empirically how health improvement helps increase productivity and growth in the economy. We hope that this volume will contribute to this field of knowledge.

The first part of the volume addresses conceptual and definitional issues. The paper by Correa attempts to provide a basis for a testable model of community health. Correa reviews the various definitions of health and concludes that they have in common the idea of the balance between the factors internal and external to an organism. He discusses the utility and the practical limitations of using life expectancy at birth as an index of health perceived as a system of internal and external balances. Correa then proceeds to identify and develop indices of those factors that can be considered to be determinants of health and disease. Peter Newman provides an imaginative framework for the planning of "medicine" in developing countries. His central focus is how does one arrive at a systematic program of health related projects in any developing country. First, the role of health in individual and social welfare is examined. The "state of health" is then defined and included as an explicit argument in the individual and society utility functions. The planning model that underlies the discussion is presented formally in the Appendix. Not only did Newman provide a reasoned argument for developing a working model for the planning of medicine in developing countries, but has also taken the profession an important step forward in that drection. The future challenge is to show how far the Newman framework could be made operational.

Gish's discussion of "health development" concludes the first part of the volume dealing with conceptual issues. Gish focuses more on specific issues related to health sector development, namely, the environment, basic needs, health planning, primary health care (PHC), and categorical disease control programs. The starting point of Gish's discussion, which can also be taken as a general conclusion based on the author's useful review and discussion, is that "the primary determinants of human welfare in any society lie within its social order and political economy, and not in the type or level of development of any particular services as such." To a large extent, these three papers represent different perspectives on the role of health in development planning.

The second part of the volume deals with the relationship between health and human capital formation. It includes four papers and a summary/discussion by Salkever. The paper by Edwards and Grossman examines the relationship between white adolescent health and preventive medical care in the USA in the context of the nature (heredity)-nurture (home and school environment) controversy. The next two papers by Chernichovsky and Coate and by Lewit deal with the economic determinants of the diet, growth and health of young children and with the demand for prenatal care and the production of healthy infants in the

Introduction xi

United States, respectively. The last paper, by Boardman and Inman, returns to an examination of adult health.

The last part of the volume deals with "health in development." Grosse and Perry investigate several hypotheses concerning the multiple determinants of levels of life expectancy in developing countries in recent decades. Malenbaum, on the other hand, illustrates the possibility of new roles for health in the wider societal objectives, e.g., educational planning. Some research directions are illustrated to improve the assessment of the "power of health" in social and economic development. The two studies by Vogel and Greenspan and by McGrath are more specific. They deal with two illustrative cases of important issues in health planning, namely the case of Sudan and that of South Africa, respectively. The last paper by Fülöp and Reinke provides an otherwise not accessible comprehensive and analytical review of the World Health Organization experience with health manpower utilization and health outcomes. Sorkin provides summary/discussion of the five papers presented in this part of the book.

Much effort has gone into the production of this volume. Aside from the contributors and numerous colleagues who provided suggestions to improve the content of this volume, we would like to thank Susan Anderson and Ann Elwan for superb editorial assistance, Beth Gerfin for early assistance and support, Camelia Sirageldin for assistance in preparing the index, the editorial staff at JAI Press for careful and professional support and, indeed Ruth Skarda for an excellent job of retyping the entire manuscript and for providing the most needed efficient secretarial support.

David Salkever
Ismail Sirageldin
Alan Sorkin
Series Editors

REFERENCES

1. Heller, Peter and Alan Tait (1982). "Comparing Government Expenditures Internationally," *Finance and Development*, 19:1, 26–29.
2. Hicks, Norman L. (1980). "Is there a Tradeoff between Growth and Basic Needs?" *Finance and Development* 17:2, 17–20.

PART I
CONCEPTS AND MEASURES

A CONCEPTUAL MODEL OF HEALTH

Hector Correa

I. INTRODUCTION

The object of this study is to present a conceptual model of the health of a population. Such a model should provide a basis for defining the health of a community, by specifying (a) some index to measure its level; (b) the factors which determine the values taken by the health index; (c) the characteristics of the relationships between these factors and health; and (d) some indices of the determining factors. Thus, if the required data are available, it should be possible to determine, through the use of statistical techniques, whether the hypotheses made with respect to the relationship between health and its determinants are supported by the empirical evidence, and if so, to estimate the magnitude of their impact.

In what follows, the concept of health is derived from that of life. For this purpose, the latter is presented in Section II, and the former in Section III. An index of health is presented in Section IV. The factors that determine health are studied in Section V.

II. CONCEPT OF LIFE

As already mentioned, the concept of life is used as a starting point in the presentation that follows. Later, this concept will be restricted in order to arrive at that of health. The basis for this is that life is a necessary, but not a sufficient, condition for health.

On intuitive grounds it can be said that life is characterized by the interaction of an organism with its environment. The content of the three basic terms in this definition, namely, organism, environment, and interaction, will be presented below, with special attention being paid to the human organism.

A. Organism

Currently, every living organism originates by reproduction from other living parent organisms. The characteristics of the parents are transmitted genetically to their descendents.

For the purpose of the analysis to be made here, the principal characteristics received from the parents through genetic inheritance by the descendant organisms are as follows:

1. Structure, i.e., subdivision into parts, body functions, the growth and maintenance process, and normal life expectancy;
2. The specification of elements that it needs to obtain from the environment and its abilities or capacities for securing them; and
3. The defense mechanisms that permit it to adapt to environmental changes and protect it from being used up or from contributing without retribution to the satisfaction of the needs of other organisms.

Some comments will be made below with respect to these three sets of characteristics.

It is useful to distinguish between the physicochemical (or somatic) and the psychosocial (or mental) components of the structure of human beings. No attempt will be made to define precisely each of these two components. It should also be observed that they are further divided into subcomponents, each with its own specific characteristics, manner of functioning, growth, maintenance, and normal life expectancy. The subdivision of the mental component into id, ego, and superego can be mentioned as an example. All the different components of an organism are interdependent.

Organisms need to obtain resources from the environment in order to maintain their lives, reproduce, etc. This need can be considered as the first immediate consequence of their physicochemical and psychosocial structure and functioning.

Following the basic structural subdivision, the needs of human organisms can be classified as physicochemical and psychosocial. The first group includes the need to receive from the environment elements such as air, water, food, clothing, shelter, etc. and the need to deliver into the environment waste materials. The psychosocial group includes the need for self-esteem, self-realization, approval, achievement, etc. Both types of genetically determined needs are, in the case of human beings, culturally conditioned. This conditioning specifies what can be used to satisfy the needs, how they should be satisfied, when, etc. The influence of culture on human needs will be discussed in more detail later.

Directly corresponding to needs and environmental characteristics, and also genetically determined, are the capabilities of organisms to obtain the resources required to satisfy these needs.

Among these capabilities is the one to feel the needs and their satisfaction, which can be considered the basis for pain and pleasure. Observe that, although the origin of pleasure might be the satisfaction of needs, it is not always, and not only, associated with them. The following are two examples of this:

1. The pleasure associated with the use of resources that satisfy need even beyond the level of satisfaction, such as the case of excessive eating; and
2. The short-term pleasure produced by drugs such as alcohol, tobacco, coffee, heroin, etc., with the risk of pain in the long run.

These exceptions suggest that pain and pleasure are immediate responses that do not take into consideration long-term consequences.

Next, other purely biological capabilities such as the ability to breathe, eat, move, produce articulated sound, see, hear, etc. can be mentioned. Intellectual or psychological capabilities should also be considered. Among the most important consequences of somatic and mental capabilities are those of interacting with other people and of working. These two capabilities are the basis for the satisfaction of social needs. For instance, communication with other people is required in order to develop the sense of belonging to a group that satisfies the need for affiliation. The capability to work is a final result or combination of physicochemical and psychosocial capabilities, and is particularly important because income from work is perhaps the most common socially accepted means of providing for the satisfaction of needs.

The capabilities mentioned above are, like the needs, culturally conditioned. For instance, culture determines the types of work that can be done, human relationships within work, etc.

The last characteristic of an organism to be considered here is the defense mechanisms it uses to protect its physicochemical and psychosocial components. A first type of these is used by an organism to maintain and restore balance among its different components, i.e., as a defense mechanism against internal disruption. A second type is used by the organism to protect itself from its environment. This includes its reactions to compensate for environmental changes, in order to avoid the damage they can produce. Examples of this type of defense mechanism are reactions to excessive cold and heat, denial of reality, construction of fantasies, etc.

The second type also includes defense mechanisms that prevent an organism from becoming a resource used up physically and socially by other organisms. As examples, natural immunity against disease and aggressive behavior can be mentioned. These genetically specified defense mechanisms are conditioned and extended, in the case of human beings, by society and culture—for instance, by the preventive and curative methods of somatic and mental health services. Also, culturally determined defense mechanisms regulate human interaction. The rights of a person as recognized by a society can be considered cultural defense mechanisms that are the social counterparts of those genetically determined for the psychosocial components. Social institutions such as police forces, whose purpose is to guarantee human rights, can also be considered as social and cultural extensions of the genetically specified defense mechanisms.

B. Environment

The environment of an organism is the set of elements not specified by the organism's genetic makeup, but having some actual or potential influence on it. In this sense, the environment is outside an organism, yet close to it. However, some elements of the environment must be internalized, i.e., must become part of the organism in order to influence it. This is obviously true in the case of, for example, food; but it is also true in the case of the cultural aspects that influence human behaviour. It is in the environment that the resources needed to satisfy physicochemical and psychosocial needs are obtained, where wastes are disposed of, where the physicochemical components and the other living organisms and human beings from which a human being has to defend him/herself are found. Finally, it is in the environment that cultural guidance about the way to satisfy needs and to improve defense mechanisms are found. More specifically, it is useful to distinguish among the following:

1. Inanimate or physicochemical environment;
2. Organic environment formed by living organisms, separated for purpose of analysis from human beings;
3. Social environment formed by human beings and their interaction; and
4. Cultural environment, including knowledge of inanimate, organic, and social environment and rules and norms of behavior.

Each of these four components of the environment will be considered below. Before doing so, it should be observed that in each of these subdivisions there are elements that contribute to the satisfaction of the needs of a human being, but also elements from which a human being has to defend himself. It can be said that the environment can be seen as a continuum from amiability to hostility toward human beings.

No attempt will be made here to present a list of the different components of the *inanimate environment,* beyond saying that weather and minerals are its more evident elements. However, it should be observed that many of its components are by-products of the interaction between the social and the inanimate environment. This is true both for useful elements, such as clean water, and for the damaging ones classified under the generic name of pollution.

For the analysis of health, the most important point to consider is the density of the different components of the inanimate environment in a specific location. This density and that of the social groups determine the likelihood of contact between human beings and favorable and unfavorable components of the inanimate environment, and it is only through this contact that benefits are obtained or damages are caused.

The *organic environment* includes everything from viruses and other microorganisms to the great trees and the great mammals. Again, the most important aspect to be considered is that of the likelihood of contact with human beings that is determined by both the density of the organisms and that of human beings. This likelihood is influenced by characteristics of the inanimate environment such as the weather and by the activities of social groups.

The *social environment,* composed of other human beings, is where a person finds the resources to satisfy his/her needs. The importance of the social environment to the reproduction process and to the satisfaction of the psychosocial needs of approval, affiliation, etc. does not need to be emphasized. It is also in the social environment where a person finds many of the conflicts and frustrations that challenge the psychosocial defense mechanisms. In addition, it is only through the association of human beings that the resources to satisfy the needs are obtained. Perhaps the associations for work and in the market are the most obvious examples.

The question of the distribution of resources for survival should also be mentioned at this point, since it appears only when the social environment is considered.

As in the case of the inanimate and the organic environment, density is a particularly important characteristic of the social environment. Clearly, the greater the density, the greater the likelihood of encountering both the advantages and the disadvantages that the social environment brings about. This is one of the reasons why levels of urbanization are relevant to the study of health.

Two main components of the *cultural environment* can be specified: (a) knowledge and (b) values and norms of behavior. The types of knowledge relevant to the study of health are (a) knowledge of alternative ways to satisfy needs; (b) knowledge used in productive work; and (c) knowledge of alternative ways to reduce the hostility of the environment, modify the probability of coming in contact with its detrimental elements, or to strengthen genetic defense mechanisms. This third type of knowledge modifies and extends the genetically specified defense mechanisms. Medical knowledge can be mentioned as an example. It is useful to distinguish between traditional and modern knowledge. Their difference is in the emphasis given to the experimental verification of their content. Knowledge in the cultural environment does not have to be the modern one.

Values specify the socially acceptable ways to satisfy needs, to obtain the means to satisfy needs, the aspects of environmental hostility that should be reduced, the forms of medical knowledge that can be used, etc. Some values are institutionalized in rights that are culturally specified defense mechanisms protecting each person from others or from certain types of organized social actions.

Acculturation is the process through which culture is transmitted from one generation to the next. Formal education is an important component of this process, but it includes many other forms of educational interactions, such as education received at home, in peer groups, etc.

C. Interaction

As mentioned, life is a process of interaction between an organism and its environment. The interaction with the social and cultural environments is the basis for satisfaction of psychosocial needs. Also, in the interaction with the inanimate and the organic environments, an organism uses its capabilities alone or in cooperation with other organisms to obtain what it requires to satisfy mainly its physicochemical needs, discharges into the environment its wastes, and uses its defense mechanisms to avoid prejudicial influences of the inanimate environment and to avoid being exploited by other organisms. These observations suggest three forms of interaction with the environment:

(a) The first form can be called interaction only in a very restricted sense. It includes the process of obtaining useful elements from the inanimate environment. In this case, the only limiting factors are the capabilities of each organism and, in the case of human beings, of human societies.

(b) The second form is cooperative interaction for the mutual or better satisfaction of needs. One example is symbiotic interaction, on the one extreme, with lower species such as microorganisms in the digestive tract, and, on the other, the case of systematic exploitation of cattle and other animals. Most social interactions among human beings can be classified as cooperative interaction, such as interaction for reproduction, education, work, pleasure, utilization of health services, etc. It should be observed that this type of interaction is the basis for the satisfaction of psychosocial needs.

(c) The third type of interaction is competitive interaction in which organisms tend to use for themselves the elements needed for the survival of others, or even, in a more extreme case, they may use each other for survival. Their defense mechanisms protect the organisms in this type of interaction.

It should be clear that, quite frequently, even human interaction has elements of both cooperative and competitive interaction.

III. CONCEPT OF HEALTH

A. Introduction

In this section a concept of individual health will be logically derived from the systemic view of organism and environment presented in Section II. This concept is presented in Section B, compared with other concepts available in the literature in Section C, and used as a basis to define the level of health in a social group in Section D.

B. Concept of Health of an Individual

It was stated that life is a necessary, but not a sufficient, condition for health. The concept of life is restricted by that of health; that is, an organism is healthy when its current or short-run internal function and interaction with the environment are normal, balanced, in equilibrium. More explicitly, a living organism is healthy if its internal function and interaction with its environment are in a self-perpetuating condition; i.e., where there is no immediate possibility of life-threatening disruptions. This idea applies equally well to the physicochemical and the psychosocial components.

The concept of individual health presented above encompasses that used in medicine. One of the purposes of a medical examination is to determine whether the internal balance, or the normality of the interaction with the environment, has been broken, and, if so, the relative importance of the deviation. However, the concept of health above refers to all deviations from balance or equilibrium, while the medical definition refers only to the more systematic or lasting deviations that can be classified as diseases.

Despite the fact that the term *disease* in medicine refers to only one type of deviation from the health status balance, *disease* below will be considered to mean the opposite of health; that is, *disease* will refer to any form of imbalance.

To achieve health, that is, the balance mentioned above, the following are necessary:

1. The internal mechanisms of defense must maintain the balance among the different components of an organism (homeostasis); and
2. The capabilities and mechanisms of defense that protect an organism from external stress must be well adapted to the conditions of the environment—i.e., have capabilities sufficient to insure the acquisition of the environmental elements to satisfy needs—and the mechanisms of defense must be sufficient to compensate for environmental changes and to pre-

vent the organism from being used up or exploited by other organisms to satisfy their needs.

To clarify the concept of health being presented, it should be observed that, although every imbalance is a deficiency in health, not every imbalance is the cause of a health deficiency. Some imbalances appear as consequences of the interdependence of all the components of an organism. For instance, bed disability brings about an imbalance between the needs of a person and his/her capacity to satisfy them. However, in most cases, bed disability is not a cause of a health deficiency, but an effect of imbalances in other components of the system.

Again, to clarify the concept of health it should be observed that pain, pleasure, and the sense of overall well-being that were mentioned in relation to the capacity to feel needs and their satisfaction actually have a more general objective. They reflect the internal and external balance of an organism as a whole, or of some of its components.

Assuming environmental stability over time, the concept of health above is equivalent to perfect adaptation, first, to the structural characteristics within an organism, and second, to the external environment. Since adaptation is the result of evolutionary genetic mutation, this means that, in a stable environment, lack of health always has a genetic origin, i.e., it is the result of defects in the evolutionary process. These defects generate needs that cannot be satisfied with available capabilities and environmental resources or fail to provide the organism with mechanisms of defense sufficient to meet the challenges of the environment.

Despite the truth of the previous observations, internal and external balance will not be considered here as equivalent to perfect adaptation to internal characteristics and to the external environment. The reason for this is that the environment is not stable, and even if it were, the genetic adaptation is only a very long-term determinant of health. Here, a short-term point of view is adopted. This means that the current needs, capabilities, and defense mechanisms of human beings are considered, rather than the potential ones that could be developed through evolution, or possibly by genetic engineering.

This concept of health can be applied to components of the system formed by an organism and its environment. No attempt will be made here to describe all the forms that this application of the concept could take. However, it should be observed that these forms include the dimensions of health considered, say, by WHO and other authors. More specifically, considering the structural components of an organism, it is possible to speak of physical and of mental health. The concept of social health is obtained by considering the interaction of an organism with its social environment. The concept of health could also be applied to more detailed subdivisions of the system, considering, for instance, the health of each of the organs in the body and so on up to individual cells if necessary.

The balance in the components of a living system is closer to the immediate experience than the balance of the system as a whole. This explains why health is frequently called multidimensional, meaning that the idea of health is a conceptual construction from a set of numerous independent components. However, this point of view is not intellectually satisfactory. If it is accepted as stated, it implies that the idea of health is a purely artificial construction without an empirical counterpart. This implies that its meaning and usefulness should be questioned. However, if this is not the case, and there is some content to the idea of health as a whole, then there is some structure and interdependence among its numerous components, i.e., health is not multidimensional. The importance of these ideas will be seen while discussing the construction of health indices.

C. Other Concepts of Health

To complete the presentation of the concept of health, the one presented before will be compared with others available in the literature.

Perhaps oversimplifying, it can be said that all the available concepts of health can be subdivided into four groups that present health:

1. As the internal and external balance of an organism;
2. As absence of disease,
3. As capacity to perform; and
4. As well-being.

Frequently, the various authors present as a concept of health a combination of these four definitions. As a consequence, several attempts to define health belong to several of the groups mentioned above.

The definition presented in Section B is an example of those considering health as the internal and external balance of an organism. For this reason, this type of concept will not be discussed in detail. It is enough to observe that it seems to have been used by Greek physicians such as Hippocrates as mentioned in Reference 1 and philosophers such as Plato as mentioned in Reference 2. Examples of more modern uses are in References 3, 4 and 5.

The idea of a specific disease is quite close to the immediate experience of everyone, as well as the idea of regaining health by recuperating from a specific disease. Probably, based on this intuitive appeal, numerous attempts to define health as absence of disease, frequently with some additional characteristics, have been made. Examples of these approaches appear in References 1 and 6. However, these attempts cannot be considered satisfactory, since they leave unexplained the meaning of disease in general, as opposed to a particular disease. Any attempt to do this makes it necessary to refer to the internal and external balance of an organism; i.e., a logical completion of the definition of health as absence of disease makes it an example of the first group of definitions mentioned above.

A typical example of the definitions in the third group is the one in Reference 7. In it health is considered to be the capacity for effective performance of roles and tasks for which an individual has been socialized. This approach can be criticized on at least three grounds. First, specific capacities to perform, as specific diseases, are quite close to our experience. However, a generalized idea of capacity to perform seems as hard to define as the generalized idea of disease. Carrying the analysis to an extreme, the point is reached in which generalized capacity to perform is health, and health is generalized capacity to perform; i.e., nothing has been defined. Second, implicit in the previous observation is the fact that the generalized idea of capacity to perform cannot be made equivalent to internal and external balance of an organism. The reason for this is that capacity to perform cannot be judged by itself. One possible way to evaluate capacity to perform is to compare it with some average in a social group; however, this only shifts the problem to that of determining the reference group. For instance, regardless of how average is the capacity to perform of a member of a malnourished group, he/she will not be considered healthy. To establish a term of reference for evaluation, it is necessary to consider that the capacity to perform of an organism is an instrument to satisfy its needs and protect itself from internal imbalance and external threats. With this, capacity to perform can be evaluated with respect to its function. However, this means that the basic problem of the attempts to define health as capacity to perform is that they present only a partial view of the system being defined. A third problem of this type of definition is that, in many cases, the conditions of the capacity to perform of an individual are effects of its internal and external balance, brought about by the interdependences of the system, and not causes of it. For instance, bed disability days are a consequence of a specific disease, and not its cause. These defects also affect the definitions of health as well-being and will be analyzed while discussing them.

One of the best-known, and perhaps one of the most controversial, definitions of health is the one suggested by WHO in Reference 6. It says that health is a state of complete physical, mental, and social well-being, and not merely the absence of disease and infirmity. This definition can be considered a characteristic example of the fourth group mentioned above. The analysis of the components of this group show that they do not indicate what health is, but rather describe the result of the integration of health with other factors or processes. When dealing with the health determinants of well-being the definitions frequently refer to the internal and external balances used to characterize health in the definitions in the first group. In this sense, the definitions become a form of the definitions in the first group. On the other hand, if these definitions do not specify the health determinants of well-being, they would include educational, economic, political, religious, and other conditions as components of health.

As a conclusion of the review of definitions of health just presented, it can be said that all of them, implicitly or explicitly, have to use as term of reference the idea of internal and external balance of an organism. However, it should be added that, in order to bring forth the idea of internal and external balance of the organism with specific examples, the definitions of health as absence of disease and as capacity to perform are also useful. Also, the definitions of health as well-being, as long as it is accepted that health is only one of the determinants of well-being, also serve to give intuitive meaning to the idea that health is internal and external balance.

D. Level of Health of a Social Group

The analysis to this point refers to the health of one organism. The health of a social group will be defined to be some average of the health of all its members. When an index of individual health is specified, it will become evident that it is possible to compute averages of the values of the individual indices of the members of a social group.

IV. HEALTH INDICES

A. Introduction

It seems intuitively obvious that a concept of health is always, implicitly or explicitly, used a as starting point in any attempt to construct an index of health. On a more normal level, it can be said that an explicit concept of health is required as a term of reference to judge the validity of any index of health constructed. Below, reference will be made to the indices constructed using the disease, capacity to perform, and balance definitions of health as starting points. A brief review of the literature suggests that the well-being definitions, perhaps as a consequence of their exaggerated scope, have not been used for this purpose.

The problems of constructing health indices will not be discussed in the presentation to be made. They are extensively covered in the literature. Useful studies appear in References 8, 9, and 10.

B. Indices of Health as Absence of Disease

The definitions of health as absence of disease are the conceptual framework that leads to its evaluation through medical examinations. This is particularly the case of examinations to determine whether an individual is affected by a specific disease and the severity of the affection. Despite this, most of the indices derived from medical tests such as pulse rates, blood counts, performance in personality tests, etc. are not disease-specific but reflect response of the complete system, indicating whether it is in balance. Using, say, the techniques of factor analysis, these indices might be integrated into an overall index for an individual. However, this integrated index does not indicate whether a person has a specific disease, but rather evaluates his/her state of general internal and external balance. An average of the integrated indices of all the members of a social group could be an indicator of its level of health.

C. Indices of Health as Capacity to Perform

As a term of reference for the comments to be presented below, the presentation in References 11 and 12 will be used. These references summarize 17 indices developed in the last 30 years and could be considered as representative of the state of the art from the point of view they adopt.

In the references being studied, the subdivision into physical, mental, and social health, suggested in the WHO definition of health, is accepted. Then, following the recommendations of several authors, physical and social health, in principal, are evaluated, using as a measuring device the performance of certain activities such as walking, running, visiting friends, etc. Frequently, a relationship is sought between the lack of ability to perform these activities with symptoms of disease, or with the presence of diagnosed diseases. This approach is also used to evaluate mental health. The measurements taken for each of the components are aggregated into overall indices of health using standard statistical procedures.

The defects of the definitions of health as capacity to perform reach a crisis point when they are used as a basis to specify indices of health as explained above. The definitions themselves do not provide any idea of which aspect or component of the capacities to perform should be measured, or how the measurements of the components should be aggregated into an overall index. As mentioned before, attempts to give meaning to the aggregated indices make it necessary to consider health as internal and external balance.

D. Indices of Health as Internal and External Balance

The observations above show that the indices of health as absence of disease or as capacity to perform evaluate whether a person is in a state of internal and external balance and, if this is not the case, evaluate the magnitude of the deviations. Particularly in medicine, it is also assumed that large deviations are an immediate threat to human life.

To define an alternative index of health it can be assumed that any deviation from internal and external balance not only could be an immediate threat to life, but that it *always* tends to reduce its length. This implies that a complete recovery from any disease to the point where there is no difference between a person who suffered the disease and one who did not is impossible, that there is always some difference reflected in a variation in the length of life. This assumption is very difficult, if not impossible, to test.

As stated, the assumption also implies that mental imbalances tend to reduce length of life. This is likely to be controversial. In Reference 13 it is stated,

A Conceptual Model of Health

"Making abstraction of suicide, the psychiatric patients died of the same causes as the general population. Death is brought about by the same pathological processes but it occurs earlier. The calendar of mortality is, as a consequence, modified." Similar conclusions are reached in References 14 and 15.

Accepting that the length of life of a person can be considered as the most appropriate index of his/her health, it follows that the average number of years of life of all the members of a community can be used as an index of community health conditions. This average is what is known in demography and in the mathematics of life insurance as life expectancy at birth. These sciences also employ methods to estimate life expectancy at birth of the members of a group before the length of the life of each of its members is known, i.e., before all its members are dead.

One limitation of this index is that the basic information needed for its computation is not available for many social groups. However, as shown in the Appendix to this paper, under certain simplifying assumptions it is true that

$$E = \frac{100}{r}, \tag{1}$$

where E = life expectancy at birth and r = 100 (number of deaths)/number of persons in the population); i.e., r is the crude death rate for 100 persons in the population.

With the formula above, r can be used as an index of the health conditions of a population.

One limitation of life expectancy at birth and of death rate as indices of health follows from the basic assumption used to link them with the concept of health as internal and external balance, namely, that every deviation from the state of balance tends to reduce the number of years of life. If this is not the case, the impact of some deviations from balance, i.e., health deficiencies, would not be measured by life expectancy or death rate. This indicates that, if such deviations from balance exist, it is necessary to modify the index above to consider morbidity conditions also. Examples of methods to do this are presented in References 16 (which also includes a brief summary of the literature) and 17. These methods are based on equivalences between morbidity and mortality, arbitrarily defined that make it meaningful to add, for instance, days lost due to disease and to death. Since the basis mentioned above seems weak from an intuitive point of view, the indices obtained are of doubtful value.

It seems natural that the definition of health as internal and external balance, which sees health as a unit, would lead to an aggregated index of health. However, the aggregated index presented above might not be satisfactory in many practical applications. Indices of health of components of the system are also necessary. The problems of developing partial indices within the framework of the aggregated index suggested above will be considered below.

The concept of health as internal and external balance can be applied to each of the different components of the organism considered in Section 2. Using this idea as a starting point, the index of health above can be easily decomposed at least at a conceptual level. The simplest way to do so follows from Formula (1) and the fact that the total death rate can be expressed as the sum of the death rates by cause. However, at a practical level, the approach described is not likely to be useful. One reason for this is that the immediate causes of death receive attention, while the internal or external imbalances that bring about the conditions required for the immediate causes of death to operate might even be unknown. This is particularly true for the causes of death that might be considered related to deficiences in mental and social health, such as suicides, homicides, accidents, etc. It is also true that disease may reduce the capacity to perform while not constituting an immediate threat to life. The effect of this type of disease on the length of life, despite the fact that this might be important, is not known in many cases. Hence, particularly for studies of the health status of components of the organisms, it is necessary to fall back on indices of health based on the definitions of health as absence of disease or as capacity to perform. However, when using these indices, it is important to keep in mind that they reflect imbalances in components of the system.

V. DETERMINANTS OF HEALTH AND DISEASE

A. Introduction

In the previous sections, several factors that can be considered to be determinants of health and disease have been suggested. The object of this section is to identify explicitly these factors and the indices that can be used to measure them. When this is done, it will be possible to test whether the conclusions obtained from the analysis above are supported by statistical evidence and to evaluate the impact of the factors identified on the level or health of a community.

In the presentation to be made, the concept of health as internal and external balance will be used as the term of reference. A justification for this is that, as already observed, this concept seems to be an unavoidable term of reference in any attempt to make the other concepts logically consistent.

To introduce the study of the determinants of health and disease, it is necessary to take into consideration that, in agreement with the observations made regarding the basic characteristics of an organism, and with the definition of health being used, it can be said that any organism with normal genetic characteristics, normal satisfaction of needs, successful defense mechanisms, and components used up by the passage of time only in accordance with the normal duration as genetically determined is healthy. This suggests that the causes of variations of the level of health must be looked for in genetic abnormalities, defects in the satisfaction of needs, unsuccessful defense mechanisms, or unusual wearing out due to the passage of time. These are considered below.

Before proceeding, it should be observed that the four aspects just mentioned are not mutually independent. Actually, they interact in the sense that, say, a genetic abnormality can produce a structural defect that reduces the capacity of a person to eat or digest food (i.e., to satisfy his/her needs). Another example is the fact that deficient food intake makes a person more susceptible to infection. However, the opposite is also true; i.e., certain types of infections reduce the capacity to absorb or utilize the nutrients obtained from food.

For simplification, the influence on health and disease of the interactions mentioned above will not be considered in the analysis that follows. This means that attention will be concentrated on the initial causes of diseases, leaving aside causes that might be more immediate, but that are themselves consequences of the interdependence mentioned.

B. Determinants of Genetic Diseases

Genetic abnormalities can be transmitted from parents to their children, or they can be the result of genetic change within a specific individual. Below, only the case of genetic changes that can be considered as the starting point of any genetic abnormality will be studied. The hereditary transmission of abnormalities, that is, the transmission of the result of genetic changes in the ancestors of a person, will not be studied, since the transmission itself is a normal process applied to a defect initiated in another person. However, as mentioned before, some observations will be made on the initiation of the defect.

The specific causes of genetic changes within the reproduction of cells are not known. On the other hand, it is known that some environmental influences, particularly radiation, tend to increase the frequency of genetic changes. This could be considered as an example of a genetic change produced by environmental factors that overwhelm the defense mechanisms. These observations suggest that, for the analysis being made here, it can be stated that the causes of genetic changes that bring about diseases are defects in the satisfaction of needs or in the defense mechanisms. As a consequence, the study of the determinants of genetic diseases can be reduced to a study of the deficiencies in the satisfaction of needs or the lack of success of the defense mechanisms that brings about the genetic changes.

A point that needs to be mentioned is that, at least in principle, genetic changes that produce unfavorable consequences tend to disappear from a population. However, this natural tendency can be checked with the cultural extension of the biological defense mechanism, i.e., by medical services. This example shows that short-term benefits of these services can produce long-term undesirable consequences. This point will be studied as part of the analysis of defense mechanisms.

C. Influence of the Satisfaction of Needs on Health and Disease

It was mentioned before that human needs can be classified as physicochemical or somatic and psychosocial or mental. The first group includes, for example, the needs for air, water, food, clothing, shelter, reproduction, etc. Psychosocial needs are not as well specified as the physicochemical ones, in the sense that it is more difficult to describe the feelings that are associated with them, or the ways that they are satisfied. Descriptive names attached to some of them are need of self-esteem, of self-realization, of approval, of achievement, of affiliation, etc. Perhaps marriage and family life, participation in productive activities, affiliation in professional, religious, political, etc. groups are the most commonly used approaches to the satisfaction of these needs.

It should be stated again that deficiencies in the satisfaction of needs could be by defect or by excess. Perhaps the most obvious example of this is the satisfaction of the need for food. In this case, the two extremes are often found in the same society, with people dying from starvation at the one extreme and, at the other, people dying as a consequence of obesity. The ideas of deficiency and excess in the satisfaction of psychosocial needs are less well defined.

A first step in the analysis of needs in relation to health and disease is to specify their levels of satisfaction. Several indices can be defined for this purpose. At the physicochemical level, indices of the intake of nutrients, the type of clothing, of housing, etc. can be used. It is more difficult to specify indices dealing with the level of satisfaction of psychological needs. However, from the observations above, it follows that the proportions of married or divorced, of employed or of unemployed persons could be the most appropriate indices. For a survey of the literature on this point, see Reference 18.

The study of the level of satisfaction of needs is not sufficient. It is also important to study the determinants of these levels of satisfaction. This question relates to the capability of satisfaction of needs.

It was mentioned before that the defects in the capabilities to satisfy needs due to genetic abnormalities, lack of balance both within and between somatic and mental components, inability to gain consciousness of the needs, in using environmental resources to satisfy them, such as in the case of absorption diseases, etc. will not be considered here as part of the influences of the satisfaction of needs on health and disease. It follows that only persons with normal capabilities will be considered here, and, as a consequence, the different levels of satisfaction of needs are determined by environmental factors. For the analysis below, these factors will be classified into two groups:

1. Determinants of the availability of the resources required to satisfy the needs; and
2. Determinants of the utilization of these resources.

A Conceptual Model of Health

The determinants of availability of resources to satisfy both the physicochemical and psychosocial needs can be classified into four groups:

1. Social characteristics;
2. Values;
3. Knowledge; and
4. Resources.

Each of these four groups will be studied below.

Social characteristics and political and economic organization specify the conditions of exploitation of resources, and, to a large extent, the way the outputs obtained are distributed among the members of the society.

Despite the statements above, it should be observed that there is little or no scientific information on the influence of different forms of political organization on the availability of resources required for the satisfaction of needs. The information available about the influence of economic organization is also somewhat limited. On the one hand, there are hypotheses suggesting that concentration of the factors of production could lead to additional expenditures on research and development—i.e., a larger output of resources—and, on the other, results in economic theory show that monopolistic concentration of the factors of production tends to reduce the volume of goods and services produced. This suggests that indices of the degree of monopoly could be useful in studying the determinants of the satisfaction of human needs. However, despite the fact that such indices have been defined and computed, they are available for only a few countries.

Several examples of the influence of values on the availability of the resources required to satisfy needs can be mentioned. A first one is the rejection as socially unacceptable of certain forms of association for the satisfaction of certain psychosocial needs. A special case of this might be the rejection of concubinage. A second example is the conditioning of the processes of production of goods and services. An instance of this is the exclusion of the use of slave labor, while other values justify forms of discrimination against some social groups. It should be observed that here the expression "goods and services" is used with a generalized meaning to include not only the resources needed to satisfy physicochemical, but also psychosocial, needs.

The internalization of values depends on the level of acculturation of the population. Despite the fact that these levels of acculturation are not associated only with the education obtained through the formal systems, indices of formal education can be used as indices of the level of acculturation. The number of years of education of a person seems, on intuitive grounds, to be a useful possibility.

The average number of years of education for members of a social group is an immediate generalization of the index proposed above for a single person. However, this generalization does not take into consideration the fact that the level of education differs from person to person, and as a consequence, capacity to use resources also differs from person to person. Thus, it is necessary to complement the average number of years of education with indices of the spread and of inequality of the distribution. The variance and the Gini or other indices of concentration of the distribution of years of education could be used for this purpose.

The main aspect of knowledge that determines the availability of the resources required to satisfy needs refers to the results of social and physical sciences used to produce goods and services.

It seems that methods for constructing indices of the stock of knowledge available to a social group have not been developed. However, number of technical publications and of patents could be indices of the growth of knowledge. Number of persons obtaining advanced degrees could be an index of the availability of persons that could put new knowledge to use in the production processes.

Resources are the final determinant of the availability of the elements required to satisfy needs. The resources used to obtain the elements needed for the satisfaction of the physicochemical needs can be indentified with the factors of production considered in economics, i.e., natural resources, labor or human resources, and physical capital. Well-known indices are available to measure at least two of these, labor and capital.

By extending the usual meaning of the term, it is possible to talk about resources for the satisfaction of psychosocial needs. An important component of these resources is population density, which influences the condition of the social and cultural environment in which these needs are satisfied. Several indices of population density can easily be constructed, such as density per square mile, per square mile of arable land, etc. Indices of urbanization can also be used. However, these indices do not seem to give an appropriate reflection of the influence of density on the availability of resources to satisfy psychosocial needs.

The determinants of the actual utilization of the resources required to satisfy needs can be classified into groups similar to those of their availability, i.e., knowledge, values, and resources.

The satisfaction of needs depends on knowledge of them, and on the methods that can be used for their satisfaction. It is possible that the level of education of the population as a whole provides an index, if not of knowledge itself, at least of the extent to which accepted ideas have been diffused in the population. The observations made previously with respect to the indices of educational achievement can also be applied in this case.

Cultural values influence the utilization of resources because they modify the priorities attached to the satisfaction of different needs and the acceptability to the social group of the different methods for satisfying needs. For instance,

A Conceptual Model of Health

cultural values might specify that, for women, the need for affiliation satisfied with a family is more important than the need for achievement satisfied with productive work. Also, cultural values might specify that a normal marriage is a better way to satisfy the need for affiliation than a consensual union.

The utilization of the elements that satisfy needs in a social group as a whole depends also on the distribution of income and status in it.

Income can be used to purchase the elements required particularly, but not exclusively, for the satisfaction of mainly physicochemical needs. It should be observed here that, at least in accounting terms, it can be shown that the total resources produced for the (mainly) physicochemical needs is equal to the total income of the population. This suggests that income per capita can be used to represent both availability of resources and the capacity to use them.

This index, like the one used for education, and for the same reasons, should be complemented with indices of the dispersion and concentration of per capita income.

Social status can be considered a social counterpart of economic wealth that offers the possibility of utilizing certain social "resources" for the satisfaction of mainly psychosocial needs.

Again in this case, the utilization of resources to satisfy needs in a social group as a whole depends also on the distribution of status in it.

D. Defense Mechanisms as Determinants of Health and Disease

A point that should be observed is that there is no clear distinction between lack of satisfaction of needs and failure of a normal defense mechanism at either the somatic or the mental level. For instance, it might be impossible to determine whether the death of a person from exposure to freezing temperature is due to lack of satisfaction of needs for clothing and shelter or to the fact that his/her mechanism to keep temperature at survival levels was overwhelmed by these conditions. For the purpose of this study, the distinction is not required.

In the analysis to be made below, attention will be paid to the defense mechanisms determined by genetic inheritance and by the social and cultural environment.

In the analysis of genetic defense mechanisms it is useful to distinguish between their deficiencies and their failure. The deficiencies of the genetic defense mechanisms can be produced by genetic abnormalities, lack of satisfaction of needs, or the direct or indirect effect of the failure of other defense mechanisms. The deficiencies associated with genetic abnormalities, since they are due to the interaction of different components of an organism, will not be considered here. All that has been said before with respect to deficiencies in the satisfaction of needs can be applied without modification to the case of the genetic defense mechanisms, and for this reason, the topic will not be analyzed below. Finally, for the analysis being made in this section, the point of interest is not the deficiencies in some defense mechanisms caused by failure in other defense mechanisms, but the original failures.

The observations above can be used as a basis for reducing the present analysis to the study of failure of normal defense mechanisms, i.e., of defense mechanisms that are not deficient due to the direct or indirect consequences of genetic abnormalities, inappropriate satisfaction of needs, or failure of other defense mechanisms. It follows that the only type of failure of defense mechanism to be considered here is that brought about by environmental factors.

In general, it can be said that environmental factors bring about the failure of normal defense mechanisms because they overwhelm them. An example of this, dealing with the physicochemical component, is the case of a person who gets an infection simply because of exposure to a large number of infective agents. The mental health problems associated with urban life are examples of failures of the psychosocial defense mechanisms. These examples show that the intensity, concentration, or high density of the damaging environmental factors is the determinant of the failure of a normal defense mechanism.

The degree of concentration or the density of environmental factors is what makes them capable of overwhelming normal genetic defense mechanisms. Characteristics of the weather such as temperature and humidity, of the terrain, such as slope that determines flow of water, etc. influence the density of several agents and vectors of disease that form a part of the organic environment. Indices of the density itself, or of the characteristics of the weather and of the terrain, can be used to study the influence of inanimate and organic environments on health.

Reference has already been made to the influence of the density of human populations on the satisfaction of psychosocial needs. At this point it can be added that a high density might overwhelm the mental defense mechanisms and bring about mental disease. The point at which density moves from advantageous to disadvantageous is not known.

The sociocultural extensions of the genetic defense mechanisms will be called *sociocultural defense mechanisms*. Several examples can be presented, such as human rights, laws, mental health services, and police services that extend the defense mechanisms of the psychosocial component of human beings; while somatic medicine extends mainly those of the physicochemical components.

In the presentation below, attention will be concentrated mainly on one example of the sociocultural defense mechanisms: the medical services. This is an effort to keep the present study within the boundaries traditionally given to health. However, it should be observed that most of the observations below are also valid for the other examples of sociocultural defense mechanisms mentioned before.

The sociocultural defense mechanisms can have three effects:

(a) Modification of the characteristics of the elements of the environment to make them less detrimental to human beings; results of this could go from the modification by means of dams of the flow of rivers to the generation of less dangerous types of useful animals such as bees.

A Conceptual Model of Health

(b) Modification of the probability of the contact of human beings with detrimental elements of the environment; pure water supply and sewage disposal are defense mechanisms characterized by this effect.

(c) Strengthening of the genetically determined defense mechanisms so that they are able to overcome environmental stresses acting on human beings; medical methods of immunization and cure are perhaps the best examples of defense mechanisms with this effect.

The analysis below is based on two characteristics of the sociocultural defense mechanism that are not shared with the genetically determined ones: (a) they are not automatically available, and (b) if available, they might not necessarily be used.

Below, a study will be made of the determinants of availability and utilization of sociocultural defense mechanisms. For this, it should be observed that, with respect to these determinants, the sociocultural defense mechanisms can be considered as a particular case of goods and services. The same is true of the elements that satisfy needs, whose availability and utilization were studied before. As a consequence of this, some degree of repetition cannot be avoided in the presentation that follows.

Little can be added to what has been said about the influence of political and social organization on the availability of goods and services to satisfy needs when the availability of sociocultural defense mechanisms is considered. For this reason, this point will not be discussed here.

The second determinant of the availability of sociocultural defense mechanisms is the value attached to health and to the different types of defense mechanisms. These values determine the priorities attached to health services in comparison with other goods and services, such as education, and the priority attached to different types of health services such as environmental sanitation and preventive and curative medical services.

The only way to define indices of the values attached to health and the different types of health services is based on the visible behavior of the persons in a community that can be attributed to these values. Once this is accepted, it becomes clear that total or per capita expenditure on health, as compared with expenditures on other goods and on different types of health services, can be used as indices of the value attached to them. Some problems associated with the use of expenditures as an index of the value attached to health will be mentioned later.

Knowledge is the third determinant of the availability of sociocultural defense mechanisms. Using the classification of knowledge presented previously, it can be said that modern knowledge, i.e., knowledge based on empirical verification, is more likely to bring about the desired effects of the defense mechanisms than traditional knowledge.

Attempts to define indices of the level of medical knowledge available to a social group are likely to result in its identification with different types of health personnel and facilities. The reason why such indices have important limitations will become clear later. On the other hand, expenditures on health research, number of patents for pharmaceuticals and medical instruments, and numbers of degrees awarded in medical sciences, whether total or per capita, could be used as indices of the efforts made to increase or improve the knowledge needed to provide health services.

The fourth determinant of availability of sociocultural defense mechanisms is whether the human, physical, and financial resources to be used for environmental sanitation and preventive and curative medical services are available. The four determinants mentioned above specify the upper bounds for the resources available and their effectiveness.

It can be said that there are natural indices of availability of human, physical, and financial resources for health. These indices could be, respectively, number of doctors, nurses, and auxiliary personnel; number and characteristics of hospitals and of their equipment; and finally, the monetary value of capital and current expenditures. A limitation of these indices is their lack of sensitivity to qualitative differences. At this point, it should be clear that the indices above also represent the knowledge available with respect to health and the value attached to health. In summary, it can be said that they are indices of the availability of health services, but it is not completely clear which of the determinants of this availability they represent.

As already observed, the determinants of the actual utilization of the sociocultural defense mechanisms can be classified under headings similar to those used for the determinants of their availability, i.e., knowledge, values, and resources.

The knowledge that influences the utilization of the sociocultural defense mechanisms is that of the need for and the advantage of using them. The principal case in which the need is known is when the patient is conscious of his/her disease and can benefit from medical services. Examples of lack of knowledge of the need exist in cases of undetected cancer and unrecognized alcoholism. Another defect in the knowledge of the need for environmental health services is the common case in advanced societies of persons who medicate themselves despite the fact that there is no medical reason to take drugs.

The importance of knowing the advantages of using the environmental health services as a determinant of their utilization seems to be more intuitively clear, so no comment on this point will be made here.

The average and dispersion of the number of years of education in a population can be used as indices of the knowledge of the advantage of sociocultural defense mechanisms.

The values attached to the different methods to reduce environmental hostility and to prevent or cure disease determine the methods that are preferred from among all those available. As before, the actual expenditures on the different methods could be used as indices of the value attached to them. The level and distribution of education in the population are also factors that influence the value attached to health and to the different defense mechanisms.

Income and status are the resources that the members of a population can use to acquire sociocultural defense mechanisms. Income is likely to be more important in the case of the acquisition of medical services; however, the relative influence of income can be greatly reduced when some form of insurance scheme takes over all or part of the costs for health services. In this case, other nonfinancial costs such as waiting time might take the place of the financial ones. These observations suggest that the level and distribution of income and the proportion of a population covered by health insurance can be indices of the capacity to utilize health services.

Status is more likely to influence the utilization of nonmedical sociocultural defense mechanisms such as police services.

VI. SUMMARY AND CONCLUSIONS

As mentioned in the Introduction, the object of this paper is to provide a basis for a testable model of community health. In summary, such a model could be presented as follows:

The dependent variable should be an index of health, such as life expectancy at birth or crude mortality rate.

In the analysis of the determinants of health, attention should be paid to the somatic and mental components of human beings.

Two main sets of explanatory variables should be considered for each of these two components: satisfaction of needs and defense mechanisms.

A first set of variables among those referring to the satisfaction of needs should simply be indices of the level of satisfaction achieved. Examples of such indices could be, in the case of physicochemical needs, those of quality of drinking water, food consumption, clothing, quality of housing, etc. Examples of indices for psychosocial needs could be those dealing with marriage, employment, mental morbidity, crime, and drug addiction.

One further step away, but still dealing with the satisfaction of needs, indices of the determinants of the availability and utilization of resources should be considered. Perhaps the most immediate indices of these determinants are levels of income and education.

The variables dealing with defense mechanisms can be subdivided into two groups, those dealing with the genetic and with the sociocultural defense mechanisms.

Since the reason for failure of the genetic defense mechanisms is found in environmental conditions, the point of interest in this case is the specification of indices that represent selected elements of these conditions. It is likely that temperature, humidity, and rainfall, mainly due to their influence on the proliferation of agents and vectors of disease, are the most appropriate indices of the inanimate and organic environments. Population density, measured, say, by the level of urbanization, could be the easiest index to obtain of the principal characteristics of the social and cultural environments.

In the analysis presented, medical services were selected as the most relevant case of sociocultural defense mechanisms. In studying them, attention should be paid to the determinants of their availability and of their utilization.

In dealing with the availability of medical services, the usual indices of population/doctors, population/hospital beds, etc. can be used. A point of particular interest refers to medical knowledge. It does not seem to be possible to specify indices of the stock of knowledge available; however, its increments could be represented by number of scientific journals published, number of patents for pharmaceuticals and equipment, number of new Ph.D.'s awarded in specified fields, etc.

The actual utilization of the medical services available could be measured, for instance, by number of doctor visits per person per year. Indices of the determinants of utilization are, as in the case of the utilization of the services to satisfy needs, levels of income and education.

It should be observed that all the indices mentioned are averages. For the study of the characteristics of health in a community, they should be completed with indices of dispersion and concentration, giving some idea of the number of persons for whom the values of the indices used are above and below the average.

APPENDIX

The object of this appendix is to show that:

$$E = \frac{100}{r}, \qquad (1)$$

where E and r were defined in the text.

To verify Eq. (1) let us consider a population of size N. Then assuming that the age-specific death rates are all equal to s, with

$$s = \frac{r}{100}, \qquad (2)$$

and that all deaths occur at the end of each year, it is possible to construct the following table:

No. of years of life (1)	No. of persons who live years in column 1 (2)	No. of persons who survive (3)
1	N*s	N − s*N = N*(1 − s)
2	N*(1 − s)*s	N*(1 − s) − N*(1 − s)*s = N*(1 − s)**2

No. of years of life (1)	No. of persons who live years in column 1 (2)	No. of persons who survive (3)
3	$N*(1-s)**2*s$	$N*(1-s)**3$
⋮	⋮	⋮
n	$N*(1-s)**n - 1*s$	$N*(1-s)**n$

From this table it follows that the total number of years of life (T) of the N persons in the group is

$$T = \text{Sum } N*s*i(1-s)**(i-1)$$
$$= \frac{N*s}{1-s} * \text{Sum } i(1-s)**1 \qquad (3)$$

and the life expectancy at birth, i.e., the average number of years of life of the person in the group, is

$$E = \frac{T}{N} = \frac{s}{1-s} * \text{Sum } i(1-s)**i. \qquad (4)$$

Finally, in Ref. 19 it is shown that

$$\text{Sum } i(1-s)**i = \frac{1-s}{s**2}. \qquad (5)$$

Using Eq. (5) in Eq. (4) and with Eq. (2), one obtains Eq. (1).

REFERENCES

1. Ahmad, P. I., A. Kolker, and G. V. Coelho. (1979) "Towards a New Definition of Health: An Overview," in *Toward a New Definition of Health*, (eds.), P. I. Ahmad and G. V. Coelho, New York: Plenum Press, p. 7.
2. Brook, R. H., J. E. Ware, Jr., A. Davies-Avery, A. L. Stewart, C. A. Donald, W. H. Rogers, K. N. Williams, and S. A. Johnston. (1979) "Overview of Adult Health Status Measures Fielded in Rand's Health Insurance Study," *Medical Care* 17:7, Supplement.
3. Cardus, D., and R. M. Thrall. (1977) "Overview: Health and the Planning of Health Care Systems," *Preventive Medicine* 6:134.
4. Cheng, M. K. (1973) "The G Index for Program Priority," in *Health Status Indices*, (ed.), R. L. Berg, Chicago: Hospital Research and Educational Trust.
5. Chlang, C. L. (1965) *An Index of Health: Mathematical Models*, Washington, D. C.: Government Printing Office.
6. "Constitution of the World Health Organization." (1947) *Chronicle of the WHO* 3.
7. Dubos, R. (1965) *Man Adapting*, New Haven: Yale University Press.
8. Dunn, H. (1961) *High Level Wellness*, Virginia: R. W. Bearry, Ltd.

9. Fried, M. (1969) "Social differences in mental health," in *Poverty and Health: A Sociological Analysis,* (eds.), J. Koda, A. Antonovsky and I. K. Zola, Cambridge: Harvard University Press, p. 113.
10. Hall, H. S., and S. R. Knight. (1948) *Algebra Superior,* Chapters V and XXI, Utena, Mexico City.
11. Hofer, B., and L. Raymond. (1977) "La mortalite d'une cohorte de patients psychiatriques traites ambulatoirement a Geneve," *Medicine Sociale et Preventive* 22:161.
12. Jazairi, N. T. (1976) *Approaches to the Development of Health Indicators,* Organization for Economic Cooperation and Development, Paris.
13. Koranyi, E. K. (1977) "Fatalities in 2,070 psychiatric outpatients," *Archives General Psychiatry* 34:1137.
14. Levy, E. (1977) "The Search for Health Indicators," *International Social Science Journal* 29(3):433–463.
15. Parsons, T. (1972) "Definitions of Health and Illness in the Light of American Values and Social Structure," in *Patients, Physicians and Illness; a Source Book in Behavioral Science and Health,* 2nd ed., (ed.), E. G. Jaco, New York: Free Press, p. 107.
16. Purola, T. (1972) "A system approach to health and health policy," *Medical Care* 10:373.
17. Stewart, A. L., J. E. Ware, Jr., R. N. Brook, and A. Davies-Avery. (1978) *Conception and Measurement of Health for Adults in the Health Insurance Study: Vol. II, Physical Health in Terms of Functioning* R-1981/2- HEW, Santa Monica, California: The Rand Corporation.
18. Tefft, B. M., A. M. Pederson, and H. M. Babigian. (1977) "Patterns of death among suicide attempters, a psychiatric population, and a general population," *Archives General Psychiatry* 34:1155.
19. World Health Organization. Consejo Ejecutivo 67a Reunion *Preparaçion de Indicoadores para Vigilar los Adiantos Hechos en el Lorgo de la Salud para todos en el Ano 2000* Informe del Comite del Programa del Consejo Ejecutivo EB67/13/Add.1 12 de Diciembre de 1980 (mimeo).

CONCEPTUAL FRAMEWORK FOR THE PLANNING OF MEDICINE IN DEVELOPING COUNTRIES

Peter Newman

I. INTRODUCTION

In a paper that is a paragon of common sense and wide experience applied to problems of medical planning for underdeveloped countries, Abel-Smith (1) has recommended that: "The economist can best assist in health planning and the training of health planners at this moment in time not by the elaboration of intricate theories and the development of **models**, but by inculcating his way of thinking into day-to-day administration." (p. 10.) This advice can only command the warmest agreement if it is taken to mean that all but a small percentage of economists' time in this field should be taken up with trying to improve existing patterns of resource allocation to medicine (increasing the growth of this activity, reducing the growth of that one) rather than in attempts to force the whole allocation into a quite new pattern imposed by an optimization model that is perhaps compounded in equal parts of poor data, simple-minded specification and inappropriate computation.

Such a worthy task of day-to-day improvement not only fails to preclude but in fact requires the investment of some hard analytical labor in the construction of planning schemes that are sufficiently sophisticated to take account of the peculiarities and opportunities offered by problems of health. For in trying to improve the existing resource allocation it is implicit that the suggested change really is an improvement, which implies in turn that there are some acceptable criteria by means of which the change is judged to be for the better. Such criteria can only be formulated properly, however, in the light of what the society is trying to achieve and of the mechanisms and resources it has, the constraints it faces, in trying to do so. The careful articulation of all these aspects is just what is meant by a framework for planning.

The problem posed is that of how to arrive at a systematic program of health-related projects in any developing country. Here *systematic* must be taken to mean within the context of the economy and society as a whole, not just within the health (still less the medical) sector itself. A country may have a well-designed set of projects for urban curative medicine which, taken on their own terms, constitute a model of systematic, interrelated health care delivery modules; yet within the context of the whole country's needs they may be so inappropriate as to seem surrealistic. So the problem comes down to how health and its attendant services fit into the general framework of economic and social planning; and that is the subject of this paper.

It is well known that in usual planning practice health is made to fit either by being placed toward the end of the queue or being treated as a thing apart from the more easily analyzable sectors, its allocations of resources being governed by such forces as past tradition or the political strength of its ministry. In planning documents the section devoted to health frequently opens with conventional pieties about its importance for well-being and then launches into detailed descriptions of the various projects, usually without much discussion of why they and not others were chosen, nor of why they were designed in the way and at the levels that appear. Of course there are exceptions, in which genuine efforts are made to achieve some kind of explicit rationality (e.g., Taiwan, Tanzania) but this is usually within the field of health itself, rather than presented as part of the general economic and social plan.

A fairly careful review of the literature on health economics in developing countries [aided considerably by the annotated bibliography compiled by Williams (17)] shows that little constructive work has yet been done on the incorporation of health factors as an integral part of economic planning. In part this must have been due to the difficulty of treating "health" as an operational concept, in the same way as, say, gross national product is an operational concept. There have been several attempts to achieve such operationalism [e.g., Sullivan (16) and Fanshel and Bush (7), which also contain good reviews of the relevant literature] but none of them has proved satisfactory, at least in this context. In the present paper a definition of *state of health* is offered which seems to have some promise as a workable idea. The main innovation here is to give up the premise that *health* can be satisfactorily captured by means of a one-dimensional index (as GNP tries to capture the idea of total output), and to be content instead with a multidimensional representation. This representation is of course more complicated to work with than a single number, but for the uses to which it is put it proves quite tractable.

The structure of the paper is as follows: In the section following this Introduction an attempt is made to explain the role which health is assumed to play in individual and in social welfare. The early subsections here may irritate in their stress on the appropriate meanings of words, but some common understanding is necessary if analysis is to proceed. A central feature of this section is its insistence that the state of health must be included as an explicit argument in the utility functions of both individual and society. This runs counter to much current practice (if not current theory) which usually attempts to measure only the production benefits of health programs.

The third section sets out the analytical framework itself, first giving an extended discussion of the various definitions of *state of health* that are used. The planning model which underlies this section is presented in full only in the Technical Appendix, but the main text and the Appendix are in fact dependent on each other, the "literary" discussion of the text being just as important to the logical structure of the ideas as the more mathematical treatment of the Appendix.

Throughout, a serious attempt has been made to make the argument intelligible to those persons interested in the planning of medicine who have not received much specialized training in economics. At times the resulting draft may be tedious to economists, but such passages are hopefully few, short, and far between. Space limitations and the desire to present a closely reasoned argument have kept illustrative examples down to a minimum, and for similar reasons the references have been limited in number.

It would have been possible to use the analytical framework developed here to present a synthesis of the existing literature on the planning of medicine in developing countries. If the framework holds against criticism a review of the literature organized along these lines might prove useful; but at present it seemed presumptuous, especially since there is no lack of surveys of work in medical economics related to the present problem [e.g., Baker (3), Klarman (8), Mushkin (10), Williams (17)].

A final disclaimer: It could be argued that problems of allocating resources to family planning also belong in the category discussed here—it is often difficult in practice to distinguish such expenditures from those on maternal and child health. There seem to be no difficulties which in principle would prevent the extension of the ideas presented here to resource allocation in fertility control but such an extension is not attempted, in order that the argument may be kept confined to what is already a wide and difficult subject.

II. THE PLACE OF HEALTH IN THE GENERAL WELFARE

A. Problems of Definition

Poor Richard's recipe for how to become "healthy, wealthy, and wise" does not convince everyone, but at least his objectives were internally consistent. Today's usage of "health, education, and welfare" is not, even when welfare is interpreted not as doles for the poor, but as general well-being. Each word of Franklin's phrase refers to a condition or state for the individual while the modern version contains one word, *education,* which unlike the other two refers not to a state but to a process, a learning process, by which transition is made from a state of lesser knowledge to one of greater.

One guess as to why this anomaly occurs is that in the study of educational planning it has been found fruitful not to discuss the attainment of wisdom in general, but instead to concentrate on the achievement of specified levels of skill in specified areas. In the same way it is generally thought useful to identify education with the more formal processes of learning, as typified in schools and colleges, even though we know that a large part of learning occurs wholly outside such formalized procedures.

The scaling down in aim and scope implied by the semantic change from *wisdom* to *education* means that some of the larger philosophical issues concerning the purposes of education get left out; such restrictions are not made without cost. But at least they help in formulating problems of policy tight enough to stand some chance of being answered with the data and the skills at hand. Provided there are always critics available, ready and willing to remind the policymaker of the gaps between being educated and attaining wisdom, little is lost by such methodological restraint.

In the planning of medical care there appears to be no such general agreement on the usefulness of restricting the domain of analysis. Many studies in this area apparently still find it needful to preface their substantive work by consideration of the meaning of *health,* even though that concept is just as loaded with difficulty, just as shot through with cultural and technological relativism, just as wide open to the smuggling in of value judgments, as the concept of wisdom itself. Which is not to say that the meaning of health should not be discussed, only to plead for sharp separation of the levels of discourse involved.

It is widely and explicitly recognized that transitions from one state of health to another can be and often are brought about by means—such as changes in the provision of food, water, clothing, and shelter—other than the application or withholding of medical care. But whereas in educational planning the distinction between (formal) education and learning is usually kept reasonably clear and constant, in medical planning the corresponding distinction between the process of medical care and transitions between states of health frequently is not. Indeed, it is odd that ordinary English usage lacks any word, analogous to *learning,* to refer to transitions between states of health. One may speculate why; perhaps it has to do with the fact that over the span of human life knowledge tends to increase, health to decrease. Whatever the reasons the absence of such a word makes it easier to slip into error in considering changes in medical care as necessarily the major, perhaps the only, cause of changes in the state of health.

This enquiry follows the pattern set in educational planning and refers to states of health (to be suitably defined later) and transitions between them, rather than to "health" itself; and is careful to regard medical care as only one among many processes, all consuming scarce resources, by which such transitions are made. In the interests of compact and unemotive terminology the value-laden term *medical care* is not used (who can be against care?) and the shorter, more neutral word *medicine* use instead.[1] The word *health* by itself is employed as little as are the words *wisdom* and *knowledge* in educational planning. Such words are too valuable to be emptied of meaning by being used to cover too many situations, too many meanings.

B. Health and Wealth

Turn now to correspondences between health and wealth. *Wealth* needs definition. For any individual some of his assets can be sold on more or less organized markets while his other assets, because of natural or man-made restrictions (such as the prohibition against selling oneself into slavery), cannot be sold outright but only leased. The first type consists predominantly of physical assets and net claims on the assets of other people, and may be identified as those items that are reckoned into an individual's estate at death. The second type consists essentially of those assets that inhere in the individual as such and which vanish at death.

Wealth normally refers only to the first, marketable kind of asset and even then has two distinct meanings, denoting either the whole inventory of goods and claims which constitute the individual's holding of such assets or the monetary value of that inventory, each item being entered at its market price. The latter meaning will not be used here, *wealth* referring only to the list of physical amounts of the various commodities and claims comprising marketable assets.[2] The other asset type, namely those—such as labor skills of many kinds—whose services can be sold but not the assets themselves, will be called *human wealth.* Like wealth, it is to be thought of as a vector of certain kinds of assets, each measured in physical and not monetary terms. Given the genetic makeup of the individual and his general environment, his human wealth at any time is dependent mainly on his states of knowledge and of health at that time, these two complex entities affecting the levels of the various human skills that he can achieve. The education and medicine he has received, accumulating in complex ways over his lifetime up to the present, are two (but only two) of the factors which determine his current states of knowledge and health. There are many others factors involved. For example, it is well known that early nutrition influences not only current states of health, but also current levels of intelligence, and hence current states of knowledge.

Moreover, since the individual's human wealth vector is partly determined by his state of health, his supply of the services of that wealth will be dependent not only on the market prices of those services (such as wage rates of various kinds) but also on his state of health. In turn, the funds obtained by the sale of those services, together with funds secured by selling items from his wealth vector, are used to purchase goods as well as other services. These purchases can then be allocated to personal consumption, to the accumulation of wealth, and to the formation of human wealth.

In brief, the individual's state of health, like his state of knowledge, is a direct determinant of his human wealth and an indirect determinant, through the appropriate system of markets (which might be quite primitive), of his wealth. Patterns of causation in the reverse direction, from wealth to health, will be taken up in Section III.E.

C. Health and Welfare

The level of an individual's welfare is usually taken as measured by the value of his utility function, which is simply a representation of his scale of preferences. It is assumed that he can compare any pair of alternatives in the set of all possible choices open to him and that such pairwise comparisons are consistent from pair to pair. The utility function then serves only to reflect these preferences, in the sense that the utility attached to alternative A is greater than that attached to alternative B if and only if A is preferred to B. *Utility* is only a word, a word that is used to denote the preference labeling of each alternative and which does not in itself imply the existence of an entity such as pleasure or satisfaction, a matter the theory treats agnostically. All this is standard economics and the only question at issue here is the precise nature of the set of alternatives open to the individual.[3] Specifically, should his utility function include among its variables his state of health, along with the usual vector of goods and services?

This seemingly straightforward question conceals many subtleties and indeed contains within it several of the major issues that seem to divide economists and physicians in their approaches to the planning of medicine. So it is worth an extended answer.

A first stumbling block is that the question is meaningless unless it is possible to have quantitative measurement of the state of health. It is at best useless to put any variable into the utility function unless that variable has a well-defined cardinal measure, and the fact that practitioners of economics have not always observed this working rule does not make it any the less mandatory.[4] Measurement of the state of health does not have to take the form of a single number, for example, an index of "health status." A vector of numbers may be required to deal with the complexities inherent in the concept, and although necessarily more complicated to work with than a single number it may still be quite tractable. Such representations of the state of health will be discussed in the next section; for the moment it is simply assumed that unambiguous measurement is possible, at least in principle. Whether in a particular situation the data available are sufficient to permit actual measurement to be made is a different question, about which little that is useful can be said a priori.

Suppose that an individual had his state of health improved without suffering any reduction in his command over goods and services. Surely such a person would consider that his welfare had improved as a consequence, which is to say that his level of utility had increased. To argue otherwise would seem to carry not only the obvious implication that he does not value a better state of health for its own sake but also the less obvious implication that his preferences as between vectors of commodities are independent of his state of health. But since that state obviously affects his judgments of some commodities (such as food and drink and active recreation) more than it does others (such as books and music and television), his preference ranking of commodity vectors will be conditional upon his state of health, which must therefore enter into his utility function in some form.

Yet precisely this conclusion is denied, at least implicitly, by those cost/benefit studies of medical programs that focus on the "production" aspects of medicine to the exclusion of its "consumption" aspects, i.e., on the consequences of improvement in the state of health for individuals' human wealth, rather than its direct impact on their welfare. These studies typically estimate the effects of such an improvement on mortality, morbidity, and debility, and thence the resulting changes in the supply of various types of human skills and possibly of other factors of production as well. These changes in input supply are then translated into potential changes in the stream of outputs, by means of production functions whose degree of realism varies from study to study. Suitable valuation of such changes in the output stream, together with what one hopes is an appropriate discount rate to reduce them to changes in present values, usually complete the benefit side of the calculation; the cost side is normally straightforward, care being taken to calculate the relevant opportunity costs wherever possible.

Such estimates are quite useful but considered as measurements of the total benefits of an improvement in the state of health they are seriously incomplete. As Schelling [(15) p. 134] has remarked in a closely related context (that of computing how much it is worth to save the life of a child) this kind of approach is "At best, ... the way a family will deal with the loss of a cow, not the loss of a collie. Though children are not pets, ... they are more like pets than like livestock, and it is doubtful whether the interests of any *consumers* are represented in a calculation that treats a child like an unfinished building or some expensive goods in process" (italics added).

Often the preoccupation with production benefits is softened by acknowledgment that consumption benefits do in fact exist. But then as Klarman [(8) p. 164] points out, the difficulty of measuring those benefits is used as justification for ignoring them. The analytic structure of the cost/benefit methodology thus comes to be quite different from what it would have been if consumption effects had been recognized throughout as an essential part of the process.

Sometimes the emphasis on production benefits goes so far that only they are thought suitable to be called economic benefits [(1) pp. 6–7], so that economists tend to be identified with practitioners of this type of cost/benefit analysis. This is rather like rating a diva solely by her ability to act. Normally words should be servants and not masters, so whether the word *economic* is reserved for just one category of benefit or not is hardly a serious affair. But here it is necessary to insist on the *economic* validity of including the consumption benefits of medical programs, not merely as an afterthought to the "real" toughminded business of tackling their production effects, but as a central feature of the planning procedure. To do otherwise, to regard the direct welfare-improving effects of medicine as in principle secondary to its output-improving effects, is to imply that as far as the allocation of resources to medicine is concerned all agents in the economy act as profit maximizers rather than utility maximizers, a poor assumption that risks serious bias and hence inefficiency in that allocation. This elementary but fundamental point needs to be driven home because, as a scrutiny of the relevant literature reveals, failure by medically trained planners and economists alike to appreciate its full significance bedevils attempts to secure compatibility of their different approaches to the planning of medicine.

Thus, the view taken here of the role of economics in medical planning is both more modest and more ambitious than that which concentrates on production effects alone. More modest because it allows that decisions on medical expenditures are, indeed should be, made in the light of factors additional to the "economic" (i.e., production) benefits expected; more ambitious because it claims that even these problems of expanded choice are economic problems, in the sense that efficient allocation of resources is desirable in achieving the specified objectives, whatever they may be.

A more subtle argument against including the state of health explicitly in the utility function runs as follows. Suppose that apart from certain parameters, such as general environmental and genetic factors, and past states of health, one's present state of health is determined completely by the goods and services that are utilized, i.e., by the commodity vector itself.[5] Then the only genuine choice variables in the utility function are the commodity levels, since once these are chosen the state of health is determined. So the utility function can be written in "reduced form," in which both the influence of the state of health on welfare and its dependence on the commodity vector are embodied in the structure of the utility function itself. Put another way, to the extent that the individual can control his state of health by the absorption of goods and services (including medical goods and services), to include it explicitly in the utility function is unnecessary.

Unnecessary—but desirable, for at least two reasons. First, while its inclusion under these conditions would not be logically wrong, merely otiose, it would have the advantage that the dependence of the welfare level on the state of health would be shown explicitly, whereas in the reduced form (in which utility is expressed solely in terms of commodities) these welfare aspects are mixed up, perhaps inextricably mixed up, with the more technological dependence of the state of health on the input of commodities; and so information is lost. Secondly, one's state of health is often seriously affected by variables outside the control of the individual but still subject to decisions by other agents in the economy, both private and public. This is obvious in the case of communicable diseases but is also true for those environmental variables, such as water supply and sanitation, that are often the responsibility of government. Here again, although the state of health need not enter the utility function explicitly since it can always be "solved out" in terms of the commodity vector and the external variables, information is conserved by leaving it in.

So far the emphasis has been on individual welfare. Difficulties in the general concept of group or social welfare form a well-known part of economics and will not be discussed in this analysis, which is concerned only with the special role of health in social and economic planning. A first problem is how the social state of health is to be described and how it is related to the individual states. It could be argued that the social state is simply the whole list, or vector, of individual states. But this is unwieldy, quite apart from the impossible requirements for data that it implies. With such a formulation, how is one to define an improvement in the social state of health—in the Paretian terms of improvement in one or more individual states, with no other individual state being made worse? This criterion would be useless for any rearrangement of resources in medicine which reduced anyone's level of medical care, no matter how many other people were thereby made better off.

One needs to develop a concept of a social state of health whose data requirements are less mountainous and whose range of application less minimal. Just what kind of aggregate or average of the individual states this might be can hardly be fruitfully discussed until there is an adequate operational definition of the individual's state of health; and that is a problem for the next section. Whatever the definition of social state of health may be, the grounds for including it explicitly in the social welfare function (or its surrogate, the planners' objective function) are essentially similar to those already discussed for the case of the individual, and just as compelling.

III. A FRAMEWORK FOR MEDICAL PLANNING

A. States of Health: A Set of Definitions

It is high time to give concrete interpretations of the abstraction "state of health." For convenience it is easier to work not with the state of (good) health but with the complementary idea of the state of ill health. The first assumption made is that ill health, or illness, can be caused only by one or more of a finite set D of diseases of all kinds—hereditary, communicable, parasitic, psychosomatic, degenerative, and so on. To make the set D useful it must be as complete as the current state of medical science and technology can make it, although the degree to which one disease is distinguished from another may well depend on the objectives in view and the data at hand. Moreover, geographic and other restrictions on the region for which the planning is to be done may permit some diseases to be omitted completely, e.g., malaria in Sweden. Often it may be appropriate to regard different degrees of severity of the same disease (e.g., tuberculosis) as different diseases, while a group of diseases (e.g., diseases of the genitourinary tract) may sometimes be regarded as one. Although such problems of classification may be quite severe, they appear to be basically no harder than a set of problems with which economists are quite familiar, those associated with the classification of commodities.[6]

Although D may well contain diseases that are essentially syndromes, i.e., collections of symptoms whose etiology is unknown, it is not to be thought of as containing those symptoms themselves. Thus observational data such as readings of temperature, pulse rate, respiration, weight, and height are not included in D, although they play an important role which is elucidated below.

It is next assumed that given (i) any individual, say the ith, out of the population of P persons constituting the region or nation under discussion; (ii) any disease, say the jth, out of the finite number q of diseases in the set D; and (iii) any time t, measured from some arbitrary origin, it is possible to assign a probability $d_{ij}(t)$ that the individual i has disease j at time t. The vector of q such numbers $d_{ij}(t)$, where $0 \leq d_{ij}(t) \leq 1$ for each j in D, is then defined to be the *state of health of individual i at time t*, and is written $d^i(t)$.[7]

If for any disease j the numbers $d_{ij}(t)$ are summed over the whole population and then divided by P, the resulting *number*, call it $d_j(t)$, will obviously lie between zero and one, since $\Sigma_{i=1}^{i=P} d_{ij}(t) \leq P$. So for any disease j the member $d_j(t)$ may be regarded as the probability that a representative number of the population has that disease at time t. The vector of q such numbers $d_j(t)$, one for each disease in D, is then defined to be the *social state of health at time t*, written $d(t)$. Provided that P is a reasonably large number and that the judgments of individuals' probabilities embodied in the $d_{ij}(t)$ are not biased estimates of their "true" probabilities, then a good estimator of $d_j(t)$ for any j is that proportion of the population, call it $d_j^*(t)$, that actually has disease j at time t. But it is unlikely that the balance between resources available and the size of the population will permit the complete enumeration and examination required for an accurate measurement of $d_j^*(t)$, so that in practice one would have to fall back on sample estimates $\hat{d}_j^*(t)$ of $d_j^*(t)$, of varying degrees of reliability.

Although most of what follows will be concerned with social rather than individual states of health, it is important to give some discussion of the ambiguities that surround the present definition of the individual's state, both because the definition of the social state rests upon it and because in doing so some light is thrown on certain phenomena which have puzzled medical planners.

A first clarification to be made is that nowhere was it assumed that the probabilities $d_{ij}(t)$ are independent of each other, either from disease to disease, or from person to person, or from time to time. If anyone has severe malarial fever, for example, he is unlikely to have syphilis (indeed, the former used to be a "cure" for the latter). Similarly, the probability that a child has chickenpox is not independent of the probabilities that his siblings and his school fellows have the same disease. Again, the probability that a man may suffer from heart disease at age 45 (and hence at a certain time t) is not independent of the probability that he may suffer from the same disease at age 44 (time $t - 1$).

A much more serious ambiguity is that which was introduced, deliberately, when it was assumed that "it is possible to assign a probability. . . ." Because of the general statistical dependence among the $d_{ij}(t)$ this assumption would be better expressed: "it is possible to assign a probability vector $d^i(t)$. . . ." But even in this revised formulation there is still a glaring imprecision. For any individual i, just who is to "assign" the vector $d^i(t)$? There are at least two answers to this question. The first—which has been tacitly adopted already in the definition of D and hence of the individual's state of health—is that $d^i(t)$ is assigned technologically, as it were, according to the current "best practice" of medical science.

More specifically, it is assumed throughout that $d^i(t)$ represents a consensus of the judgments of the medical profession concerning the situation of individual i, given that the physicians have access to such information as (i) observations on his current physical and mental state (i.e., the symptoms mentioned in the second paragraph of this section); (ii) his past medical history, including hereditary factors; (iii) his past and present economic position; (iv) his level of education; and (v) the general social and natural environment in which he is placed. Symbolizing such pertinent information by $h_i(t)$, where $h_i(t)$ is an element of some more or less abstract information set, the probability vector $d^i(t)$ then appears as a vector of *conditional* probabilities $(d_{i1}(t), d_{i2}(t), \ldots, d_{iq}(t)|h_i(t))$, conditional on the information embodied in $h_i(t)$. Presumably the confidence with which the assignment $d^i(t)$ is made increases as the volume of information in $h_i(t)$ increases, which leads to the important applied problem in statistical decision theory, not discussed here, of how much it is worth to increase the reliability of $d^i(t)$ by using resources to increase $h_i(t)$.

This interpretation of the vectors $d^i(t)$ is obviously fuzzy at the edges (like most definitions in social science) and perhaps at the center, too; for some diseases and for some individuals it might be difficult to reach any medical consensus at all.[8] Over time $d^i(t)$ might change not only because the individual changes but also because medical opinion and practice change. Nevertheless, taken sufficiently broadly it seems an interpretation with definite meaning, indeed one which is observable under ideal conditions. For ease of reference call it the *diagnosed* individual state of health, with a corresponding meaning for the diagnosed social state. It gains sharper focus by being contrasted with the second answer to be discussed, viz.: that the assignment of the probability vector $d^i(t)$ is done by individual i himself.

B. Perceived Status of Health

To see that the theory must also make room for a second, "subjective" interpretation of the probabilities of disease observe that for social planning it cannot be the case that the diagnosed social state of health $d(t)$ can enter into the planner's objective function, for it is not directly observable except under the absurdly ideal conditions which permit everyone to achieve a diagnosed individual state of health. For practical purposes it must be the vector $d^*(t)$ of estimates of the q proportions $d_j^*(t)$ which enters as one of the ingredients in the social welfare function. In precisely the same way individual i cannot hope to achieve a diagnosed state of health without expenditure of a quantity of resources that may be quite beyond his reach; not every Indian peasant can get to the Mayo Clinic. There must therefore be another kind of "state of health" which enters into his utility function and which helps determine, among other things, his expenditure on medical goods and services; such a state must incorporate elements of personal judgment to a greater or lesser degree.

Let $d_{ij}^*(t)$ denote individual i's own judgment of the probability that he has disease j at time t, where j now belongs not to D but to a finite set D_i^* consisting of q_i diseases and syndromes (possibly including some that have no counterpart in established medicine) that the individual considers relevant to his well-being. The vector of q_i such numbers $d_{ij}^*(t)$, one for each disease in D_i^*, is then defined to be *the perceived state of health of individual i at time t*, written $d^{i*}(t)$.

Three aspects of these perceived individual states need comment, even in this preliminary enquiry: (i) How does the individual come to judge his perceived state of health? (ii) How is the perceived individual state related to the diagnosed individual state? and (iii) How does the perceived individual state enter into the individual's utility function? The discussion of these three aspects will be kept quite brief, but a more extended analysis would seem to be a promising topic for future research.

Concerning (i) it is clear that $d^{i*}(t)$, like $d^i(t)$, must be interpreted as a vector of conditional probabilities $d_{ij}^*(t)$, each conditional upon that amount and type of information in the possession of individual i at time t which he considers pertinent to his judgment. How he comes by that information, and how he determines which information is relevant and which not, are matters that are difficult to say much about *a priori*. For the individual who seldom consults medical practitioners $d^{i*}(t)$ must be formed from knowledge garnered elsewhere, from the beliefs and practices prevalent in his own culture, from personal untrained observation of what he judges to be relevant symptoms, both his own and those of people in his immediate environment, and from his own intelligence, experience, and imagination.[9] Many people may rely heavily on the estimated vector $\hat{d}*(t)$ of aggregate proportions of the various diseases in forming their own $d^{i*}(t)$. More commonly, perhaps, estimates of such proportions for relevant groups of people rather than for the whole society may be used, as when an inveterate smoker of cigarettes, computing his chances of having lung cancer, looks at the incidence of that disease among males of his own age who are heavy smokers rather than that among the population at large.

In any event such personal probabilities are likely to be enveloped by considerable uncertainty, so that $d^{i*}(t)$ may be looked on as being the mean of a multivariate probability distribution with high variances. This in turn may induce enormous personal anxiety—it is common for people to suffer more from uncertainty about what ails them than from the ailment itself. In such circumstances medical advice is sought and bought and the medical adviser's judgment about the probability vector $d^{i*}(t)$, in those cases where he chooses to reveal it, may be assumed by the individual as his own. It is unlikely, however, that the adviser will provide information on more than a few of the diseases in D_i^*, those few being mainly those which correspond to the anxieties most keenly felt by the patient, so far as the latter is aware of them.[10] Moreover, even when the information is obtained the individual is still left with uncertainty about the appropriate course of action to follow in order to reduce those probabilities that are unacceptably high (e.g., those close to 1), and so is likely to turn again to the medical adviser for help. At each stage—diagnosis, prescription, and prognosis—the individual's uncertainty is likely to be quite high and therefore a premium is placed on securing trustworthiness (if not reliability) in the information and advice dispensed by the physician; the consequences of this need for the organization of many branches of the medical profession have been explored by Arrow (2).

Concerning (ii) it is clear that an individual's perceived and diagnosed states of health will coincide only by chance. More generally, given any individual i and any disease j there is no reason to suppose that $d_{ij}*(t)$ bears any systematic relation to $d_{ij}(t)$. They may be expected to differ for two reasons. First, the nature and the volume of information on which $d_{ij}*(t)$ rests is likely to be quite different from that on which $d_{ij}(t)$ is based. Secondly, the modes of inference from any given information to the assignment of probabilities are likely to be quite different for the untrained layman than for the trained physician.

Perhaps the only general statement possible is that the relevant set D_i^* of diseases entering into the individual's perceived state is likely to be smaller than D, since many diseases known to modern medicine may be unrecognizable by the layman; of course the reverse situation is possible if there exists a number of "folk" diseases unknown—perhaps because purely imaginary—to the medical profession. But note that for analytical purposes it is not necessary for the individual to know the name and the etiology of each disease in D_i^*, only that he can distinguish one syndrome from another. Thus he may not perceive schistosomiasis as a disease, merely its symptoms as some of the natural conditions of life like being tired or feeling hungry; but he may still be able to perceive those symptoms as a separate phenomenon and to attach a probability to having them.

This leads to the third aspect of perceived states of health, namely, how they fit into the individual's utility function. In general, as the last section discussed in some detail, this function will include the individual's state of health among its argument variables. It has also been assumed that it is the perceived rather than the diagnosed state that should be included, although a case might be made for including the latter as well. Obviously, the individual's level of utility is affected not only by the magnitude of the probability $d_{ij}*(t)$ but also by the disease to which that probability refers; thus if disease 1 is the common cold and disease 2 is diabetes mellitus, most individuals would consider a $d_{i1}*(t)$ of 1 a much less unpleasant state of affairs than a $d_{i2}*(t)$ of 0.1.

Whichever probability vector one uses, perceived or diagnosed, it has been implicit so far that the whole of it should enter the utility function. But it is quite possible that some of the diseases in D_i^* (or in D) may not appear in that function. For a disease like schistosomiasis (bilharzia) under conditions of high endemicity, it may not occur to the individual that a world without its symptoms is possible so he may make no allowance for that disease in his preference ordering, even though quite aware of the relevant perceived probability.[11] If the individual's demand for goods and services follows the predictions of the theory of consumer's behavior, then such an absence of a disease from his utility function implies that he will have no demand for any medical (or other) goods and services that are aimed exclusively at alleviating that disease. This is of course not the whole story. In fact there may not exist any commodities with such highly specific properties, so whether there is demand for them or not is a superfluous question. Even if there were such goods, if they had any effectiveness (even just as placebos) then it is likely that the individual's perception both of the range of probabilities for the disease and of its importance for his utility level would be quite different. For example, to the extent that malaria can be made subject to personal control by means of chemotherapy and other measures at a cost within the range of at least some of the richer peasants and urban workers, it becomes a matter for economic decision making [cf. Barlow (4, pp. 7–8)].

So far no meaning has been given to a perceived social state of health analogous to the perceived individual state. Since the latter is intrinsically a subjective concept and since there are severe and possibly insurmountable difficulties in such concepts as the social will or a consistent social preference ordering, it is not immediately obvious what a perceived social state might mean. Who or what is to make the perceptions for such a state? Neither is it obvious that such a concept is required, either theoretically or operationally. Nevertheless it proves convenient (to put it no higher) to define *the perceived social state of health at time t* as the vector, call it $d^*(t)$, of the q empirical proportions $d_i^*(t)$ that were introduced in Section III.A., which show for each disease j the fraction of the population that has it at time t. As discussed, in practice one has to rely on sample estimates $\hat{d}^*(t)$ of $d^*(t)$ [and of the diagnosed social state $d(t)$ as well] since the complete enumeration needed for accurate measurement of $d^*(t)$ is seldom available. Note carefully that according to this definition $d^*(t)$ is in no sense a group average of the $d^{i*}(t)$, as $d(t)$ is an average of the $d^i(t)$; the perceptions involved are quite different.

The relation between the perceived social state $d^*(t)$ and the perceived individual states $d^{i*}(t)$ is indeed loosely analogous to that existing in the theory of consumer demand between the market demand functions and the individual preference orderings, or utility functions. In each of these two theories all the concepts are in principle observable but in both cases only the aggregate concepts are observable in practice. Nevertheless in each case the individual concepts, although not directly operational, are needed to construct a theory for the aggregate which is reasonably well rooted in a consistent theory of the behavior of individuals.

In most work on indices of health status, as reported and developed, e.g., by Sullivan (16) and Fanshel and Bush (7), the emphasis is placed not on developing consistent planning frameworks but more on devising practical measures by which to evaluate performance, so little attention appears to be paid to such theoretical concepts as perceived individual states of health.

An approach broadly favored by all three authors mentioned is to measure health status (usually as a single number rather than a vector) by means of the ability to perform all of a set F of certain physical functions, such as walking, doing certain kinds of manual tasks, etc., rather than by means of morbidity as done here. Hence in such a scheme inference is made either from a set of symptoms S to the set of diseases D and thence to the set of functions F; or directly from S to F; or by direct observation of F itself.

This approach also has an analogy in the theory of consumer's behavior, where Lancaster (9) has argued that the consumer's utility depends essentially not on vectors of consumption goods (which correspond to our states of health) but on vectors of characteristics of goods (corresponding to the physical functions described above). Consumption goods then only give utility to the extent that they produce characteristics, just as in the functional approach diseases are apparently disfavored to the extent that they impair functions. Although for some purposes Lancaster's theory is brilliantly right, as a general theory of consumer demand it lacks, precisely, generality; and a similar remark would seem appropriate to the situation in health.

C. Preventive versus Curative Medicine

A possible structure of personal preferences is this: suppose that in judging the importance of a disease to his well-being the individual looks not at its subjective probability $d_{ij}^*(t)$ in absolute terms, but at that probability compared with the corresponding probability for some group of people among whom he lives. For simplicity of argument suppose that such a group is the nation itself, for which the vector $d^*(t)$ is relevant. To make matters even simpler assume that $D_i^* = D$ and that the comparison takes the form of evaluating the difference between the perceived individual state $d^{i*}(t)$ and the (universally) perceived social state $d^*(t)$ so that only the vector of differences, i.e., $d^{i*}(t) - d^*(t)$, enters individual i's utility function. This structure would "explain" the absence of a disease from the utility function for if disease j were completely endemic $d_{ij}^*(t)$ would tend to coincide with $d_j^*(t)$, so that $d_{ij}^*(t) - d_j^*(t)$ would be very small; in the extreme case where the two probabilities coincide the difference—and hence disease j itself—would not enter the utility function at all.

It is not pretended that actual preferences concerning disease are as starkly simple as this. Misery may love company but it is misery nonetheless. Any individual who had any of his perceived probabilities reduced to zero through benevolent outside action would almost certainly regard himself as thereby better off, even if none of the other people in his reference group was so treated; and that result would be in contradiction to the structure of preferences just described. All the same, it seems reasonable to assume that personal judgment concerning the importance of a disease to one's well-being would include as data not only the absolute level of its perceived probability, but also its level relative to some averaged group probability.

This last hypothesis seems relevant to the explanation of a persistent and puzzling phenomenon in poor countries. Very often it is easy to show that on any criterion one selects, be it reduced mortality, or morbidity, or debility, or increased output, or some other indicator, a dollar's worth of resources spent on one or more programs of preventive medicine (defined to include supplemental nutrition, sanitation, and water supply) yields better results than a similar amount spent on curative medicine. Yet there seems to be genuine popular preference for curative medicine, in the sense that political promises of increases in the supply of hospitals and other curative services apparently appeal more than do increased programs in preventive medicine, even when the coverage of the expanded curative program is still pitifully inadequate.

Where these conditions exist it is just not helpful for advisers merely to recommend a shift of expenditures toward preventive medicine, for as long as that is not how the individuals themselves see the matter such measures have little chance to succeed. What needs to be done is to understand as thoroughly as possible the reasons for this apparent paradox and then to harness such understanding into a program for medicine as a whole that will be politically effective as well as economically efficient.

Discussion of this paradox is often combined, sometimes even mixed up, with discussion of several other propositions that are analytically distinct not only from this one but also from each other, and which will not be examined here. Examples are as follows:

1. It is difficult to persuade medically qualified persons to work in rural areas.
2. (a) There is relatively greater provision of both curative and preventive medicine, especially the former, in urban areas than in rural. (b) As a consequence, migration from country to city is encouraged. (c) This migration is undesirable.
3. (a) Redistribution of income and wealth from rich to poor and (especially) from urban to rural areas is a good thing to do. (b) Appropriate differential expenditure by government on both preventive and curative medicine is an effective way to make such transfers of real income.
4. (a) The preponderance of curative over preventive medicine in the training of doctors leads to an emphasis on the former, at all levels of medical administration. (b) This is one reason why the supply of those paramedical personnel who are principally engaged in preventive activities is often inadequate, both in number and quality.
5. (a) Local doctors often receive training in rich countries and so become especially knowledgeable about the diseases prevalent in those countries, such as degenerative ailments. (b) This leads to a misplaced emphasis on the importance of such diseases and of expertise in them, within the structure of the practice of curative medicine in the home country.

Such propositions deserve careful examination in enquiries more detailed than this one. In particular cases any one of them could be of overwhelming practical importance. But it should be clear that the issues involved refer to quite varying orders of enquiry, ranging from purely factual matters (1, 2a, 4a, 5a), to patterns of causation (2b, 4b, 5b), to theoretical assertions of the relative effectiveness of policy instruments (3b), to outright value judgments (2c, 3a). Clarity of thought and hence relevance of policy requires that separate matters be kept separate.

Return now to reasons for the original paradox. A first possibility has been hinted at already. If an individual's well-being is most affected by those diseases whose perceived probability of occurrence at any time diverges sharply from the group average, then he will tend to favor personal cure rather than communal prevention. Such ailments are not likely to be the communicable nor parasitic diseases (which by their nature tend to affect the group as a whole) but hereditary, degenerative, or psychosomatic in character, or to be acute episodes like appendicitis. Except in modernized societies the occurrence of this kind of sickness may appear to the individual as a judgment against him, unlike illness that affects the whole community and so can be taken as a fact of life. The former diseases may therefore give rise to much greater anxiety than the latter and so it becomes natural for the individual to seek the personal attention and reassurance possible with curative medicine, either in the form of modern hospitals and doctors or in the form of indigenous practitioners of herbal and psychosomatic medicine.

Secondly, even in poor countries the level of governmental provision of curative medicine for each patient is likely to be above that which the average individual can afford for himself. This is so partly because that provision takes on some aspects of a lottery, whose prizes go only to those who are "fortunate" enough to be sick; and the lottery aspects are even plainer in those almost universal situations where the supply of public curative services is quite inadequate, so that whether a sick person manages to secure them is a matter possibly of patronage or persistence, mainly of sheer luck. In such circumstances the appeal of public curative medicine takes on some of those features of income redistribution that make lotteries so attractive to so many people.

Uncertainty is present even in preventive medicine but now it takes the form not of a lottery with several substantial prizes and no obvious losers except public revenue, but of a situation where the failure of some members of the community to comply with the rules of the game (e.g., not being immunized or not maintaining sanitary facilities in adequate condition) may prejudice the effectiveness of the whole operation. If the society lacks enough cohesion to provide either the formal or the informal mechanisms required to secure sufficient compliance for proper disease control, then the individual may not be foolish to opt instead for curative medicine, whose uncertainties—he may feel—are more within his power to influence than is the case with preventive medicine; it may be easier to camp in a hospital waiting room for two days than to persuade one's neighbor to dig a proper latrine.

It is not denied that there may exist much less rational reasons for individuals to prefer curative to preventive medicine. Consumers' ignorance of how diseases are transmitted, the propaganda and influence of those (such as hospital builders and administrators, most drug companies, many doctors, and perhaps some aid donors) who perceive their interests to lie in expansion of curative medicine, the inertia of a bureaucracy which already underplays the preventive side—all these may have a part in such misallocation of resources. But one does not need these hypotheses in order to explain the seeming paradox; sufficient rationale can be given without them.

The methodological point made here is quite general and of particular importance for medical economics in poor countries. Before condemning apparent misallocations of resources as due to ignorance, naivety, "lack of development,'' or worse, strong efforts should be made to find out whether the objectionable practices have a deeper, more logical justification and then to design policy that takes this more rational basis into account, should it exist. Only then will one get improvements that run with the tide, as it were, rather than struggling, perhaps ineffectually, against it.

D. Economic Roles for Health

The preceding subsection was a slight digression from the main theme, which is that of constructing a conceptual framework for medical planning. In this preliminary analysis no attempt is made to treat problems of health in a fully dynamic, general equilibrium setting, in which one analyzes how the optimal plans of households, firms, and government are or are not made consistent by means of the market mechanism, and possibly by the political mechanism as well. This certainly needs to be done at some stage, if a firm grasp is to be had of the interplay between personal and public planning, involving so-called externalities of many kinds. The chief problems here are not the technicalities of economic or mathematical analysis, but the broad one—which also affects the simpler model that is discussed below—of breathing sufficient empirical life into the abstract scheme, in particular the functional relationships involved, to make the conditions for optimization more than mere formalism.

Instead of this general treatment a simpler view is taken, of a system whose evolution is controlled by a central planning agency, equipped with a definite set of social preferences in which the social state of health appears alongside more commonly included variables such as the consumption of goods and services, including the services of human wealth. An example of such a planning model will be found in the Technical Appendix, appearing there not because of any substantial mathematical difficulty involved (the Appendix builds the model but does not run it) but because a reasonably precise description requires a proliferation of symbols that would disrupt the main text. The Appendix should be regarded as an integral part of this paper and the analysis below will refer freely to the rather simple structure of the model it contains. However, in the text the focus is not on that structure but on problems of interpreting those of its relationships that involve the state of health in an essential way. It is primarily as a vehicle for exhibiting clearly the logical structure of these health-involved relationships that the model is best considered, and not as a direct tool for planning in its own right.

Examination of the model in the Appendix shows that the state of health enters in four places. In their order of occurrence in the list of equations, although not necessarily in order of either conceptual or empirical importance, these four relationships are as follows:

1. The dependence of human wealth on the state of health and other variables [Eq. (5)]

2. The "production function for health," i.e., the dependence of the state of health on capital stock, consumption, and other variables [Eq. (6)]
3. The dependence of the state of knowledge on the state of health and other variables [Eq. (7)]
4. The dependence of social utility on the state of health and other variables [the maximand (8)]

Each of these patterns of dependence has been the subject of lively debate in the economics of medical planning. In what follows each of 1, 3, and 4 is reviewed but only briefly, either because it has already been discussed here or because not much can be said at this level of generality. However, problems connected with 2 have not so far been analyzed and are so important that they deserve more extensive treatment, even in this preliminary enquiry.

The problems associated with 1 have already been touched upon in Section II.B. and include all those concerned with the effects of reductions in the prevalence of disease on mortality, fertility, morbidity, and debility, and hence on the availability of labor of various types. Thus the application of cost/benefit analysis to medical planning referred to in Section II.C. belongs in this category, as do those investigations that explore the consequences of lessened mortality on the level of fertility. In spite of the many studies available only the fringe of this huge area has really been explored and much remains to be done, especially for those situations, frequent in tropical countries, where the national level of prevalence of a disease is very high [e.g. Newman (11)].

It has often been pointed out that one's ability to learn is dependent on past and present states of health—for example, the influence (already cited) of early nutrition on present intelligence; but here again in spite of much research there is still much to know about 3.

With regard to 4 the importance of including his state of health in the individual's utility function (and by extension in the social utility function as well) has already been treated at some length in Section II.C. The discussion in Section III.B. of some of the ways in which his state of health may enter the individual's function shows that there may be marked divergence between the structure of the social utility function with regard to health and that of the individual. It is possible for a disease not to enter into any individual's preferences while still appearing in social preferences, perhaps with a significantly high weight. In this way, and not only through differences in the technologies available to individuals and to government, respectively, there may arise a strong case for social action.

E. The Production Function for Health

In the model it is assumed that what determines the evolution of the state of health is (i) the current level of that state; (ii) the vector of capital goods present in the economy; (iii) the current vector of consumption of all goods and services, other than the services of human wealth; (iv) the vector of human wealth; (v) that part of the vector of human wealth allocated to be used in consumption activities; (vi) the state of knowledge; and (vii) time. It is convenient to refer to this gargantuan function by the name "the production function for health," a title which is only slightly misleading and better than any others that suggest themselves. Each of the seven variables in the function will be discussed, though not in order and not in the same degree of detail.

The last two variables (vi) and (vii) are inserted to take account of the obvious facts that levels of health depend in part on knowledge of such things as personal hygiene, and that medical science, sanitary engineering, and other knowledge pertinent to health is likely to evolve over time. At this level of generality only banalities can be said about these patterns of dependence and they will not be said here.

The assumption that the change in the social state of health depends in part on the present level of that state [variable (i)] is obvious enough, but needs comment in two respects. First, in many cases the dependence is likely to be not only on the current social state but also on how that social state has evolved in the past; this extension makes the model much harder to analyze and should be introduced only when the facts of the situation make it absolutely necessary to do so. Secondly, the present assumption allows for those cases where the probability of having one disease is dependent upon the presence of other diseases, since each element in the vector of rates of change of the social state is (in principle) dependent on the whole of the vector representing the current social state.

The dependence on variable (ii) is again reasonable, especially when it is realized that the vector of capital goods can be disaggregated down to the level of different types of structure and machinery, such as hospitals, clinics, radiographic equipment, and so on. The chief difficulty here is that much of the equipment and buildings is not specific to one disease but can be used in the treatment of several ailments. Hence it is necessary to specify the level of prevalence of the other diseases before being able to say how the addition of specified items of capital will affect that of the remaining one. The analytical problems involved are familiar from the economic theory of the multiproduct firm.

The use of the word *consumption* to describe the third variable is in part a misnomer, since it includes such things as medical and educational goods and services. Notice that the vector of consumption goods is *not* further divided into that part destined for personal consumption, that part destined for medical purposes and that part destined for education. This is for two reasons: first, a sufficiently fine classification of commodities can accomplish much the same objective (for example, the consumer item "gloves" can be subdivided so that surgical gloves appears as a separate element in the consumption vector); secondly, it is often the case that the whole of the consumption of a particular item affects the state of health, and not merely that part which is labeled "for medical purposes" (for example, how is one to disentangle that part of the consumption of milk which is for consumption purposes from that for medical purposes?).

Similar remarks apply to the fifth variable, that part of human wealth used in consumption activities. Included in this category would be general leisure, together with all use of labor by medicine and education. The reasons for including these are obvious. Perhaps slightly less obvious is the inclusion of variable (iv), the total vector of human wealth. The reason for this is that the amounts of human wealth employed in production may, through fatigue and other such elements, affect the state of health; and one can calculate these amounts from the total amounts available less those used in consumption.

This review of the variables in the production function for health is not meant to give the impression that it is as yet a well-defined practical concept. Indeed, the ordinary aggregate production function [an example of which is Eq. (1) in the Appendix] is itself far from being empirically well defined, even in the disaggregated form shown there. It is, for example, an all too common experience in economic planning, and not merely in underdeveloped countries, for an expected capital/output ratio to turn out in practice to be quite wrong. So it is only to be expected that the state of the art in medicine is not likely to permit us now to write down functional relationships like Eq. (6) with any degree of precision. It is only in recent years, for example, that systematic studies have been carried out on economically efficient ways of reducing disease [e.g., Piot and Sundaresan (12), Cvjetanovic, Grab, and Uemura (6)] and such valuable work is still in its infancy. But plans are made and actions carried out all the time in medicine (as in similar fields) which imply *some* assumptions, however crude or implicit, about the causal nexus postulated by Eq. (6). It would help rational planning if such assumptions were made explicit and the empirical basis for making them greatly improved.

IV. CONCLUSION

With this discussion of "the production function for health" the last piece of the formal structure outlined in the Appendix has been slotted into place. But the completed structure is only a scheme, a framework, and not a working model. In fact there appear to be no such comprehensive working models for the planning of medicine in developing countries, although there are tentative beginnings in that direction [Barlow and Davies (5), Ray *et al*. (14)]. One hesitates to say how far the present framework could be made operational. In principle there seems no reason why not—that, after all, was a chief motivation for adopting the definition of state of health. But much experience with planning models for the general economy leads one to be quite cautious about the extent to which the present closely knit structure could be preserved through all the grim struggles and compromises forced by lack of data, problems of specification, and difficulties of estimation. The emperor may well have believed that he was having his coat cut according to his cloth, but by the end of the story he was left with neither.

Even were one to have such a fully operational model one could not expect it to do most of the work involved in the appraisal of particular health projects, any more than the most carefully articulated model for the industrial sector can of itself decide the merits of embarking on a particular textile mill or automobile plant. The most that can reasonably be asked of such models is to explore how a particular program of medical projects can help to achieve, along with other programs in the economy, the stipulated set of national health targets, and how far that set of health targets is itself appropriate, both in its direct (consumption) and indirect (production) aspects, for the economy and society as a whole. But even that task is ambition enough.

TECHNICAL APPENDIX

Introduction

The aim of this Appendix is to set out in straightforward fashion a model of optimization involving states of health for the economy as a whole. The chief concern is to show how the state of health, affecting and being affected by the other variables of the system, fits logically into it. No attempt is made actually to solve the optimization problem, nor even to give conditions that are necessary or sufficient for an optimum to be achieved, since that would not be immediately relevant to the present enquiry and would involve considerable mathematical detail.

The social state of knowledge will be included as a variable in this bare abstract model, since that can be done with little complication and contributes substantially to its realism and completeness. Assume first, and without arguing the matter in detail, that for any individual his state of knowledge at any time t can be adequately represented by a vector $s_i(t)$ of *skills*, taken from a finite set S of skills analogous to the finite set D of diseases. This implies that for each skill in S, say the kth, there exists a cardinal measure $s_{ik}(t)$ showing the ith individual's level of achievement of skill k at time t. The existence of an enormous body of literature on the testing of educational and occupational skills suggests that this assumption is not nonsense *a priori*, although it is readily admitted that the vector $s_i(t)$ may not capture all the complexity of the idea of a well-educated human being, any more than $d_i(t)$ captures the idea of being in excellent health.

It is next postulated that a sufficient description of the social state of knowledge is provided by a vector s(t), whose kth component (for any skill in S) is the aggregate of the kth components of all the individual states of knowledge, divided by the total population. In the same way that a distinction was drawn between perceived and diagnosed states of health so here a distinction between perceived and *tested* states of knowledge can be made. As in the case of health, it is the former, perceived state which is principally relevant for individual behavior.

The variables used here follow broadly the meanings of those introduced in the main text. Human wealth enters the model as a vector representing the aggregate stocks of various types and qualities of labor, measured in numbers of persons of the various types.

In reading the dictionary below, the following points should be observed: (i) For simplicity of notation the superscript * on both the perceived social state of health and of knowledge has been eliminated; (ii) the fact that each variable in the list is a function of time is not explicitly noted, although it is so noted in the description of the model.

Notation

c	Consumption, a vector of m components
d	Perceived social state of health, a vector of q components
i	Gross investment, a vector of m components
k	Wealth, a vector of m components
ℓ	Human wealth, a vector of n components
ℓ_c	Amount of human wealth used in consumption, a vector of n components
ℓ_p	Amount of human wealth used in production, a vector of n components
o_t	Depreciation rates at time t, a vector of m components
s	Perceived social state of knowledge, a vector of r components
t	Time
u_t	Planners' social utility function at time t
v	Planners' valuation function of terminal stocks
y	Output of commodities, a vector of m components

All variables are assumed to be nonnegative. A dot placed above any variable, e.g., $\dot{s}(t)$, denotes the derivative of that (vector) variable with respect to time. Note that all variables are endogenous, except o_t and (of course) t.

Description of Equations in the Model

1. The first set of equations is the production function for commodities, which asserts that social output is a function of the stock of capital goods (wealth), of the amount of human wealth devoted to production, and of time (to allow for technical progress). Here "capital goods" in fact means all goods and all services other than those of human wealth; but obviously for perishable commodities and for services the corresponding entry in the k(t) vector must be zero.

$$y(t) = F_1(k(t), \ell_p(t), t). \tag{1}$$

2. The next set says simply that total output of commodities can either be used to add to capital stocks, or be used in consumption. *Consumption* here is not quite an adequate word, since it includes also purchases of such things as medical and educational goods and services. These equations have the additional implication that the economy is closed to international trade.

$$y(t) = i(t) + c(t). \tag{2}$$

3. The third set is an obvious identity connecting the human wealth variables.

$$\ell(t) = \ell_c(t) + \ell_p(t). \tag{3}$$

4. The next set asserts that the rate of change in the capital goods vector is equal to gross investment minus depreciation. The form of the depreciation function assumed is rather simple, but if the device is adopted of treating capital goods at different dates as different goods, the present simple form has considerable flexibility.

$$\dot{k}(t) = i(t) - o_t k(t). \tag{4}$$

5. The fifth set assumes that the rate of change in human wealth depends on the current level of that wealth, on how healthy and how knowledgeable the society is, and on time (to allow for changes in population size and composition, here kept in the background). The variables c and k could also be included without much complication, but are omitted here.

$$\dot{\ell}(t) = F_2(d(t), \ell(t), s(t), t). \tag{5}$$

6. The sixth set of equations, "the production function for health," is of central interest to this investigation. Its rationale has been discussed in the main text, as have some of the severe problems connected with it. The equations represented by (6) are in quite general form but not entirely vacuous. Thus the possibility that d(t) might be affected not only by the present social state of health d(t) but also by past states is not included, an omission which could be serious in practical cases [and is likely to be so for the corresponding omission in Eq. (7)]. The possibility of technical progress in medicine has been allowed for by the inclusion both of s(t) and of t, although s(t) also covers improvement in the public's awareness of existing medical knowledge.

$$\dot{d}(t) = F_3(c(t), d(t), k(t), \ell(t), \ell_c(t), s(t), t). \tag{6}$$

7. This last set is similar to the previous one, only applied now to $\dot{s}(t)$ rather than $\dot{d}(t)$.

$$\dot{s}(t) = F_4(c(t), d(t), k(t), \ell(t), \ell_c(t), s(t), t). \tag{7}$$

Optimization in the Model

It is assumed that for each time t the planners assign a social utility which is a function of the variables c(t), d(t), k(t), ℓ(t), ℓ_c(t), and s(t); the inclusion of k(t) and ℓ(t) is for completeness' sake, and indeed most growth models would omit them. The planners' optimization problem is then to maximize some appropriate aggregate of this utility over some appropriate period of time, subject to the constraints in the problem. The vexing questions posed by what kind of aggregate and what period of time are well known but unsolved, and will not be discussed here. For simplicity's sake assume merely that the planners fix a finite horizon T years hence, and then decide to maximize social utility over that time interval, together with a valuation of the terminal wealth and human wealth and the social states of health and knowledge. Thus the planners' objective is to

$$\max[(\int_0^T u_t(c(t), d(t), k(t), \ell(t), \ell_c(t), s(t))\, dt) + u(d(T), k(T), \ell(T), s(T))] \tag{8}$$

subject to the constraints of the problem.

Two things are not yet clear. First, just what are the decision variables that the planners must choose in order to solve (8)? Secondly, just what are the constraints on the problem? It is best to begin with the second problem. The model will now be stripped down to its basic form by eliminating the inessential variables $i(t)$, $\ell_p(t)$ and $y(t)$.

Recall that

$$\dot{k}(t) = i(t) - o_t\, k(t). \tag{4}$$

From (2) this may be rewritten

$$\dot{k}(t) = y(t) - c(t) - o_t\, k(t),$$

which from (1) takes the form

$$\dot{k}(t) = F_1(k(t), \ell_p(t), t) - c(t) - o_t\, k(t)$$

and finally from (3)

$$\dot{k}(t) = F_1(k(t), \ell(t) - \ell_c(t), t) - c(t) - o_t\, k(t). \tag{4a}$$

For convenience, repeat (5)–(7):

$$\dot{\ell}(t) = F_2(d(t), \ell(t), s(t), t); \tag{5}$$

$$\dot{d}(t) = F_3(c(t), d(t), k(t), \ell(t), \ell_c(t), s(t), t); \tag{6}$$

$$\dot{s}(t) = F_4(c(t), d(t), k(t), \ell(t), \ell_c(t), s(t), t). \tag{7}$$

Then Eqs. (4a) and (5)–(7) constitute a system of first-order differential equations in the four *state* variables $d(t)$, $k(t)$, $\ell(t)$, $s(t)$; the two *control* variables $c(t)$, $\ell_c(t)$; and t. For it is readily verified that only those seven variables appear on the right-hand sides of these equations, apart from the known (exogenous) variable o_t. Moreover, it is precisely the four state variables and two control variables which are the arguments of the utility functions in (8); thus (4a) and (5)–(8) form a complete system, subject to control by $c(t)$ and $\ell_c(t)$.

It follows that the planners' optimization problem may be expressed: Find time paths of the control variables $c(t)$ and $\ell_c(t)$ from 0 to T which solve (8), subject to: (1) the transition equations (4a) and (5)–(7); (2) initial values $d(0)$, $k(0)$, $\ell(0)$,

Conceptual Framework for the Planning of Medicine in Developing Countries 55

s(0) for the four state variables; (3) non-negativity conditions on all variables. This is a well-defined problem in the theory of optimal control and may be attacked with the techniques of that theory. Until the various functions F_1, F_2, F_3, and F_4 are invested with some empirical content, however, the conditions for optimization are unlikely to be of more than formal interest.

NOTES

1. Even this word has its difficulties, of which three are mentioned here. First, there is danger of confusion between medicine considered as the process of medical care and considered as the medical care industry itself; the context usually makes clear which is meant, as with the similar double meaning of the word *education*. Secondly, *medicine* is sometimes used in a much more restricted sense, to contrast with surgery and obstetrics, a usage never employed here. Finally, it can be used as synonymous with therapeutic drugs; but will not be so in this analysis.

2. Such a list, written algebraically $(a_1, a_2, \ldots, a_i, \ldots)$ with a_i the physical amount of the ith asset, will usually be called a wealth vector. The wealth vector for any group of people, for example a nation, is obtained by summing all the individual wealth vectors, care being taken to net out all intragroup claims.

3. Problems of defining group welfare are more complicated and will be touched on later, but only to the extent necessary for this enquiry. The argument that follows could in its essentials be repeated for the analogous problem of including the state of knowledge in the utility function, along with or instead of the state of health. This would have bearing on what are sometimes called the "consumption" benefits of education, as distinct from its "investment" aspects. But since medicine and not education is the subject under discussion, this parallel enquiry is not pursued in detail here, although it is briefly included in the Appendix.

4. There are many problems (though by no means all) for which it is enough that some of the variables be measurable only in integers. This possibility enables one to enter into the utility function variables, essentially qualitative, that are capable of only a finite number of states, such as the so-called (0, 1) variables. In such cases, however, even more care than usual must be taken in framing conditions of continuity, differentiability, and convexity of the utility surface.

5. In a more general model it would be reasonable to include the state of knowledge (e.g., of hygienic practices) as a further determining factor. But if it is then assumed that the state of knowledge is itself determined in the same way by the commodity vector (and possibly the state of health) the conclusions reached in the text still apply, everything being expressible in terms of commodities alone.

6. However, the existence of markets and market prices provides a useful framework for the classification of commodities, a framework which seems to have no counterpart in the field of health; I owe this point to Jurg Niehans.

7. The whole of this analysis could be carried through essentially unchanged using not $d_{ij}(t)$, but the probability that individual i does not have disease j at time t, i.e., the probability $1 - d_{ij}(t)$. This would correspond more closely to the intuitive idea of state of health than the present definition, which should perhaps more accurately be called the state of ill health. But the usage in the text seems more natural and direct than the alternative.

8. It could be argued that given complete information about any individual i it would always be possible to know for certain whether he has any disease or not at time t, so that $d^i(t)$ would become simply a vector of zeros and ones. This argument, which in its essentials is as old as the debate on the meaning of probability itself, runs into difficulty over the problem of giving a noncircular definition of *complete information*. It is not pursued here.

9. At this point one feels more acutely than usual the loss of analytic power caused by the

decision not to include states of knowledge along with states of health in the main part of the investigation.

10. One thinks here of those not uncommon cases where, say, a middle-aged man is discharged from his annual hospital checkup with a clean bill of health, only to be dead within 6 months from a previously unsuspected brain tumor. Moral: One seldom finds what one is not looking for.

11. Cf. Prothero (13), p. 15: "people come to accept malaria as part of life: among many illiterate communities it is scarcely even recognized as a disease.... Endemic malaria is insidious in the way it manifests itself and its effects on people, and it is frequently difficult to convince those who are affected by it of the need for measures to control the disease or to eradicate it altogether."

REFERENCES

1. Abel-Smith, B. (1972) "Health Priorities in Developing Countries: The Economist's Contribution," *International Journal of Health Services* 2:5–12.
2. Arrow, K. J. (1963) "Uncertainty and the Welfare Economics of Medical Care," *American Economic Review* 53:941–973.
3. Baker, T. D. (1972) "Human Capital and Cost Benefit," Background Paper No. 1, *Research on the Relationship Between Health and Development: A Feasibility Study*. Final Report, Grant No. AID/CSD-3320, Baltimore: The Johns Hopkins University, School of Hygiene and Public Health, Department of International Health.
4. Barlow, R. (1968) *The Economic Effects of Malaria Eradication*, Bureau of Public Health Economics, Research Series No. 15, Ann Arbor, Michigan: School of Public Health, University of Michigan.
5. Barlow, R., and G. W. Davies. (1971) "Project Evaluation with a Detailed Macro-economic Model," Department of Economics, University of Micigan, 30 pp.
6. Cvjetanovic, B., B. Grab, and K. Uemura. (1971) "Epidemiological Model of Typhoid Fever and its Use in the Planning and Evaluation of Antityphoid Immunization and Sanitation Programmes," *Bulletin of the World Health Organization* 45:53–75.
7. Fanshel, S., and J. W. Bush. (1970) "Health-Status Index and its Application to Health-Services Outcomes," *Operations Research* 18:1021–1066.
8. Klarman, H. E. (1965) *The Economics of Health*, New York: Columbia University Press.
9. Lancaster, K. J. (1971) *Consumer Demand: A New Approach*, New York: Columbia University Press.
10. Mushkin, S. (1962) "Health as an Investment," *Journal of Political Economy* 70:129–157.
11. Newman, P. (1970) "Malaria Control and Population Growth," *Journal of Development Studies* 6:133–158.
12. Piot, M., and T. K. Sundaresan. (1967) "A Linear Programme Decision Model for Tuberculosis Control," WHO/TB/Techn. Information/67.55, mimeo.
13. Prothero, R. Mansell. (1965) *Migrants and Malaria*, London.
14. Ray, D., B. Christian, T. Tanahashi, D. Fisher, and P. Diethelm. (1972) "A Dynamic Model of Health and Socio-Economic Development: An Intersectoral Model," WHO, mimeo, 85 pp.
15. Schelling, T. (1968) "The Life You Save May Be Your Own," in *Problems in Public Expenditure Analysis*, (ed.), Samuel, Washington, D. C.: Brookings Institution.
16. Sullivan, D. F. (1966) "Conceptual Problems in Developing an Index of Health," *Report Series of National Center for Health Statistics*, Series 2, No. 17, U.S. Public Health Service Publication No. 1000, 18 pp.
17. Williams, K. N. (1972) *Health and Development: An Annotated, Indexed Bibliography*, Baltimore: The Johns Hopkins University, School of Hygiene and Public Health, Department of International Health, mimeo, 160 pp.

HEALTH DEVELOPMENT:
A DISCUSSION OF SOME ISSUES

Oscar Gish

This paper is concerned with some selected issues of interest to general development planners, as well as those more directly involved with the health sector. After touching upon a few background questions the paper will turn to more specific issues in the areas of health and health sector development, namely, the environment, basic needs, health planning, primary health care (PHC), and categorical disease control programs. The paper will not rely upon specifically cited materials, but rather attempts to synthesize experience within the areas under discussion.

I. HEALTH AND DEVELOPMENT

The terms *health* and *development* are neither unambiguous nor easily defined, in fact, it is easier to state what they are not than what they are. Thus, health is not just an absence of medically defined bodily malfunction usually having some specific etiology, and development is no longer to be equated merely with increases in the output of goods and services. What, then, are health and development? The WHO definition of health refers to a "a state of complete physical, mental, and social well-being." All right, and almost certainly to be preferred to only "the absence of disease" but, for functional purposes anyway, a particularly difficult concept. With regard to development, there is no one new definition which alone satisfactorily replaces that of growth of national product. Nonetheless, the level of agreement, at least in academic and international circles, about many of the components of that replacement definition is very high. These would include income redistribution between countries through such mechanisms as the New International Economic Order, and within countries through the creation/provision of basic needs—most particularly as the result of land reform and employment creation, universal access to social services, and popular participation in the processes of development. Although the widespread acceptance of these broader concepts of the nature of health and development is relatively recent, the ideas themselves are not new and many can be traced back even to the precolonial period.

The new "centrist consensus" about the nature of the development process described above has its critics on both the left and the right. Those on the left see the central and continuing issue as being the struggle between classes contending for state power and control over the means of production, and that only a victory in that struggle by the laboring classes would create the conditions under which could be met the true health and development needs of the masses—and further, that although a New International Economic Order might result in the transfer of some productive capacities from richer to poorer countries, this might only further strengthen the hand of the oppressive class of "lumpen bourgeoisie" that rules in much of the Third World (i.e., underdeveloped capitalist countries) in the interest of the transnational corporations. With regard to the basic needs strategy, this is seen at best as nothing more than a "soup kitchen" approach to development and the needs of the masses, and at worst as an attempt to provide still another reformist support to an unjust and dying social system. On the right, most of the arguments against the new consensus are based on the postulates of neoclassical economics and the historical experience of today's industrialized capitalist countries. It is argued that output is best maximized by the workings of the free market, that developing countries need to be modernized in keeping with the experiences of Western Europe and North America, and that no more resources should be invested (or wasted) in health and other social services than is compatible with the requirements of capital being invested in production. While debate over these issues rages in international forums, parliaments (where they exist), and universities, as is usual, the issues themselves are being resolved mostly within the context of the constant interaction and struggles of the contending groups and classes.

Table 1. Life Chances in the 1840s: Preston, Lancashire, England

	Gentry	Trades	Operatives
Born	1000	1000	1000
Remaining alive at end of:			
1st Year	908	796	682
5th Year	824	618	446
20th Year	763	516	315
40th Year	634	375	204
60th Year	451	205	112

Source: J. N. Morris, *Uses of Epidemiology*, Third Edition, Churchill Livingstone, London, Copyright © 1975. Reprinted with permission.

A. Determinants of Health

It is quite remarkable how the link between *some* specific medical care interventions and *some* individual cures have been confused in the popular and even professional mind with more general determinants of individual good health and positive national/community health indices. Although it may not yet be possible to quantify precisely the components of individual good health or improved overall health indices, there is sufficient evidence to show quite clearly the limited contribution to such indices that has been made in the past by medicine and the health care services. Historical experience offers important evidence with regard to this question.

As can be seen in Table 1, 140 years ago in industrializing England the lowest social class grouping had an infant mortality rate (first year of life) more than three times greater than that of the highest social class grouping. This disparity was based not on differences in access to medical care (of course, in any event, medicine had relatively little to offer at this time), but rather on the differing social and economic conditions of the classes concerned.

To take a somewhat more recent example, between 1900 and 1930 infant mortality rates in New York City fell from 140 to around 55 per 1000, an overall fall of over 60 percent. Of that fall, two-thirds occurred within the so-called diarrhea-pneumonia complex of childhood diseases. The most striking aspect of this rapid fall of infant mortality in New York City is that it occurred before there were any antimicrobial drugs or vaccines with which to treat this particular disease complex. This striking decrease in mortality was accomplished through a series of primarily social and political interventions. These interventions took place at a time of rapid economic growth and social change that was occurring in a society which had already reached a relatively advanced state of economic development and political participation by the mass of the population. Some of the specific public health measures of the period included stronger control over the distribution of more and better foodstuffs, an improved water supply, major campaigns against illiteracy, and the creation of visiting nurse services and well-baby clinics. In the years since 1930 infant mortality has fallen in New York City from around 55 per 1000 to near 20. Current studies of the effects of modern medical care on death rates indicate that even during the recent years in which powerful modern drugs and improved medical procedures have become available, they have played only a relatively minor role in the further decreases in infant (and more general) death rates experienced over the last half-century or so.

One of the most important changes to be seen in the Third World in recent years has been the rapid growth of populations, mostly due to falling infant and child mortality rates. Although specific reasons for this fall are not precisely known, it has often been argued to be primarily due to public health measures such as the international smallpox and malaria campaigns, the increasing availability of supplies of clean water, and improved nutritional status. Although death rates may have been affected by the smallpox, malaria, and other campaigns, the large falls that actually took place do not appear to be adequately explained by these alone. With regard to clean water, waste disposal and other aspects of sanitation, little has changed for the bulk of the population of the Third World which remains primarily rural, although those who have migrated to the towns may have substantially improved their position in this respect. With regard to the nutrition factor, it may be that its important contribution to falling (especially) infant mortality rates has come about primarily through the more rapid availability of at least minimum quantities of foodstuffs at times of extreme food shortage and famine. Although famines still occur in the Third World, they are no longer so regular as they were during the colonial era. There is little evidence that average nutritional standards within much of the Third World are rising, within the context of relatively wide variations, but the very availability of national and international food stocks and the transport systems to move them quickly make it less possible under conditions of independent sovereignty to allow starvation to the point of immediate death. However, although fewer people may die outright from starvation, many survive only at lowered nutritional and energy levels; thus, undernutrition becomes a chronic process rather than an acute event.

A considerable amount of traditional and current health research is directed toward quantification of the components of health, mostly defined in terms of lessened mortality and, sometimes, morbidity. This research equates health with the diminution of mortality and morbidity, usually by a process of reductionism. Among other things its purpose is to offer, to those responsible for the use of resources, policy proposals directed toward an optimal allocation of those resources toward interventions assumed to positively influence health status. These interventions are generally perceived to be purely technical and usually discrete in character. Thus, in its more vulgar forms, the choice falls to "one more health center" or "one more water well." Such research accepts the basic medical-technical model of good health, to the exclusion of contrary historic and contemporary experience.

II. THE ENVIRONMENT AND HEALTH[1]

The relationship between man and the physical environment in which he lives is a complex and delicate one. In the absence of an appropriate relationship, the health of both will suffer. Although appreciation of this relationship is rooted deeply in human experience, it has often been blunted under the pressures of "development," and most particularly development as it takes place under market pressures in conditions of extreme social and economic inequality. This section will touch upon some of these issues.

A. Conceptual Aspects

Environmental analysis has mostly constituted an assessment of environmental impacts and the consequent proposal of acceptable pollution standards and regulations intended to achieve those standards. Now, however, under the pressure of widened environmental understanding, it has become necessary (at the very least in developing countries) for such assessments to become part of an integrated development perspective and overall development planning. This means that analyses and actions intended for the protection of the environment cannot exist simply as "add-ons," but must become intrinsic to both overall and specific socioeconomic policies. For those persons and organizations that always have been specifically concerned with health, the logic of this position will be obvious. The most appropriate health promoting policies have always been perceived by (the best of) those engaged in public health work as being holistic in both conception and application.

It has become common for economists to consider solutions to environmental problems in two ways, both "marginalist." One requires that the industrial producer of pollution pay for the damage, thus bridging the gap between the private and social costs of industrial production. The other approach is to treat the environment as any other economic good and expect particular consumers to pay for it. Although certain problems in connection with these approaches have always been recognized, it is only recently that a major conceptual change in economic thinking regarding environmental protection has taken place. This change can be traced to the massive expansion in technological and industrial development as well as human populations, and their consequent environmental impacts, during the post–World War II period. The earlier marginalist approaches to the environment are perceived as inadequate for coping with the greatly expanded scale and complexity of potential environmental damages and their control. Increasingly, questions and criticisms are raised about growth-oriented development and wasteful life styles.

Until very recently the initial explosion of interest in the environment has tended to view the question almost entirely from the perspective of the already industrialized countries. In addition, many of the specific ways in which the problem was being discussed in the rich countries appeared as frivolous or even threatening to the countries of the Third World, e.g., zero-growth strategies. The rich countries, having already reached astonishingly high overall levels of output and consumption, were in the enviable position of being able to view the environment as a separate good that had to be preserved for its own sake, virtually independent of the "purposes of men." In fact, it was man per se, in some biological sense, who came to be perceived as the villain of the piece. This rather simplistic view is now rapidly being overtaken by a perspective which puts man and his (sensible) needs at the center, rather than the environment as such.

It should be recognized that the early environmental missionaries did in fact make an enormous contribution to the opening of discussion about these issues. The problem now is to fit these earlier purist views about the environment into a wider perspective which incorporates into itself the economic and social needs of man, while still giving proper respect to the environment in which man lives. Obviously man will always have to live in that environment and so its destruction would run counter to a perspective which, in fact, begins with the "needs of man" rather than the "needs of the environment." It is also obvious that at least some changes in the environment will take place as a result of that fact that "man is." Nonetheless, such changes should not come about by default, but rather in ways that are consciously understood and acted upon.

As touched upon earlier, between the end of World War II and the later 1960s the central view of development was that it could be equated, more or less, with growth of national product. Over the last decade a number of themes have emerged as replacement, or enlargement, of the "GNP equals development" view. Some of these have been discussed earlier. The "new development wisdom" brings to the fore the need to be concerned much more directly than in the past, and in many ways differently, with such issues as nutritional status, health and health care, and environmental impacts on human health. When carried to their logical ends the current discussions about health, the environment, and development become almost indistinguishable. It is especially important to note that an appropriate form of development need not be antienvironmental. However, to the degree development is based only on reproduction of the resource wasting patterns of socioeconomic organization and behavior now current in the industrialized countries—to that degree may it, in fact, actually turn out to be antienvironmental. The satisfaction of basic human needs, if carried out in the context of increasing social and economic equality and the appropriate utilization of technologies, does not require destruction of the environment. As an example, there are estimates which suggest that achieving a minimally acceptable standard of life for everyone under existing patterns of income distribution will require three to five times more natural resources than with a more equitable distribution of income.

B. Technical Aspects

The basis of (neoclassical) economic thought about environmental problems resides in its treatment of them as externalities, that is, factors external to the normal workings of a freely competitive marketplace in which buyers and sellers have equal capacity to express, through the price mechanism, their preferences with regard to any particular level or type of environmental impact/damage. This (pure) market mechanism can be helped to become more perfect through incorporation into its workings of relevant education for all those concerned, appropriate taxes and subsidies, and various legal frameworks. However, for the following reasons, which are only listed here, the traditional neoclassical framework of analysis is now found to be seriously deficient: (1) the very rapid expansion of economic activity; (2) the vastly greater capacities and impact of science and technology; (3) increasing doubt about the purity of the free marketplace; (4) the long-term effects of large environmental projects; (5) the fact that most large environmental impact projects are now public sector undertakings; and (6) the complex interrelatedness of large environmental impact activities. Rather than relying only on traditional neoclassical economics, the basis for appropriate economic analysis of environmental projects now must incorporate the kinds of wider considerations that have been indicated earlier.

Environmental assessment as it is now mostly practiced has its origins in the United States Environmental Protection Act. As conducted in the United States, environmental impact statements are primarily a set of techniques for specifying environmental problems. Generally they do not offer an evaluation in social terms either of particular environmental projects or the effects on the environment of more general socioeconomic policies. Significant progress is being made in the development of technical tools for the better specification of environmental damage. These tools include various kinds of checklist and ranking procedures, input/output matrix analysis applied to energy or other materials accounting, compatibility analysis, technology assessment, and network impact analysis. Such instruments constitute an absolute necessity for consideration of environmental problems; however, they are not sufficient by themselves to offer policy guidance to those concerned with appropriate utilization of the environment in keeping with the overall shorter- and longer-term needs of different population groupings.

Socioeconomic evaluation attempts to understand the longer- and shorter-term effects of various policies and decisions on society and its constituent parts. Acts affecting the relationship between populations and the environment fall squarely into this area. The economic perspectives most suited for such evaluation, especially (but not exclusively) within the context of developing countries, or in other disciplines for that matter, would not be primarily directed toward the technical specification of environmental impacts. The basic tools of economics allow for the monetary quantification of gains and losses in production resulting from particular decisions affecting the volume and type of production, e.g., the diversion of a given quantity of river water from agricultural production to mining operations. Beyond this it is necessary to assess effects on the specifically concerned populations which may result from particular decisions. For example, in the case given above (agricultural or mining production) it would be necessary to know, in addition to the impact on the environment itself of a changed pattern of water use, its effect on different social groups in terms of food consumption, income distribution, the volume of employment created or destroyed, the effect on rural/urban migration patterns, etc. Further, and particularly difficult to analyze, there is the transference into terms of human health of the particular decision under discussion.

In many ways the concept of ecodevelopment, in its various forms, comes closest to matching the kind of economic thinking that would be most appropriate for those concerned more directly and specifically with the health of populations. Probably the three most common interpretations of ecodevelopment are as follows: (1) an appropriate overall development strategy incorporating within itself a sound respect for the environment; (2) environmentally sound planning at the local/regional level; and (3) the incorporation of ecological ideas within development strategies.

The most common theoretical technique for evaluating the social effects of alternative patterns of resource use is cost/benefit analysis. To the degree that the basic assumptions of neoclassical economics are accepted, cost/benefit analysis can be accepted as a value-free methodology. Conversely, to the degree that the marketplace is rejected as an adequate determinant of social welfare (including health), cost/benefit analysis will be rejected as a value-free methodology. One of the most significant problems of the methodology is that it will place too small a value on the revealed environmental preferences of the poor who must inevitably opt for (say) continued employment, even if it is very hazardous to the environment or their health, rather than run the risk of being unemployed even if in a cleaner and safer environment. There are other problems that could be cited with regard to the application to environmental hazards of strict economic evaluation in general and cost/benefit techniques in particular—in fact, so many as to raise doubts about its ultimate usefulness. However, the problem of the best way of using resources to attain specified objectives still remains, as does the need to be reasonably consistent in the use of those resources. At the very least, cost/benefit analysis can help with the systematic ordering of variables as well as offer some overall way of weighting them, even if they cannot be quantified satisfactorily.

A variant of cost/benefit analysis that is potentially quite valuable is that of cost effectiveness. This method measures the costs of alternative ways of achieving preselected goals—for example, a specified upper level of industrial contamination. Once technically specified objectives have been determined it is possible through cost effectiveness analysis to determine the optimum allocation of a given quantity of resources in relation to the specified objective. Also, the originally specified objective can be modified later in keeping with the results of cost effectiveness analyses.

A special problem for those specially concerned with human health is that of quantifying its value. Although it is possible to quantify in money terms the trade-off between (to use the earlier example) the diversion of river water from agricultural production to mining operations and to consider in various ways the effects of this diversion on income distribution or employment, there is no convenient way of incorporating into the economic calculation any resulting changes in morbidity or mortality. Although some calculations are possible, e.g., the cost of treating illness, there is no generally acceptable way of pricing the value of a human life or (human) cost of an episode of serious illness. Of course there are economic calculations that can be made, based on the value of the production or earnings lost due to the death or illness involved. However, for many it would be morally unacceptable to consciously accept a given number of deaths in exchange for a given volume of money. In any event, this procedure accepts the value of poor people's lives as being less than that of wealthy ones. In fact, in the context of many Third World countries the actual trade-off in such a calculation would be immense. This problem is not easily solved and is probably best dealt with in the context of wider discussion and analyses of the relationships between development and health.

III. PRIMARY HEALTH CARE, HEALTH PLANNING, AND BASIC NEEDS

The meeting of everyone's basic needs has replaced growth of national product alone as the key goal/indicator of development. In its more complete interpretation the concept of basic needs has been defined as (1) goods (food, clothing, etc.); (2) universal access to social services (education, clean water, health services, etc.); (3) employment (including access to land—the major determinant of nutrition in poor countries); (4) a growing national industrial capacity; and (5) popular participation in and/or control over the processes of development. However, in some discussions it appears that "basic needs" is to be equated with nothing more than the provision of some particular volume of economic goods and services.

A. Basic Needs and Planning

It would be good indeed if the successful meeting of everyone's basic needs could be accomplished by no more than the provision of "stuff" to a given, predetermined basic level. In that case, in principle anyway, it would seem to be a relatively straightforward matter to calculate the volume of resources required, which in any case would not be vast relative to ongoing expenditures in such areas as education or health or as a percentage of new development expenditures for clean water or improved agricultural outputs, and to get on with the task of making them available for the provision of these so-called basic needs. There are at least two basic problems involved, however. One, which ought to be at least somewhat less difficult to deal with, is that of squeezing very much (or sometimes anything at all) out of the better-off and powerful who now control the allocation of resources and directing it toward the needs of the poor and less powerful. It might be, though, that the better-off will become wiser in the future than they have tended to be in the past and will actually give up or forego—or have taken from them, as has happened before—enough so as to make possible the provision of basic needs for all. The other still more fundamental issue lies within the concept of the "provision of basic needs." Can, in fact, basic needs be "provided," or does it also require "participation"?

The current discussion of basic needs, in its best form, does recognize the participatory element in the "provision of basic needs." What is not so often explicitly addressed is the relationship between participation and democratic control and the nature of the wider national context required to create an environment in which community participation and community control become virtually interchangeable realities. A strong case can be made to the effect that basic needs are unlikely to be satisfactorily provided in the absence of participation and that genuine participation is highly unlikely in the absence of democratic controls. The provision of increasing amounts of basic needs "stuff" in Europe, North America, and elsewhere has certainly increased life expectancy and improved the quality of life considerably for the majority in those countries. However, these changes took place under highly favorable economic and demographic conditions, at least when compared with those of Africa and Asia today, and in any event under conditions of increasing political and social participation and control by the middle and working classes. Nonetheless, it is also the case that recent years have seen increasing skepticism over the basic nature and value of the provisions being made under welfare state conditions. In sum, then, although it may be economically and technically possible to provide basic needs for all, it is highly unlikely to happen in the absence of increasing popular participation and democratic control at all levels of government and other institutional life.

In most countries formal plans bear little relationship to actual happenings. The rhetoric of the plans is usually about the poor and the need for rural-based activities, but the spending programs mainly benefit the better-off in the urban centers. The so-called implementation gap in planning has at least two major bases. One is the general political difficulty of redistributing income and power from the rich to the poor, and thus the "implementation gap problem" actually may be more of a "false rhetoric problem." The other, related issue is the framework and ways in which planning is usually conducted in Third World countries. This planning generally begins from the national center which is heavily influenced by international agencies. In its turn, the national center dominates the state or region, which plays the same role with regard to the district and other local authorities. These authorities are generally in the hands of local vested interests having no great interest in the basic needs of the bulk of the population. In practice, then, formal planning tends to be a far less significant activity than is sometimes indicated; in fact, it is often no more than an exercise in rhetoric, having the intention of giving the appearance of activity and progress.

B. Health Planning and PHC

It would be especially useful for health planners to distinguish between and clearly define actions—formally called plans or not—directed toward (1) improved health; (2) the health sector; and (3) the activities of the Ministry of Health or other agencies responsible for the provision of health care services. In this connection it is important to keep closely in mind the fact that the great advances in the state of human health to be seen in the industrialized countries of the world have stemmed largely from factors other than the provision of medical services, and especially curative ones. As touched upon earlier, the provision of clean water supplies, generally improved hygienic standards, and, perhaps particularly, increased incomes for the lower social classes leading to better nutritional standards, improved educational possibilities, better housing, etc., during the late nineteenth and early twentieth centuries all made their important contributions to improved standards of human health. The very drastic fall in the levels of morbidity and mortality in the technologically advanced parts of the world which preceded the development of potent modern drugs bears ample testimony to this fact.

From the above it follows that the most important aspects of health planning are those directed toward improving the capacity of the poorest in the population to care for their own most basic needs. Thus a more equitable distribution of land and the fruits of that land—or an industrial policy which emphasized the creation of employment, especially for those with few formal skills—would in many countries have the most profoundly positive effect on health status. In sum, it is clear that the overeaching character of good health, as distinct also from only the absence of any specific disease, means that government policy on virtually any question will have important effects on the health of the population, and especially on those in the population with least access to curative health services.

More specifically, the "health sector," as such, is generally taken to include not only preventive and curative health care services, but discrete nutrition programs, community water supply and sanitation activities, aspects of housing and educational programs, and many elements of planned rural and community development. These various programs and activities need not be governmental in character, but would also include various semigovernmental and private sector developments.

With regard to the activities of most ministries of health, these are concentrated on the provision of curative health care services and the management of certain "vertical" categorical disease control programs. In some countries water supply programs may be a component of health ministry budgets. Virtually all countries also have some element of a general preventive health program built into their work, even if often only to a limited degree. In any event, in reality most ministries of health in most of the world might more correctly be termed ministries of disease. This can be said because health ministries as we mainly know them are doing relatively little in relation to health itself; instead, they primarily provide services for those already suffering from disease. It is probably the case that ministries of health could do considerably more for health, as opposed to disease, than they are now doing, but it is also probably the case that the bulk of health inputs, as opposed to antidisease inputs, are in fact beyond the scope of their conventional activities. There are two main reasons for this. One is that the major factors influencing human health are more related to the types of services provided by such ministries as agriculture or industry, rather than ministries of health. The second reason is that much of that which has to do with health is not provided (in the conventional sense) by government at all; that is, people basically "do for themselves" with regard to their own health requirements. Of course, governments and ministries may help to create the circumstances in which people can better help themselves, but they cannot substitute for the actions of people. It may be that ministries of health can best facilitate the development of the health of national populations by (1) being supportive of the physical, social, and environmental needs of (particularly rural and poor) populations as they are now being met by other ministries than health; and (2) helping to create conditions which allow people both as communities and individuals to care for themselves in a health-promoting way.

Primary health care fits in with the concept of universal access to social services and is understood to be the key health sector component of basic needs. Planning for health care, including primary health care, must be very different in developing countries from what we now see in the industrialized ones, for the following reasons: (1) a very different resource base; (2) a very different demographic base (structure and location of the population); (3) a very different disease pattern; and (4) a very different environmental base (both physical and social). In most Third World countries we see few links between the factors just listed and the type of health care systems being developed. The systems that exist mostly try to reproduce the high-technology-based medical care systems to be seen in the wealthy countries. These systems are based on the needs of those who are most ill and have relatively little bearing on the medical care needs of the mass of the population. Such systems begin with some abstract clinical concept of "needs," mostly unrelated to the overall requirements of communities and certainly to the level of available resources, which, for health care, is often no more than several U.S. dollars per head per year.

C. Allocating Health Sector Resources

Appropriate resource allocation is central to appropriate health care planning. Health care systems must be specific with regard to the "spread effects" of resources so as to enable coverage of the whole of the population in relationship to their most basic health care needs. The central problem for the planner is not so much the volume of available resources as the ways in which they are being spent. Even if more resources were to become available, if they were to continue to be spent in the ways they are at present there would be little reason to expect much improvement in the coverage and effectiveness of the health care services. The central issues are the distribution of resources and the technical composition of the health care system. This technical composition is determined by the type of health care personnel being utilized, the type of equipment, the complexity of the pharmaceuticals, etc. All these must be appropriately adjusted so as to make it possible with limited resources to cover the whole of a population in keeping with its most basic health care needs. In keeping with this perspective it is necessary to examine the flow of resources through the whole of the health care system. It is necessary to know not only the sources of financing, but the flows between sources and providers, and then on to health care for the populations making use of the resources.

The making of health care plans must begin with the needs of the whole of the population and the possibilities for full coverage of that population. Any particular activity, project, pilot, demonstration, etc. must have built into it the possibility of replicability of its activities on a national or at least regional/district scale. As discussed above, it is necessary to change the technology being utilized in the delivery of health care; that is, the "production functions" for the outputs of units of health care must be altered. To do this would require a reorientation of the entire health care system in favor of support to primary health care. Primary health care (PHC) cannot be seen simply as another "pillar" within the health care system, but rather as the central aspect of the entire system, and all other parts of that system must be built around that central aspect. Given the limited resources available it will be necessary for the community to participate in this primary health care system. Of course, the issue of community participation/control is very complex and will be impossible to accomplish in very many, if not most, countries which tend to be run very strictly from the top down and in which the bottom is expected merely to respond to directives from the top.

There are three basic sources of finance for an expanded basic needs strategy (and its PHC component). The most obvious is increased budgetary support by ministries of finance and planning for the basic needs elements of ministerial activities. The second source, which might be considered essential for the first, is a redistribution of existing ministerial budgets toward their basic needs activities. This course of action would be most easy to follow when combined with growing ministerial budgets. The third source of finance and (in the longer term, at least) the most significant is based on the creation of new resources from within local communities themselves. These resources would take the shape of community workers, local administrative schemes, community labor for the construction of facilities, etc. With regard to the potential of community resources it is important to note that such resources would be most unlikely either to be forthcoming or to be socially useful in the absence of a reorganized national effort capable of offering appropriate support to community level efforts. In some ways this is analogous to increased ministry of finance support for ministry budgets being predicated on the redirection of ministry resources toward basic needs activities.

There are several different ways (and jargons) of formulating the fundamental requirements needed for the successful creation/planning of basic needs. One (rather crude) way of recapitulating these requirements is as follows:

1. Central planning policies and decisions directed toward the needs of the whole population, beginning with the poorest
2. Individual ministries and other government bodies offering maximum support to the basic needs components of their respective areas, e.g., agriculture, education, health, etc.
3. Appropriate decentralization of planning, implementation, and administrative decisions and activities
4. Popular participation/control and democratization of the nation's institutions

IV. PRIMARY HEALTH CARE AND CATEGORICAL DISEASE PROGRAMS

A. The Issue

Recent events, most notably the first International Conference on Primary Health Care, have tended to highlight the differences between two classic approaches to many disease control activities. In somewhat oversimplified terms, there appear to be, on the one hand, the "verticalists," favoring categorically specific, hierarchically organized, discrete disease control programs, and, on the other, those favoring integrated, "horizontal" health care delivery systems as the basis of a mixed group of disease control/health-promoting activities. In practice, these two ideal type tendencies are often brought together, to a greater or lesser degree.

The verticalists are accused by the integrationists of being overly narrow, of not appreciating the social causation of most disease and hence essentially social nature of its cure, of seeking only technological ("magic bullet") solutions to problems that are better approached through improved forms of human organization, of attempting to impose external technological hierarchies on peoples rather than working through organized communities, and, finally, of having failed too often in their past efforts (even their successes are said to have been mostly unique events, e.g., smallpox) and in their continuing zeal for the vertical campaign approach to be blocking the pathways leading to improved (integrated) health care systems. On the other hand, the integrationists are accused by the verticalists of being woolly-minded and unscientific, of trying to impose vague concepts in the social sciences on very real disease vectors, of romanticism and lack of appreciation of hierarchial discipline, and, finally, of failing to appreciate the progress that has already been made through specific disease control campaign-type activities.

The International Conference on Primary Health Care held in Alma Ata in September 1978 reflects, within the health sector, the important shift in thinking about the nature of Third World underdevelopment discussed earlier, e.g., from GNP to basic needs. Although there is little doubt of the conference's importance, at least potentially, for the industrialized countries, the more immediate impact is likely to be felt in the less industrialized parts of the world. There are many reasons for this; two will be cited here. One is the obviously pressing need to narrow the very wide gap in health status that currently exists between the richer and poorer countries. The other is the fact that the very limited health resources available to Third World countries appear to make a primary health care approach not only more relevant, but perhaps also uniquely essential. The movement away from growth of national product per se as the key component of development means that the improvement of people's health need no longer be perceived as primarily a resultant of growth, nor need it wait upon that growth, but rather should be and indeed can be accomplished within the framework of existing resource constraints. In fact, it is argued that a healthy, educated, socially involved population is a necessity for any true development. The lesson of "economic growth without development" has been learned by many development theorists (if not by all that many governments).

The primary health care approach, typically as developed under the guidance of WHO, represents the key health services/sector component of the basic needs strategy. The Alma Ata Conference on PHC represented a major international organizational effort to stimulate understanding and adherence on the part of governments to the ideas and practices of primary health care. However, despite the universal rhetorical support being given to the idea of PHC (although very basically different definitions of PHC are being offered), many governments and agencies remain tied to more traditional views of the causes of disease and the best ways of organizing scarce resources within existing and proposed disease control programs. These more traditional views are often expressed in the context of continuing support for categorical disease control programs and opposition to the integration of such programs into more generalized PHC activities. Not surprisingly, at least some of the apparent intellectual struggles between verticalists and integrationists are based more on considerations of empire protection and/or building than anything else. It is also worth noting that a number of programs originally conceived in more or less vertical terms are exploring new PHC related strategies for the accomplishment of their goals, e.g., malaria control.

B. The Case of Malaria

The malaria eradication campaign was probably the largest single disease control program ever attempted. In technical terms its problems were reasonably representative of those encountered by other such efforts (in the apparently successful smallpox effort the disease had no animal reservoir). After considerable early success the malaria eradication effort had to be replaced by one directed at "control," and even this is now tending to break down. In conjunction with the end of the eradication effort and limitations on control, a massive resurgence of malaria has occurred over large parts of the Third World. This massive resurgence has created a pressing need for new approaches to the control of the disease.

Professor D. Banerji, one of the leading public health researchers in India, has summarized the history of the vertical campaign approach to malaria in India as follows:

> The fact that despite their obvious over-riding importance, preventive services have received a much lower priority in the development of the health service system of India provides an insight into the value system of the colonels of the Indian Medical Service, the British trained bureaucrats of the Indian Civil Service and, above all, the value system of the political leadership of free India. The colonels did not appear to relish the prospects of dirtying their hands—getting involved in problems which required mobilisation of vast masses of people living in rural areas. The rural population raised in the minds of these decision makers the spectre of difficult accessibility, dust and dirt and superstitions, ignorant, ill-mannered and illiterate people. Therefore, when they were impelled to do some preventive work in rural areas, characteristically, they chose to launch military style campaigns against some specific health problems.
>
> Undoubtedly, because of the enormous devastation caused by malaria till the early fifties, this disease deserved a very high priority. But the program became a special favorite of the colonels not only because it required relatively much less community mobilisation, but it also provided them with an opportunity to build up an administrative framework to launch an all out assault on the disease in a military style—in developing preparatory, attack, consolidation and maintenance phases in having "unity of command," and surprise checks and inspections and in having authority to "hire and fire." Significantly, some of the followers of the colonels went so far as to compare the malaria campaign with a military campaign (1, p. 10).

In practice, the vertical campaign approach has proven to be antithetical to the development of technically integrated, nonbureaucratic, community-controlled health care systems. It need hardly be said that the possibility of such systems in turn are dependent on appropriate social relationships in the larger society. The campaign approach has been fostered by those who seek the bases of ill health in the characteristics of individual diseases rather than the social order and the solution to those disease problems primarily in the availability of more funding and more efficient bureaucratic controls.

The goal of complete eradication and the setting up of totally separate and vertical malaria control organizations often meant their taking a very large proportion of all public sector health spending. It might very well be argued that the public sector share of all spending has been far too small in many countries, both generally and specifically in the health sector, or that much of it was being spent badly, but the fact remains that malaria was frequently taking an exceptionally large part of total public sector health spending (in Ethiopia in 1970 the Malaria Eradication Service had a budget that was more than three times greater than that for the entire rural health service, and in Bangladesh in 1975–76 about one-fifth of recurrent health sector expenditure went to the malaria service, and it had been even higher at an earlier date).

The goal of eradication and the consequent vertical campaign approach to malaria was justified by the argument that with sufficient spraying it would be possible to interrupt transmission of the malaria parasite and thus see the complete disappearance of the disease within some particular time frame. It is important to note that, in contrast, a control strategy does not presuppose specific interruption of the transmission or an end to the disease, only its control at some particular level. This distinction is critical to an understanding of the political economy of "malaria campaigning." The basis for the creation of independent malaria organizations and the input of a very large volume of resources (at least as measured in Asia and Africa) into these bureaucracies was justified by the expectation that malaria would be done away with *everywhere and forever*. However, if control was to have been the target, then it would have been necessary to weigh expenditures directed against malaria in keeping with other priority areas within the health sector. The problems which prevented the eradication of malaria are too extensive to be gone into here, but one important technical consideration has been rapidly growing vector (mosquito) resistance so that "a switch by antimalaria campaigns from DDT to malathion would increase by five times the cost per head per annum while a switch to propoxur would increase it by 20 times" (4, p. 11).

It is unfortunate that so much of the current discussion about disease control is perceived in terms of either a "PHC approach" or another vertical campaign. Most disease control activities should and most likely will contain elements of both. The problem is to determine the most relevant specific national historical experiences and possibilities. It is clear that to the degree there is a primary health care infrastructure capable of carrying appropriate specific disease control activities, such programs become more feasible and less subject to the cost and other constraints which have led to the defeat of so many programs in the past. This is an area in which much fruitful research remains to be done.

V. CONCLUSION

This paper does not appear to call for any particular set of conclusions, as most of the discussion has been relatively self-contained. It is only necessary to restate the major underlying assumption of the paper, namely, that the primary determinants of human welfare in any society lie within its social order and political economy, and not in the type or level of development of any particular services as such. The bases of poverty among Third World populations are the maldistribution of power and resources between the industrialized and developing countries and within the less developed countries themselves. Without a fundamental redistribution of resources and power, basic needs and other such related programs cannot be expected to have any significant and lasting positive effects. It may be, though, that the struggle for greater equity, as expressed in redistributive programs, can be made part of wider effort to affect still more meaningful political and social change.

NOTE

1. This section relies very heavily on the ideas contained in two outstanding papers written by researchers based at the University of Sussex, England. One, done for United Nations Environment Programme (UNEP), is by B. Dasgupta, B. Johnson, and H. Singer; the other is by C. M. Cooper and R. Otto (see references 2 and 3).

REFERENCES

1. Banerji, D. *Social and Cultural Foundations of the Health Services Systems of India,* Centre of Social Medicine and Community Health, Jawaharlal Nehru University, New Delhi, 1974..
2. Cooper, C. M., & R. Otto. *Social and Economic Evaluation of Environmental Impact on Third World Countries—A Methodological Discussion,* mimeo, 1977.
3. Dasgupta, B., B. Johnson, and H. Singer. *Environment and Development: A Conceptual Overview,* mimeo, 1978.
4. World Health Organization. (1976) *Vector Resistance to Insecticides: A Review of its Operational Significance in Malaria Eradication and Control Programs,* WHO/MAL 76.833, WHO/VBC 76.634.

PART II
HEALTH IN HUMAN CAPITAL FORMATION

ADOLESCENT HEALTH, FAMILY BACKGROUND, AND PREVENTIVE MEDICAL CARE

Linda N. Edwards and Michael Grossman

This paper investigates the health of white adolescents, focusing particularly on the roles of family background and preventive medical care. This emphasis is motivated in part by our desire to study adolescent health in the context of the nature-nurture controversy. Despite the existence of a massive literature on the relative importance of heredity (nature) and the home and school environment (nurture) in the determination of cognitive development,[1] the corresponding issue has not been directly addressed by researchers in child and adolescent health. This is partly because much of the health research is limited either to poverty or to minority populations [Hu (25); Kessner (31); Inman (26); Dutton (15); Dutton and Silber (16)], and partly because researchers who use representative samples do not adopt the multivariate context necessary for distinguishing between genetic and environmental influences [Douglas (13); Douglas and

Bloomfield (14); Kellmer-Pringle, Butler, and Davie (28); Haggerty, Roghmann, and Pless (24); Zimmer (52)]. Our research uses multivariate statistical techniques to provide some evidence of the degree to which nurture—that is, the family and local environment—acts in determining the health levels of a representative sample of white adolescents.

One aspect of the adolescent's environment, medical care, has been recognized as the logical vehicle for public policy aimed at improving adolescent health. For example, Newberger, Newberger, and Richmond (42), Keniston and the Carngeie Council on Children (30), and Marmor (35) all have proposed that national health insurance should provide coverage of prenatal care, pediatric care, and dental care. Bills with this aim have been introduced in Congress by Senator Jacob K. Javits and Congressman James H. Scheuer, both of New York. To cite another illustration, recently enacted federal legislation has attempted to increase the availability of pediatricians and dentists in medically underserved areas to expand the use of preventive care in such areas. The Emergency Health Personnel Act of 1970 (PL 91-623) created the National Health Service Corps., whose members are assigned to health manpower shortage areas. The Health Professions Assistance Act of 1976 (PL 94-484) encourages new graduates of medical and dental schools to locate in urban ghettos and rural regions by forgiving their medical education loan obligations. Further, the Health Maintenance Organization (HMO) Act of 1974 (PL 93-222) gives priority for developmental funding of HMOs in medically deprived areas. One objective of our research is to provide estimates of the potential payoffs to national health insurance and medical manpower policies directed at improving youths' health.

The specific health indicators we study are oral health, obesity, anemia, and corrected distance vision. These four are chosen not only because they represent health problems that create discomfort for the teenager, but more importantly, because they may be good predictors of subsequent adult health. Indeed, they all partly reflect poor health habits that are likely to persist into adulthood. With the growing evidence that adults' choice of life styles and health behaviors can have important impacts on their health [Breslow and Klein (7); Fuchs (19), (20); Grossman (22); Manheim (33)], it is natural to look into adolescence to understand the formation of these habits. A second motivation for choosing these indicators is that they represent health problems that are capable of being affected by family decisions concerning diet and other forms of at-home health care, as well as by pediatric and dental care. This is in contrast to many adolescent health problems that are either self-limiting, such as morbidity from acute conditions, or irreversible, such as congenital abnormalities of the neurological system.

To analyze these health problems we use data from Cycle III of the U.S. Health Examination Survey (HES), an exceptional source of information about a national sample of 6768 noninstitutionalized youths aged 12 to 17 years in the 1966–70 period.[2] The data comprise complete medical histories of each youth provided by the parent, information on family socioeconomic characteristics, and birth certificate information. Most important, there are objective measures of health from detailed physical examinations given to the youths by pediatricians and dentists employed by the Public Health Service. These data are supplemented by two medical resource inputs specific to the youth's county of residence (the number of pediatricians per capita and the number of dentists per capita) and information on the presence of controlled or natural fluorides in the water supply system that services the youth's community. The last piece of information enables us to evaluate the impact of a collective, as opposed to an individual, preventive dental practice.

These data are used to estimate two types of relations: a health production function and a derived demand function for preventive care. The resulting estimates permit us to answer the following four questions. What is the size of the home environmental effect on adolescent oral and physical (obesity, anemia, corrected distance vision) health outcomes? How important is the home environment as a determinant of the demand for preventive dental and pediatric care? How large are the effects of dentists, preventive dental care, and fluoridation on oral health outcomes? How large are the effects of pediatricians and preventive pediatric care on physical health outcomes? In addressing the last two questions, we recognize explicitly the commonsense proposition that an increase in a community's physician or dental manpower will not increase health outcomes unless it encourages more utilization of medical care services. Previous empirical work on the impact of physicians or dentists on health has not taken account of this restriction [for example, Newhouse and Friedlander (43)].

Our findings indicate first, that family characteristics do have a significant impact on adolescent health, and second, that preventive care is an important vehicle for this impact in the case of dental health but not in the case of the three physical health measures. Similarly, the greater availability of dentists has a positive impact on dental health, but greater availability of pediatricians does not alter the physical health measures. On the basis of these results we predict that government efforts to improve the dental health of adolescents with policies to lower the cost of dental care or increase the availability of dentists are much more likely to be successful than similar policies directed at improving their physical health.

I. ANALYTICAL FRAMEWORK

In a previous paper [Edwards and Grossman (17)], we have argued that offsprings' health can be examined fruitfully within the context of the economic models of fertility developed by Becker and Lewis (4), Willis (51), and Ben Porath and Welch (5). In these models the parents' utility function depends on their own consumption, their family size, and the "quality" of each child. Child "quality" refers to those characteristics of the child that generate utility for the parents: his health, sex, wealth, social adjustment, intellectual development, etc.

Therefore, when parents choose their optimal family composition, they choose not only how many children they will have but also what portion of the family's resources will be devoted to each child. This choice is made in the usual way: parents choose the number and quality of children, as well as of other consumption goods, so as to maximize their utility subject to the constraints imposed by their wealth (their potential earned and nonearned income) and the various prices they face. In the case of children, there is a further constraint in the form of children's genetic endowments which in part determine their quality. Genetic endowments act as a constraint because they are largely outside the family's control.

The prices of children and of the various components of their quality are determined by a fundamental insight embedded in the household production function approach to consumer behavior: consumers produce their basic objects of choice with inputs of goods and services purchased in the market and their own time [Becker (2)]. This insight is of particular relevance in dealing with children and their health because parents obviously do not buy these objects of choice directly in the market; both a child's home environment and his genetic endowment are important determinants of his ultimate health level. Therefore, the price of health depends on the cost of the parents' or other caretakers' time and the prices of medical care, nutrition, and any other purchased inputs used to improve children's health. It also depends on the number of children in the family because the more children there are in the family, the more costly it is to raise their average health level. In addition, to the extent that there are systematic differences in the ability of families to produce children's health with given inputs, these differences in efficiency are also relevant. For example, more educated parents are more likely to be able to follow doctors' instructions, to have general information about nutrition, and to be willing and able to acquire medical information from published materials. Consequently, one would expect more educated parents to be more efficient in producing healthy children.

Given these considerations, the following factors are expected to influence children's health levels: the child's exogenous (genetic) health endowment, family wealth, parents' wage rates, family size, parents' educational attainment and other measures of their efficiency in household production, and the direct and indirect costs of medical care and other market health inputs (vitamins, sanitation, etc.).[3] (The indirect costs of medical care are generated by the time spent in traveling, waiting, and obtaining information about this care.[4]) The relationship between the child's ultimate health and this set of factors may be termed a demand function for the output of health. In this demand function a positive association between children's health and family wealth is predicted (assuming that child health is a normal good). Similarly, a positive association is expected between both parents' education and children's endowed health status and children's ultimate health status. Negative associations would be anticipated between all of the prices of health inputs and children's health, and between family size and children's health. Parents' wage rates may have negative or positive effects on children's health levels depending on whether the household production of children's health is more or less time-intensive than the production of other aspects of child quality and/or other types of parents' consumption commodities. In this framework a child's health is treated as a single datum—his permanent health measured, say, at the beginning of adulthood or as an average over his childhood and adolescence. This type of model is not formulated to explain variations in health over childhood or to examine the child's contribution to his own health.[5]

The above model provides a useful setting within which to view adolescent health, but empirical estimation of the resultant "demand for health" function would not yield answers to the questions posed in the Introduction. Such estimates would only yield information about the total impact of family characteristics or medical input prices on children's health. To determine the effect of preventive care on health we need estimates of the health production function. Similarly, to determine whether families with specific characteristics are more efficient at producing healthy children also requires estimates of this production function. Alternatively, to assess the role of family characteristics in determining the amount of preventive care received by adolescents, an estimate of a derived demand function for medical care is needed. Finally, a computation of the impact of health manpower availability on adolescent health requires not only the above functions but also a set of market demand and supply for health manpower functions. In the latter case, we employ a simplified approach which yields rough estimates of these manpower availability effects on health.

A. The Health Production Function

A simple, linear health production function[6] is represented by

$$H = \beta_0 + \beta_1 E + \beta_2 G + \beta_3 M + \beta_4 X + \beta_5 R + u_1. \tag{1}$$

Here H is a health measure, E is a vector of family efficiency characteristics, G is a vector of the adolescent's endowed health characteristics, M is a medical or dental care input, X is a vector of other family inputs (nutrition, parents' time, etc.), R is a vector of relevant regional characteristics (city size, region of the country, and whether or not the water supply is fluoridated), and u_1 is a random error term with the usual properties.

The health production function actually estimated in Section III does not correspond exactly to Eq. (1) because of inadequate data. First, data on the amount of "other" inputs (X) are not available. Therefore, we include the following proxy measures for X: family income, family size, and the mother's labor force status. Family income is positively related and family size is negatively related to nutrition and other unmeasured market health inputs. Family size and mother's labor force status are proxies for the amount of time the mother spends with each of her offspring. Women who work full-time or part-time in the labor market and women with many offspring have less time to spend with each one. In addition, our data do not include good information about curative care. Consequently, M represents only preventive care. This is not a serious deficiency because we have chosen health measures for which the impact of preventive care (with the associated remedial treatment) is relatively large. (By focusing on health problems for which the medical input is primarily preventive, we also avoid the necessity of modeling the simultaneous determination of health levels and curative care utilization.)

B. The Derived Demand for Preventive Care

The derived demand function for medical care depends on the same set of variables as the demand function for health:

$$M = \gamma_0 + \gamma_1 F + \gamma_2 G + \gamma_3 P + \gamma_4 R + u_2. \tag{2}$$

F represents family income, education, family size, and other family characteristics affecting either the demand for health or the family's efficiency in producing healthy children; G and R are the same as in Eq. (1); P represents a vector of relevant direct and indirect input prices (wage rates, the cost of a doctor or dental visit, etc.); and u_2 is the usual random error term.

We cannot estimate this derived demand curve exactly as stated because data on P are not available. Inclusion of variables representing the mother's labor force status helps control for variations in the mother's wage rate. Other input prices are partially controlled for by the region and city-size variables in R. Finally, physician or dentist availability measures are included to represent differences in the direct and indirect costs of medical or dental care.[7] Thus, rather than Eq. (2), we estimate the following:

$$M = \alpha_0 + \alpha_1 F + \alpha_2 G + \alpha_3 D + \alpha_4 R + u_3, \tag{3}$$

where the vector F now includes the mother's labor force status and D represents the number of pediatricians or dentists per capita in the adolescent's county of residence.

C. The Role of Health Manpower Availability

It is the inclusion of manpower availability measures in the derived demand for preventive care functions that permits us to obtain a rough assessment of the impact of health manpower on the demand for preventive care, and consequently on adolescent health. Only a rough assessment is possible because to get precise estimates it is necessary to have, first, data on the direct and indirect costs of medical care and second, measures of the price elasticity of supply of physicians or dentists. Good estimates of the supply elasticities do not exist, and it is almost impossible to measure all of the indirect costs of medical care. Although data on direct costs do exist, they are not usually found in conjunction with the detailed health and family background data used here. Thus, our estimate of the impact of health manpower on health is the best that can be obtained given the limitations of existing data sets. The coefficients of the health manpower variables in the derived demand equations embody both the relationship between health manpower availability and direct and indirect medical care prices and the relationship between medical care prices and the demand for preventive care.

Implicit in the above discussion is the assumption that an increase in a community's health manpower will not improve the health of adolescents unless it encourages a greater utilization of preventive care services. This assumption is explicitly incorporated in Eqs. (1) and (3): D is assumed to have no direct effect on health in Eq. (1) but alters health only via its impact on M in Eq. (3). Substituting Eq. (3) into Eq. (1) yields estimates of the total impact of doctor or dentist availability on health:

$$H = \beta_0 + \beta_1 E + \beta_3\alpha_1 F + (\beta_2 + \beta_3\alpha_2)G + \beta_3\alpha_3 D + \beta_4 X + (\beta_5 + \beta_3\alpha_4)R + u_1 + \beta_3 u_3. \quad (4)$$

The total impact of pediatrician or dentist availability on health is given by $\beta_3\alpha_3$. Note that an estimate of the total impact computed from individual estimates of α_3 and β_3 differs from that obtained from direct estimation of Eq. (4) because the latter does not incorporate the restriction that D does not appear in Eq. (1).[8]

D. The Role of Family Background Variables

To the extent that there are family background variables common to both the set E and the set F (parents' educational attainment is a good example of one), the substitution in Eq. (4) provides an additional insight. Parents' education is clearly seen to have two effects on adolescent health: a direct or "efficiency" effect given by β_1 and an indirect or "allocative" effect given by $\beta_3\alpha_1$. The latter refers to the ability of parents with greater schooling levels to select a better input mix in the production function.[9]

II. EMPIRICAL IMPLEMENTATION

Equations (1) and (3) are estimated using Cycle III data for white adolescents who live with either both of their parents or with their mothers only. Black adolescents are excluded from the empirical analysis. Preliminary results revealed significant race differences in slope coefficients so that pooling blacks and whites for estimation was inappropriate. Separate estimates for black adolescents are not presented because the black sample is too small to allow for reliable coefficient estimates. Observations are also deleted if there are missing data. The final sample size is 4121. Table 1 contains definitions, means, and standard deviations of all of the dependent and independent variables. It also contains a notation concerning the source of each variable.

Table 1. Definitions of Variables

Variable Name	Sample[a] Mean	Sample Standard Deviation	Definition	Source[b]
A. Health Measures				
APERI[c]	−.114	.857	Periodontal index, standardized by the mean and standard deviation of one-year age-sex cohorts	2
IDECAY[c]	−.146	.839	Number of decayed permanent teeth, standardized by the mean and standard deviation of one-year age-sex cohorts	2
OBESE	.103	.305	Dummy variable that equals one if the physician rates the youth as obese or very obese	2
PVIS	.042	.201	Dummy variable that equals one if youth wears glasses and his corrected binocular distance vision is 20/40 or worse or if youth does not wear glasses and his uncorrected binocular distance vision is 20/40 or worse	2
ANEMIA	.023	.149	Dummy variable that equals one if youth is a female whose hematocrit level is more than two standard deviations below the mean for females 12 to 17 years of age *or* if youth is a male whose hematocrit level is more than two standard deviations below the mean for his stage of sexual maturity	2
B. Preventive Medical Care Measures				
DTPREV	.697	.460	Dummy variable that equals one if youth saw a dentist for a checkup within the past year	1
DRPREV	.588	.492	Dummy variable that equals one if youth saw a doctor for a check-up within the past year	1
FLUOR	.584	.493	Dummy variable that equals one if the community in which the youth lives uses naturally fluoridated or controlled fluoridated water	See text
C. Other Variables				
FINC	9.614	5.112	Continuous family income (in thousands of dollars) computed by assigning midpoints to the following closed income intervals, $250 to the lowest interval, and $20,000 to the highest interval. The closed income classes are:	1

Variable Name	Sample[a] Mean	Sample Standard Deviation	Definition	Source[b]
			$500– $999	
			$1,000– $1,999	
			$2,000– $2,999	
			$3,000– $3,999	
			$4,000– $4,999	
			$5,000– $6,999	
			$7,000– $9,999	
			$10,000–$14,999	
FEDUCAT[d]	11.327	3.227	Years of formal schooling completed by father	1
MEDUCAT	11.142	2.843	Years of formal schooling completed by mother	1
NOFATH	.099	.297	Dummy variable that equals one if youth lives with mother only	1
FLANG	.139	.346	Dummy variable that equals one if a foreign language is spoken in the home	1
LESS20	3.360	1.853	Number of persons in the household 20 years of age or less	1
MWORKFT	.268	.443	Dummy variable that equals one if the mother works full-time or part-time, respectively; omitted class is mother does not work	1
MWORKPT	.154	.361		
DENT	.584	.216	Number of dentists per thousand population in community of residence of youth	See text
PED	.051	.027	Number of pediatricians per thousand population in community of residence of youth	See text
NEAST	.253	.435	Dummy variables that equal one if youth lives in Northeast, Midwest, or South, respectively; omitted class is residence in West	1
MWEST	.291	.454		
SOUTH	.203	.402		
URB1	.193	.395	Dummy variables that equal one if youth lives in an urban area with a population of 3 million or more (URB1); in an urban area with a population between 1 million and 3 million (URB2); in an urban area with a population less than 1 million (URB3); or in a non-rural and non-urbanized area (NURB); omitted class is residence in a rural area	1
URB2	.132	.339		
URB3	.194	.396		
NURB	.146	.353		
LMAG	.077	.267	Dummy variable that equals one if the mother was less than 20 years-old at birth of youth	1
HMAG	.096	.294	Dummy variable that equals one if mother was more than 35 years-old at birth of youth	1

(*continued*)

Table 1 (Continued)

Variable Name	Sample[a] Mean	Sample Standard Deviation	Definition	Source[b]
LIGHT1	.010	.098	Dummy variable that equals one if youth's birth weight was under 2,000 grams (under 4.4 pounds)	3
LIGHT2	.032	.177	Dummy variable that equals one if youth's birth weight was equal to or greater than 2,000 grams but under 2,500 grams (under 5.5 pounds)	3
BWUK	.245	.430	Dummy variable that equals one if youth's birth weight is unknown	3
FYPH	.117	.321	Dummy variable that equals one if there was a medical difficulty with youth before the age of one year	1
ABN	.200	.400	Dummy variable that equals one if the diagnostic impression of the physician was that the youth had a significant abnormality	2
TWIN	.023	.150	Dummy variable that equals one if youth is a twin	1
FIRST	.497	.500	Dummy variable that equals one if youth is the first born in the family	1
AGE	14.335	1.661	Age of youth	1
MALE	.528	.499	Dummy variable that equals one if youth is a male	1

Notes:
[a] The means and standard deviations are computed for the sample of 4121 white youths described in the text.
[b] The sources are 1 = parents, 2 = examination, 3 = birth certificate. See text for sources of FLUOR, PED, and DENT.
[c] The mean of this variable is not zero because standardization was done using the entire Cycle III sample rather than the subsample reported on in this paper. In particular, the negative mean reflects the better oral health of white youths compared to black youths.
[d] For youths who were not currently living with their father, father's education was coded at the mean of the sample for which father's education was reported.

A. Measurement of Adolescent Health

In the Introduction to this paper, we expressed an intention to study physiological measures of adolescent health that (1) reflect detrimental health behaviors or life styles that may persist and create more serious problems in adulthood; and (2) relate to problems that can be modified by endogenous inputs in the health production function such as proper diet, parents' time, and especially preventive medical care.[10] Based on these criteria, we focus on two correlates of poor oral health: the periodontal index and the number of decayed permanent teeth; and on three correlates of poor physical health: obesity, abnormal corrected distance vision, and anemia as reflected by low hematocrit levels. All five measures clearly relate to conditions that can carry on into adulthood, and all can be modified by appropriate care. Dental care provided by dentists has a direct impact on tooth decay and periodontal disease. The prescription of eyeglasses by an ophthalmologist or an optometrist can remedy abnormal distance vision. Pediatricians also play an important role in eye care because they often are responsible for examining a youth's eyes initially and referring his parents to an eye specialist if necessary. Finally, all the health measures excluding vision reflect basic nutritional factors that can be modified by the appropriate diet. These measures are described in detail below.

The periodontal index (APERI) is a good overall indicator of oral health as well as a positive correlate of nutrition [Russell (45)]. Kelly and Sanchez (27, pp. 1–2) describe the periodontal index as follows:

> Every tooth in the mouth . . . is scored according to the presence or absence of manifest signs of periodontal disease. When a portion of the free gingiva is inflamed, a score of 1 is recorded. When completely circumscribed by inflammation, teeth are scored 2. Teeth with frank periodontal pockets are scored 6 when their masticatory function is unimpaired and 8 when it is impaired. The arithmetic average of all scores is the individual's [periodontal index], which ranges from a low of 0.0 (no inflammation or periodontal pockets) to a high of 8.0 (all teeth with pockets and impaired function).

It is clear from this description that higher values of the periodontal index correspond to poorer dental health. Our measure, APERI, is scaled somewhat differently from that described above in order to remove the well-known age and sex trends in the periodontal index. APERI is computed as the difference between the adolescent's actual periodontal index and the mean index for his or her age-sex group divided by the standard deviation for that age-sex group.[11] A similar method of age and sex standardization is used for our other measure of oral health, the number of decayed permanent teeth (IDECAY). We employ two measures of dental health because it is one of the few health problems for which well-defined continuous health measures have been developed.

Obesity is represented by a dichotomous variable that equals 1 if the physician rates the youth as obese or very obese (OBESE). The physician presumably takes account of the youth's height, age, and sex in making his evaluation.

Anemia is represented by a dichotomous variable that equals 1 if the youth's hematocrit level is "excessively" low (ANEMIA).[12] The hematocrit level of a female youth is considered to be excessively low if it is more than two standard deviations below the mean for all females 12 to 17 years of age. The hematocrit level of a male youth is considered to be excessively low if it is more than two standard deviations below the mean for all males in his stage of sexual maturity. This procedure is based on Daniel's (10) findings that (1) hematocrit values differ by sex; (2) these values depend on sexual maturity rather than age for male adolescents; and (3) hematocrit levels are independent of age and sexual maturity for female adolescents.[13]

Abnormal corrected distance vision is denoted by a dichotomous variable that equals 1 if a youth wears glasses and his corrected binocular distance vision is 20/40 or worse *or* if a youth does not wear glasses and his uncorrected binocular distance vision is 20/40 or worse (PVIS). This standard of abnormal distance vision is the one used by National Center for Health Statistics (38).

It is instructive to consider measures of adolescent health that are excluded by our selection criteria. Abnormal hearing is subject to medical intervention, but the prevalence rate of this condition is less than 1 percent in the HES. Hence, it is far too rare to pose a threat to the future lifetime well-being of a significant percentage of adolescents. High blood pressure is not studied because there is a lack of consensus among pediatricians concerning the importance of this condition in adolescence and the appropriate treatment [National Heart, Lung, and Blood Institute's Task Force (41)]. Moreover, the measures of high blood pressure in Cycle III are somewhat suspect (National Center for Health Statistics (39)]. Congenital abnormalities are a source of current and future difficulties, but we do not study them because to a large extent they are irreversible. Parental ratings of adolescent health and other subjective indicators are avoided because of the possibility that responses depend on the parents' socioeconomic status. Parents with low levels of income and schooling are likely to be dissatisfied with many aspects of their life including the health of their offspring. Finally, we do not include measures relating to the "new morbidity" such as "learning difficulties and school problems, behavioral disturbances, . . . and the problems of adolescents in coping and adjusting . . . " [see Haggarty, Roghmann, and Pless (24), p. 316]. While such measures may well reflect life styles that have serious health consequences, they are unlikely to be revealed in a physical exam. Nor are they likely to be easily altered by preventive medical care. Although examination of these and other excluded health measures would be necessary to paint a complete picture of the health of this adolescent cohort, it is not relevant to the objectives of this paper.

B. Measurement of Preventive Dental and Medical Care

Preventive dental care is measured by a dichotomous variable that equals 1 if the youth saw a dentist for a checkup within the past year (DTPREV). Similarly, preventive pediatric care is measured by a dichotomous variable that equals 1 if the youth saw a doctor for a checkup within the past year (DRPREV). These variables distinguish between two groups of adolescents: (1) those who received preventive care; and (2) those who received no care at all or only curative care. These two measures of preventive care are preferred to alternatives like the number of dental or physician visits or the receipt of curative care alone because our measures are less likely to reflect reverse causality from poor health to more medical care. Of course, our measures reflect the possibility that adolescents received treatment as well as an examination, but the appropriate treatment of problems revealed by an annual checkup is an integral component of preventive care.

Fluoridation is indicated by a dichotomous variable that is equal to 1 if the community in which the youth resides uses naturally fluoridated or controlled fluoridated water (FLUOR). Naturally fluroridated communities are serviced by a water supply system that contains a natural fluoride content of 0.7 part per million or higher. They are identified by the Division of Dental Health of the National Institutes of Health (12). Controlled fluoridated communities are those that have adjusted the fluoride content of their water supply systems to the optimum level. They are identified by the Division of Dental Health of the National Institutes of Health (11). For youths who reside in controlled communities, the fluoridation variable equals 1 only if the date on which that youth was examined in the HES succeeds the date on which the community adjusted the fluoride content of its water supply system. This insures that youths in controlled communities actually were exposed to fluoridated water.[14]

C. The Pediatrician and Dentist Availability Measures

The youths in Cycle III were selected from 38 distinct primary sampling units. The primary sampling unit is a county or a group of several contiguous counties, some of which form a standard metropolitan statistical area. We obtained data on the number of dentists per capita (DENT) in each youth's primary sampling unit (hereafter termed his county or community of residence) for the year 1968 (the midyear of the Cycle III survey) from publications of the American Dental Association. The number of pediatricians is not available for the years during which the HES was conducted (1966–70). Therefore, we use the number of pediatricians per capita in the county of residence (PED) for the year 1964 from the American Medical Association [Theodore and Sutter (47)].[15] We believe that the number of pediatricians in 1964 is a good proxy for the number in 1968. Although youths receive medical care from other types of physicians—general practitioners, internists, and ophthalmologists—these physicians also service adults while pediatricians do not. Therefore, we focus on pediatricians as the most important suppliers of physicians' services to youths.[16]

D. Measurement of Other Explanatory Variables

Many of the remaining explanatory variables called for in Sec. 2 require no further elaboration. Parents' educational attainment and family income, for example, are adequately described in Table 1. Some of the other variables listed in Table 1, however, do require additional explanation.

Family size is represented by the number of people in the family who are under 20 years of age at the time of the Cycle III interview (LESS20). Consequently, it may overstate or understate actual completed family size.

Three measures of the family's efficiency in producing healthy children are included in addition to the parents' educational attainment. These are dichotomous variables that identify youths whose mothers were under age 20 when the youths were born (LMAG), youths from homes in which a foreign language is spoken (FLANG), and youths who live with their mothers only (NOFATH). Young mothers are notoriously less efficient at contracepting and may be similarly less efficient in producing healthy offspring. Foreign-born families are likely to exhibit differences in household productive efficiency. The absence of her spouse from the household is likely to hinder the mother's allocative efficiency in selecting the input mix with which to produce health. The absence of a father also impinges upon the amount of time that a mother can spend with her children.[17]

The youth's endowed health status is represented by four variables relating to his early health. The first two (LIGHT1, LIGHT2) are dummy variables identifying youths of low birth weight. Low birth weight is a typical indicator of a less healthy birth outcome [for example, Birch and Gussow (6)]. Birth weight was obtained from the youth's birth certificate. Since birth certificates are missing for approximately 25 percent of the sample and since we do not focus on the effects of birth weight, we do not delete these observations. Instead, we include a dummy variable that identifies youths with missing birth certificates (BWUK) in the regression estimates. The third endowment measure is a dummy variable identifying youths whose mothers were over 35 years old at the youth's birth (HMAG). The rationale for including this variable is that older mothers are more likely to have offspring with health defects. The last of these measures is a dummy variable which identifies youths whose parents reported a medical difficulty with the youth before the age of 1 year (FYPH). Although parents' reports of youths' medical problems before the age of 1 year are subject to recall error, the first year of a child's life is likely to stand out in his parents' minds relative to other stages in his life cycle. Therefore, we believe that the measurement error in this variable is small.

One current health indicator is used as a proxy for the child's unmeasured genetic health endowment and his health history beyond age 1. This indicator is the presence of at least one significant abnormality as reported by the HES physician who examined the youth (ABN). Abnormalities include heart disease and neurological, muscular, or joint conditions; other major diseases; and otitis media. Except for the last condition, which constitutes a relatively small percentage of all reported abnormalities, these health problems are to a large extent congenital and irreversible.

We also control for several other characteristics of the youth which are not necessarily health-related but may cause him to receive better or worse treatment within the family. They are his birth order (FIRST) and whether or not he is a twin (TWIN). First-born youths (or nontwins) will have greater access to individual parental attention because they arrived in the family first (or they arrived alone). In addition, the youth's age (AGE) and sex (MALE) are included in regressions in which the dependent variable is not adjusted for age and sex (i.e., when the dependent variable is either obesity, abnormal corrected distance vision, preventive dental care, or preventive pediatric care).

Finally, three region variables (NEAST, MWEST, SOUTH) and four sizes of place of residence variables (URB1, URB2, URB3, NURB) are included to control for regional differences that are not otherwise taken into account. We are agnostic about the nature of these differences, but want to avoid the possibility that the health manpower and fluoridation effects are biased by an omission of unmeasured regional characteristics.

III. EMPIRICAL RESULTS

In this section we present estimates of Eqs. (1) and (3) and compute the total impact of family characteristics and health manpower availability on adolescent health as given in Eq. (4). Equations (1) and (3) form a recursive system which can be estimated using single equation techniques as long as $E(u_1 u_3) = 0$. We make this assumption here. Although all the dependent variables except the two oral health measures are dichotomous, the method of estimation is ordinary least squares. Preliminary investigation revealed almost no differences between ordinary-least-squares estimates and dichotomous logit estimates obtained by the method of maximum likelihood. When the dependent variable is dichotomous the fitted equation can be interpreted as a linear probability function in which the regression coefficient of a given independent variable represents the change in the conditional probability of poor physical health or receipt of preventive care for a one-unit change in the independent variable. The resultant estimates also embody the assumption that several variables that may be considered endogenous (mother's labor force status and family size, for example) are exogenous to adolescent health.[18] Finally, our estimates cannot be unambiguously interpreted as production functions or derived demand equations because insufficient data forced us to use proxy measures for some of the explanatory variables.

Estimates of the dental health production functions and the preventive dental care demand function are discussed in the first part of this section. The physical health production functions and the preventive pediatric care demand function are discussed in the second part. Both discussions are centered on answering the questions posed in the introduction concerning the roles of the family, preventive care, and health manpower availability in determining adolescent health. In examining the results, it is important to remember that the five health measures (APERI, IDECAY, OBESE, PVIS, ANEMIA) are negative correlates of good health, so that negative effects of independent variables in the production functions reflect factors associated with *better* health outcomes. The two preventive medical care measures (DTPREV, DRPREV), on the other hand, are positive correlates of care; thus positive effects of independent variables in the demand functions reflect factors associated with higher propensities to obtain preventive care.

A. Oral Health

Estimates of the oral health production functions and the preventive care demand function are in Tables 2 and 3, respectively. When the number of decayed permanent teeth (IDECAY) is the dependent variable, two production functions are estimated. The first contains the same set of independent variables as the periodontal index (APERI) regression, while the second includes APERI as

Table 2. Ordinary-Least-Squares Estimates of Oral Health Production Functions[a]

Independent Variable	APERI Regression Coefficient	t-Ratio	IDECAY Regression Coefficient	t-Ratio	IDECAY (with APERI) Regression Coefficient	t-Ratio
FEDUCAT	−.016	−2.95	−.011	−2.10	−.007	−1.41
MEDUCAT	−.030	−4.80	−.028	−4.67	−.021	−3.57
DTPREV	−.255	−8.47	−0.27	−9.42	−.212	−7.48
FLUOR	−.082	−2.96	−.159	−5.96	−.139	−5.39
FINC	−.007	−2.36	−.015	−4.85	−.013	−4.40
LESS20	.036	4.41	.023	2.90	.014	1.86
MWORKFT	.046	1.47	.088	2.91	.077	2.63
MWORKPT	−.007	−0.20	.055	1.54	.057	1.64
NEAST	.038	1.01	.264	7.26	.255	7.24
MWEST	−.072	−1.99	.183	5.24	.200	5.92
SOUTH	−.057	−1.42	.131	3.38	.145	3.85
URB1	.003	0.10	−.071	−1.85	−.072	−1.93
URB2	.027	0.63	−.028	−0.68	−.035	−0.87
URB3	−.088	−2.36	−.153	−4.25	−.132	−3.78
NURB	−.001	−0.00	−.077	−1.96	−.077	−2.02
LMAG	−.009	−0.17	−.032	−0.67	−.030	−0.64
HMAG	.023	0.47	−.048	−1.10	−.053	−1.26
LIGHT1	.187	1.40	−.127	−0.98	−.172	−1.38
LIGHT2	.081	1.10	.123	1.73	.103	1.50
BWUK	.014	0.45	.062	2.07	.058	2.02
ABN	.173	5.39	.032	1.04	−.009	−0.32
FYPH	.023	0.57	−.102	−2.64	−.107	−2.87
NOFATH	.016	0.33	.148	3.29	.145	3.31
FLANG	−.051	−1.28	−.172	−4.45	−.159	−4.27
TWIN	−.023	−0.26	.107	1.26	.112	1.37
FIRST	−.013	−0.47	−.024	−0.88	−.021	−0.79
APERI	—	—	—	—	.240	16.47
CONSTANT	.566		.532		.396	
Adj. R^2	.086		.114		.169	
F	15.85[b]		21.37[b]		31.99[b]	

Notes:
[a] The critical t-ratios at the 5 percent level of significance are 1.64 for a one-tailed test and 1.96 for a two-tailed test.
[b] Statistically significant at the 1 percent level of significance.

Table 3. Ordinary-Least-Squares Estimate of Preventive Dental Care Demand Function[a]

Independent Variable	Regression Coefficient	t-Ratio	Independent Variable	Regression Coefficient	t-Ratio
FEDUCAT	.009	3.22	LIGHT2	.070	1.84
MEDUCAT	.023	7.27	BWUK	.005	0.32
FLUOR	.003	0.20	ABN	−.002	−0.14
DENT	.170	4.05	FYPH	.018	0.88
FINC	.011	6.99	NOFATH	−.025	−1.04
LESS20	−.033	−7.95	FLANG	−.037	−1.79
MWORKFT	−.034	−2.07	TWIN	−.002	−0.00
MWORKPT	.044	2.30	FIRST	.007	0.49
NEAST	.046	2.24	AGE	−.006	−1.53
MWEST	.041	2.17	MALE	−.022	−1.67
SOUTH	−.017	−0.83	CONSTANT	.331	
URB1	−.008	−0.37	Adj. R^2	.150	
URB2	−.016	−0.64	F^b	27.02	
URB3	.003	0.14			
NURB	−.029	−1.37			
LMAG	−.065	−2.53			
HMAG	−.018	−0.75			
LIGHT1	.054	0.79			

Notes:
[a] The critical t-ratios at the 5 percent level of significance are 1.64 for a one-tailed test and 1.96 for a two-tailed test.
[b] Statistically significant at the 1 percent level of significance.

an additional independent variable. It has been suggested by Russell (45) that variations in APERI result largely from genetic factors. If these genetic factors are correlated with the home environment and imperfectly measured by the health endowment variables, the second regression will give a more accurate estimate of the effects of the home environment on IDECAY than the first. Of course, APERI has an environmental component as well as a genetic component (as is evident from our estimate in Table 2). Therefore, the two IDECAY regressions contain upper and lower bound estimates of the impact of the environment on IDECAY.[19]

Most notable among the results are the large significant impacts of a preventive dental visit on both the periodontal index and the decay index.[20] The coefficient estimates imply that adolescents who did not have a preventive checkup within the past year have periodontal indices and decay scores that are each about .3 standard deviation worse than adolescents who received a checkup. When APERI is included in the decay equation, the decay differential between the two groups of adolescents declines to 2 but remains statistically significant. To gauge the magnitudes of these effects, recall that APERI and IDECAY have means of approximately 0 and standard deviations of approximately 1. Therefore, the oral health differentials associated with absence of preventive care are relatively large; they range from 20 to 30 percent of the standard deviations in the scores. Moreover, the differentials apply to a substantial proportion of the sample: 30 percent of the youths in the HES did not have a checkup in the past year. These findings underscore the efficacy of preventive dental care.

The results pertaining to a publicly provided form of preventive care—water fluoridation—are also strong. Youths exposed to fluoridated water (FLUOR) have significantly better oral health than other youths at all conventional levels of confidence.[21] The fluoridation differentials are smaller, however, than the corresponding preventive dental care differentials in oral health. For example, the fluoridation coefficient in the periodontal index equation is one-third as large as the preventive dental care coefficient. In the decay equations, the ratio of the two coefficients ranges from three-fifths to two-thirds. Nevertheless, given that the per-child cost of fluoridation is substantially below the cost of a preventive dental visit, this remains a cost-effective method of improving dental health.[22]

Let us turn now to the role of the family in determining adolescent dental health levels. The four characteristics of the family environment we focus on are parents' education (MEDUCAT, FEDUCAT), family income (FINC), family size (LESS20), and mother's labor force status (MWORKFT, MWORKPT). An overview of the production function estimates in Table 2 reveals that all six variables have statistically significant effects in the expected directions (with the exception of mother's labor force status in the periodontal index equation). Children of more educated parents have better oral health, as do children from families with higher income; while children whose mothers are employed full-time or who come from larger families have poorer health. The impacts of these variables on IDECAY are reduced in absolute value when APERI is held constant, but the pattern of statistical significance is not dramatically altered (only the coefficient of father's schooling becomes insignificant). It is clear, then, that these family characteristics have an important impact on adolescent dental health.

We interpret these findings as evidence that the home environment plays an important role in determining children's health. It can be argued, however, that our results do not really constitute strong evidence in favor of "nurture" because of the likelihood of positive correlations first between these family characteristics and the parents' health and second between the genetically determined components of parents' and children's health. Put differently, this argument states that family characteristics such as income or parental education largely reflect genetic health factors. For example, parents who are themselves healthy are more likely to be in the labor force and will have higher earnings. Or, parents who have had a healthy childhood and adolescence are more likely to have attained a higher level of education. Two of our findings, however, cast doubt on the applicability of this argument in our case. First, when we include APERI in the decay equation in an effort to more fully control for genetic factors, we still find that these family environment variables have significant impacts on IDECAY. This is noteworthy because the inclusion of APERI is likely to bias the coefficients of the family environment variables toward zero (see note 19).

A second and stronger reason revolves around the coefficients of the educational attainment of the two parents. If the education effect is primarily genetic, we would expect the coefficients of both mother's and father's education to be equal because both parents make an equal genetic contribution to the child. On the other hand, if the education effect is primarily environmental, we would expect the impact of the mother's education to be larger because she is the family member most concerned with the children's health care. In Table 2 we observe that in every case the coefficient of mother's education exceeds that of father's education. In addition, despite a high correlation between the two education variables (r = .61), the difference in coefficients is always statistically significant at the 10 percent level.[23] Thus, our results clearly indicate that the family environment, and in particular the mother's education, plays an important role in producing healthy children.

Besides having an important impact on the production of health, family characteristics work to improve adolescent heath by increasing the probability that an adolescent receives preventive care. In Table 3 we see that all six of the family variables have significant impacts on the probability that an adolescent received preventive care. Children from families with higher annual income, more educated parents, and in which the mother works part-time are more likely to receive preventive care, while children from larger families or families where the mother works full-time are less likely to receive preventive care. As an example of the magnitude of these effects, the probability that a child received preventive care in the previous year increases by about 2 percentage points for each additional year of education received by the mother and declines by about 3 percentage points for each additional child in the family. Once again we believe that these results reflect environmental rather than genetic influences: the mother's education coefficient is more than twice as large as the father's education coefficient and the difference between them is statistically significant (t = 2.82).

To determine the total effect of family characteristics on health—both the direct effect embodied in the production function estimates and the indirect effect that operates through the family's proclivity to obtain preventive care—we compute the total impact of these family characteristics in Table 4. The reported coefficients are analogous to the sum $(\beta_1 + \beta_3\alpha_1)$ in Eq. (4).[24] Comparison of the coefficients in Tables 3 and 4 indicates that the total impact is from 10 to 100 percent greater than the "direct" effect alone. We also observe, as before, a large and statistically significant (at the 5 percent level) difference between the impacts of father's and mother's education, again lending support to our conclusion that "nurture" matters.[25]

With regard to the role of health manpower, we see that it has a large significant effect on the family's propensity to obtain preventive care for its children

Table 4. Total Impacts (Direct and Indirect) of Selected Variables on Oral Health

Variable \ Oral Health Measure	APERI	IDECAY	IDECAY (with APERI)
FEDUCAT	−.018	−.013	−.009
MEDUCAT	−.036	−.034	−.026
DENT	−.043	−.047	−.036
FINC	−.010	−.018	−.015
LESS20	.044	.032	.021
MWORKFT	.055	.097	.084
MWORKPT	−.018	.043	.048

(Table 3). An increase of one dentist per thousand population increases the probability that adolescents visited the dentist for preventive care in the previous year by 17 percentage points. This estimate is identical to one obtained by Manning and Phelps (34) and is insensitive to the exclusion of region and size of place of residence from the equation.[26] The implied effect on adolescent health (assuming that dentist availability has no direct impact on adolescent health but operates only by increasing the family's propensity to obtain preventive care) is given in Table 4 and ranges from −.036 to −.047 standard deviation in the dental health measures.[27] Thus, an increase in the number of dentists in an area by 1 per 1000 population is equivalent in its effect on dental health to an increase in the level of the mother's education by one and one-third years.

It should be noted that the positive impact of dentists on the propensity to obtain a checkup is unlikely to reflect demand manipulation by dentists. The concept of demand manipulation refers to the ability of health personnel to shift the demand curve for their services, when all direct and indirect costs of these services are held constant. In his extensive treatment of this phenomenon, Pauly (44) shows that the demand manipulation effect should be larger in a sample of consumers with positive utilization than in a sample of all consumers. Moreover, his model gives no basis for expecting a demand manipulation effect in an equation that explains the probability of a checkup. Based on these considerations, we view the dental manpower variable as reflecting the importance of information, entry, travel, waiting, and direct costs in the parents' decision to obtain preventive dental care for their off-spring.

Most of the other results in Table 4 are consistent with our expectations and will not be discussed.[28] We do wish to point out, however, that although fluoridation does have a significant impact on dental health, it is not significantly related to the probability of obtaining preventive dental care. This is not surprising since from a theoretical point of view either a positive or negative relationship could be predicted. If fluoridation is regarded as an increase in the child's health endowment, the quantity of care demanded should fall. On the other hand, if the increased endowment also increases the marginal product of preventive care, or if it lowers the psychic costs of obtaining care by reducing the severity of the tooth decay uncovered by a preventive checkup, a positive effect on the quantity of care demanded would be predicted.[29] Both types of results have been reported in other studies. Manning and Phelps (34) report mixed effects of duration of exposure to fluoridation on the propensity to obtain preventive dental checkups for white children below the age of 15 in a 1970 health survey conducted by the National Opinion Research Center. Upton and Silverman (49) use 1966 data for 15 midwestern towns, half of which used fluoridated water, and report fewer restorations of children's permanent teeth in the fluoridated towns.

We conclude this subsection by using our results to estimate the impacts of three government programs to improve the oral health of youths. First, consider a $1000 income transfer to low-income families. As shown by the reduced-form coefficients of FINC in Table 4, the transfer would lower the periodontal index of youths from these families by .01 point and would lower their decay index by .02 point. (Such a program would naturally also have other beneficial effects on children and their families.) Next consider a program to reduce or eliminate regional differences in the number of dentists per thousand population. Dentists are more numerous in urban areas than in rural areas. To take two sites in the HES, there were 1.1 dentists per 1000 population in San Francisco, while there was .2 dentist per 1000 population in San Benito, Texas. Suppose that this difference were eliminated by raising the number of dentists in San Benito by 1 per 1000 population. Then the periodontal index of youths in San Benito would fall by .04 point, and their decay index would fall by .05 point.[30] Finally, consider an 80 percent reduction in the price of a dental checkup as a result of the enactment of a national health insurance plan for dental care with a 20 percent coinsurance rate. Based on research by Manning and Phelps on the impact of price on the propensity to obtain preventive dental care for children and youths, we estimate that such a policy would raise the probability of obtaining care by 16 percentage points. This would improve both the periodontal and the decay scores by .04 point.[31]

We view the above computations as illustrative rather than definitive. To choose among the three programs, information on the cost of each program and on the number of youths affected clearly is required. Moreover, as indicated in Section I, definitive computations of impact effects should take account of the supply elasticity of dental care and the exact nature of the relationship between dental manpower and the indirect costs of obtaining dental care.

B. Physical Health

Estimates of the physical health production functions and the preventive care demand function appear in Tables 5 and 6, respectively. Looking first at the production function estimates, we are struck by the fact that these physical health

Table 5. Ordinary-Least-Squares Estimates of Physical Health Production Functions[a]

Independent Variable	OBESE Regression Coefficient	t-Ratio	PVIS Regression Coefficient	t-Ratio	ANEMIA Regression Coefficient	t-Ratio
FEDUCAT	.001	0.46	−.002	−1.51	−.001	−0.65
MEDUCAT	−.005	−2.01	.001	0.71	−.002	−2.07
DRPREV	−.005	−0.50	−.005	−0.75	.002	0.47
FINC	.0002	0.20	−.001	−1.04	−.001	−1.39
LESS20	−.007	−2.41	.006	3.31	.004	2.60
MWORKFT	.005	0.40	.001	0.10	.002	0.37
MWORKPT	−.007	−0.54	−.00003	−0.00	.0003	0.00
NEAST	.040	2.94	−.003	−0.36	−.013	−1.93
MWEST	.036	2.80	.012	1.45	−.010	−1.63
SOUTH	−.020	−1.36	.009	0.93	.002	0.24
URB1	−.017	−1.15	−.0002	−0.00	.007	0.96
URB2	.023	1.47	.010	0.91	.010	1.33
URB3	−.011	−0.79	−.009	−1.01	.015	2.27
NURB	.016	1.09	−.008	−0.85	.014	1.94
LMAG	.023	1.27	−.011	−0.92	.005	0.54
HMAG	.017	1.01	.002	0.20	.003	0.40
LIGHT1	−.001	−0.00	.003	0.10	−.020	−0.84
LIGHT2	−.010	−0.36	−.003	−0.17	−.010	−0.73
BWUK	.0004	0.00	−.012	−1.59	−.013	−2.39
ABN	.122	10.38	.009	1.17	−.008	−1.40
FYPH	−.006	−0.40	.006	0.59	.001	0.10
NOFATH	.021	1.22	.009	0.81	.006	0.71
FLANG	.015	1.06	.021	2.13	.001	0.20
TWIN	.004	0.10	.017	0.79	−.020	−1.25
FIRST	.021	1.95	.008	1.20	.003	0.66
AGE	−.004	−1.23	−.008	−4.28	—	—
MALE	−.053	−5.65	−.024	−3.87	—	—
CONSTANT	.190		.164		.049	
Adj. R^2	.039		.013		.008	
F	7.21[b]		2.97[b]		2.35[b]	

Notes:
[a] The critical t-ratios at the 5 percent level of significance are 1.64 for a one-tailed test and 1.96 for a two-tailed test.
[b] Statistically significant at the 1 percent level of significance.

measures are much less amenable to statistical explanation than are the dental health measures. Of course, lower R^2 values would be expected for the three physical health measures because they are dichotomous rather than continuous. But many fewer explanatory variables are statistically significant in the physical health case. Clearly unmeasured genetic or "luck" factors play a much larger role in the case of these health measures.

Table 6. Ordinary-Least-Squares Estimate of Preventive Pediatric Care Demand Function[a]

Independent Variable	Regression Coefficient	t-Ratio	Independent Variable	Regression Coefficient	t-Ratio
FEDUCAT	.013	4.28	LIGHT1	−.031	−0.40
MEDUCAT	.007	1.82	LIGHT2	.063	1.47
PED	.675	1.97	BWUK	.001	0.10
FINC	.005	2.59	ABN	.094	5.00
LESS20	−.017	−3.60	FYPH	.042	1.81
MWORKFT	.015	0.80	NOFATH	−.071	−2.61
MWORKPT	.005	0.24	FLANG	.017	0.72
NEAST	.062	2.77	TWIN	−.025	−0.49
MWEST	−.031	−1.50	FIRST	.058	3.47
SOUTH	−.021	−0.91	AGE	.004	0.95
URB1	.021	0.87	MALE	.064	4.30
URB2	.036	1.34	CONSTANT	.192	
URB3	−.014	−0.61	Adj. R^2	.067	
NURB	−.037	−1.56	F	11.99[b]	
LMAG	−.032	−1.09			
HMAG	−.055	−2.09			

Notes:
[a] The critical t-ratios at the 5 percent level of significance are 1.64 for a one-tailed test and 1.96 for a two-tailed test.
[b] Statistically significant at the 1 percent level of significance.

In contrast to the results for preventive dental care, there is little evidence that preventive medical care is efficacious. Youths who saw a doctor for a checkup within the past year (DRPREV) have a one-half percentage point smaller probability of being obese or of having abnormal corrected distance vision than other youths, and a one-fifth percentage point *higher* probability of having anemia. None of these three differentials is statistically significant. One possible explanation for these findings is that there are fairly long lags between the receipt of preventive care and an improvement in physical health. Alternatively, one might argue that physicians play a minor role in the outcomes studied here relative to unmeasured endogenous inputs such as proper diet. The nonsignificant impact of preventive care also means that family characteristics operate on health only through the production function. There are no indirect effects of the various family characteristics on physical health, only direct effects. Consequently, we do not present a table of "total" effects (comparable to Table 4) in the case of the physical health measures.

The relationship between family characteristics and health is also much weaker in the case of physical health. Most of the six family characteristics variables studied are not even statistically significant in the production function; only the mother's education and family size variables have significant impacts. Children of more educated mothers are less likely to be obese or anemic, and they are more likely to have poor vision (the latter relationship is not significant). Children from larger families are more likely to have poor corrected vision or be anemic, but they are less likely to be obese. To get an impression of whether these effects can be viewed as environmental as opposed to genetic, we again look at the difference between the coefficients of the two parents' education variables. For both OBESE and ANEMIA, the mother's education coefficient is larger than the corresponding father's education coefficient, but for PVIS the opposite is true. Only in the case of obesity is the difference significant at the 10 percent level. Thus, in this case the evidence regarding a nature versus nurture interpretation of the family effects is not conclusive, but it does suggest that with respect to obesity at least one component of the family environment—mother's education—has an important impact.

We noted that the family size variable has a perverse sign in the obesity equation: children from larger families are in better rather than worse health in that they are less likely to be obese. The positive relationship between family size and the incidence of the other health problems is easy to rationalize (it may reflect a substitution away from higher "quality" children as the shadow price of quality rises),[32] but a justification for the negative relation reported for obesity is less obvious. One possible explanation for this negative family size effect (as well as for the positive income effect) is the existence of joint production among various aspects of quality. For example, families with fewer children or higher income may consume richer and higher-calorie foods. This consumption raises some aspects of quality but at the same time makes obesity more likely.

The finding of nonsignificant effects of family income in physical health outcomes has important implications. First, it suggests that policies to improve the well-being of adolescents via income transfers would have little impact on our physical health measures. Second, this finding coupled with the significance of the mother's schooling variable underscores the key role in health production of nonmarket productivity as opposed to market goods and services as measured by family income and preventive care. This result echoes our earlier findings for a group of younger children [Edwards and Grossman (17)]. In the case of obesity, we believe that the impact of mother's schooling reflects the information that highly educated mothers have acquired as part of the schooling process about the dangers of obesity and about what constitutes an appropriate diet.

Family effects in the derived demand for preventive pediatric care (Table 6) tend to be much stronger than they are in the production functions (although the R^2 in the preventive pediatric care equation is still substantially lower than in the preventive dental care equation). Among the six family variables, only the mother's labor force status variables do not have a significant impact on the family's probability of obtaining preventive care. Families with higher parental education and more income are more likely to get preventive care for their children, while larger families are less likely to. In addition, father's education has a larger impact than mother's education. It is not clear how to interpret these results, however, since we have no evidence that preventive pediatric care is efficacious.

The last result to be discussed concerns the role of pediatrician availability. Similar to the corresponding findings for preventive dental care, the number of pediatricians per 1000 population in the county of residence (PED) has a positive and statistically significant regression coefficient in the demand curve for pediatric care. This finding complements those reported by Kleinman and Wilson (29) and Colle and Grossman (8). However, the implied effects of an increase of 1 pediatrician per 1000 population are small ($-.003$, $-.003$, and $.001$ for OBESE, PVIS, and ANEMIA, respectively), primarily because the health impact of preventive care is small and not significant. Thus our findings indicate that a policy to increase pediatric manpower in medically underserved areas would not improve the physical health of adolescents—at least as represented by our three measures. Such a policy should be given a much lower priority than an analogous policy to expand dental manpower in areas characterized by shortages.[33]

IV. SUMMARY AND IMPLICATIONS

The purpose of this study has been to examine the determinants of the oral and physical health of white adolescents with special emphasis on the roles of family background and the use of preventive medical care. The main results of the study are (1) nurture plays an important role in determining oral health but less so for the other health problems studied; (2) preventive care is efficacious in the case of oral health but not for the other health problems studied; and (3) the three physical health measures are largely unexplained by the family and preventive care variables used here. Only mother's education and family size have significant impacts.

With respect to the first result, mother's schooling is singled out as a crucial component of the home environment. Although mother's schooling, father's schooling, family income, and family size all make significant contributions to oral health, mother's schooling dominates father's schooling. Moreover, mother's schooling tends to dominate both income and father's schooling in the physical health equations, especially in the case of obesity. The finding that the impact of mother's schooling almost always exceeds that of father's schooling is especially important because equal effects would be expected if the schooling variables were simply proxies for unmeasured genetic endowments.

Two additional pieces of evidence underline the robustness of the finding that nurture "matters." First, the relative magnitude of the effect of the various family background variables on the index of tooth decay is not greatly altered when the periodontal index, a proxy for genetic oral health endowment, is held constant. Second, the identification of a plausible mechanism by which family characteristics influence adolescent health—preventive care—increases our confidence that these variables reflect a behavioral effect as opposed to a genetic effect or a statistical artifact.

With regard to the role of preventive dental care, youths who received a preventive dental checkup within the past year and youths exposed to fluoridated water have much better oral health than other youths. Moreover, the probability of a preventive examination is positively related to the number of dentists per capita in a youth's county of residence. This implies that a program to increase the availability of dentists in medically deprived areas would improve the oral health of youths in these areas. Indeed, we estimate that the payoffs to increasing dental manpower by 1 per 1000 population are about the same as the payoffs to the coverage of preventive dental care under national health insurance.

The probability of obtaining a preventive checkup by a doctor is also positively related to family income and to the number of pediatricians per capita in the county of residence. But we have little evidence that preventive care delivered to youths by physicians is efficacious in terms of their physical health. Therefore, the payoffs to national health insurance for physicians' services delivered to youths or programs to increase the availability of doctors who treat youths are very small.

Our results for the physical health measures are weak, but one pair of findings does stand out. Adolescents are less likely to be obese if their mothers are highly educated, and they are more likely to be obese if they come from small families. The latter relation provides a partial explanation of the dramatic increase in obesity during recent decades, since over the same period we have seen a startling decline in family size. The former relation, on the other hand, suggests a strategy for slowing down the trend in the incidence of this health problem. What is needed is a public information program—similar to that mounted in the case of childhood immunization—directed at alerting less educated parents, and especially mothers, to the dangers associated with childhood obesity.

Overall, what our results suggest is that selective rather than general programs would be most effective in improving the health of the population under 18 years of age. For instance, instead of providing complete coverage for physicians' services delivered to persons from birth to age 18 under national health insurance, the government should direct its attention at prenatal care and physicians' services during the first year of life. It is known that appropriate prenatal and infant care can make a difference in terms of health outcomes [for example, Lewit (32)]. Conversely, our results for oral health in this paper and in our previous research [Edwards and Grossman (17)] suggest that the payoffs to the coverage of dental care from the age it is first received until age 18 or beyond would be substantial.

ACKNOWLEDGMENTS

Research for this paper was supported by PHS Grant No. 1 R01 HS 02917 from the National Center for Health Services Research to the NBER and by a grant from the Robert Wood Johnson Foundation to the NBER. We are indebted to Jody Sindelar for comments on an earlier draft and to Ann Colle for research assistance. In addition, we would like to thank Dorothy Rice, Mary Dudley, Kurt Mauer, and especially Robert Murphy, all of the National Center for Health Statistics, for their cooperation and assistance in adding county-level variables to our Cycle III Health Examination Survey data tapes. This paper was presented at the fifty-fourth annual conference of the Western Economic Association, Las Vegas, Nevada, June 1979. It has not undergone the review accorded official NBER publications; in particular, it has not yet been submitted for approval by the Board of Directors.

NOTES

1. For a partial survey of this literature, see Grossman (22) and Edwards and Grossman (18).

2. A full description of the sample, the sampling technique, and the data collection is presented in National Center for Health Statistics (40).

3. Children's health also depends on the prices of inputs used to produce other aspects of their quality and the prices of other forms of parents' consumption. The effects of these variables will not be studied here.

4. For discussions of the indirect costs of obtaining pediatric care, see Colle and Grossman (8) and Goldman and Grossman (21).

5. One possible objection to using this type of framework to analyze the health of adolescents is that the goals of parents and youths are likely to differ. For instance, cigarette smoking by a youth might increase his utility but reduce his parents' utility because it is detrimental to his current or future health. This type of conflict between parents and youths has been analyzed by Becker (3) in the context of an economic model of social and family interactions. He shows that such conflicts are important when the parents' utility function depends on particular "merit" commodities consumed by the youth rather than on his consumption of all commodities. In such a case parents have an incentive to allocate resources not only to their children's consumption, but also to policing their offsprings' consumption patterns. An explicit melding of our model with Becker's would be a difficult task, and although it would alter the interpretations of the effects of various family characteristics, it would not add to or delete from the list of relevant explanatory variables.

6. Given the essentially arbitrary scaling of all of our adolescent health measures and the general ignorance concerning the exact specification of a health production function, we believe that it is inappropriate to experiment with more sophisticated functional forms.

7. The Bureau of Labor Statistics does collect measures of the prices of various goods and services, including physician and dental office visits, for 40 cities and 4 nonmetropolitan areas. We do not take price variables from this source because they are based on small samples, and the sites in the HES survey are not identical to the sites in the Bureau of Labor Statistics (BLS) survey. On the other hand, the number of dentists and pediatricians are based on complete enumerations in all countries by the American Dental Association and the American Medical Association and can be matched easily to the HES sites. Thus the two manpower variables have little measurement error, while the price estimates from the BLS would contain a great deal of measurement error.

8. This is in contrast to the work of Newhouse and Friedlander (43), who fit an equation similar to our Eq. (4).

9. The term *allocative effect* and the decomposition of the schooling parameter into direct and allocative components is due to Welch (50). He uses this framework to study the impact of schooling on market production. Technically, schooling is a relevant determinant of the demand for medical care even if it has no allocative efficiency effect. In simple models of schooling as an efficiency variable in household production [Grossman (23); Michael (37)], schooling raises the amount of health output obtained from a given vector of inputs. In such models schooling can lower the quantity of medical care demanded at the same time as it raises the quantity of health demanded. In particular,

medical care would rise only if the income and price elasticities of demand for offspring's health exceeded unity. We stress a model that incorporates an allocative efficiency effect because schooling should increase the parents' knowledge about what constitutes an appropriate diet, when to take their children to the doctor or the dentist for a preventive checkup, how to follow the doctor's advice, and how to foster appropriate oral hygiene behavior by their children. The ability of parents with extra schooling to select a better input mix, as well as to obtain a larger health output from given inputs, is likely to encourage them to demand larger quantities of preventive care even if the income and price elasticities of health are less than one. In part the effect may reflect a substitution toward preventive care and away from curative care.

10. In adopting these two criteria for the selection of health measures, we are guided in part by Kessner's (31) tracer methodology for studying the health of children and adolescents.

11. If the actual periodontal index of each age-sex group is normally distributed, APERI could be translated directly into the youth's periodontal index percentile. We have experimented with the actual value of the periodontal index as the dependent variable in a multiple regression that includes age, the square of age, and a dummy variable for male adolescents in addition to the remaining independent variables. The results obtained (not shown) are similar to those reported in Section III.

12. Dutton (15) advocates the use of a continuous, rather than a discrete, measure of anemia. She conducts a multiple regression analysis of actual hematocrit levels of black children between the ages of 6 months and 4 years. The only statistically significant variables in this regression (at the 5 percent level) are age and sex. Therefore, it is not at all clear what we would gain by adopting her measure.

13. Similar patterns are present in the Cycle III data. Tanner (46) stresses the importance of sexual maturity in the determination of the health and cognitive development of adolescents. Preliminary analysis revealed, however, that sexual maturity does not have an effect on our health measures except in the case of hematocrit levels of females.

14. Clinical evidence suggests that exposure to fluoridated water is particularly important if it occurs during the ages at which the permanent teeth are being formed [McClure (36)]. These teeth do not appear until a child is approximately 6 years old but start to be formed a few months after birth. Therefore, it is useful to identify youths who had been exposed to fluoridated water before they reached the age of 6 years. Unfortunately, we cannot do this because the youth's current residence alone is reported in the HES. We did create a fluoridation variable that identifies youths exposed before age 6 *under the assumption of no migration,* but it had no effect on oral health in regressions that included the fluoridation variable described in the text.

15. Our measure of the number of dentists excludes those in the federal dental service. The number of pediatricians pertains to those in private practice.

16. Since pediatricians treat only children and youths, the number of pediatricians per person under a certain age (say age 18) might appear to be a more relevant measure than the number of pediatricians per capita. We did not employ such a variable for several reasons. First, the appropriate age cutoff is not obvious. Second, even if pediatricians do not treat youths beyond the age of 17, since mothers typically are responsible for taking youths to the physician, the indirect costs of obtaining pediatric care might be more related to the number of pediatricians per woman with children below the age of 18 than to the number of pediatricians per person below the age of 18. Third and most important, there is little variation in persons under age 18 as a percentage of the population or in women with children under age 18 as a percentage of the population among the 38 sites in the HES.

17. The educational attainment of absent fathers is not known. For children with absent fathers, we code FEDUCAT at the mean level of father's education in the subsample of youths who live with both parents. This coding scheme is consistent with the assumption that father's education has the same relationship with adolescent health whether or not the father is actually present. An alternative assumption is that father's education has no affect on adolescent health if he is absent. Under this assumption, the education of absent fathers would be coded at zero. Use of the alternative coding scheme would alter the regression coefficient of NOFATH but would not alter the coefficient of FEDUCAT or the coefficients of other independent variables in the regression.

18. The health endowment variables are also endogenously determined because they are affected by family choices regarding prenatal care, timing of childbearing, and resources allocated to children since birth. Despite the endogeneity of the health endowment measures, mother's labor force status, and family size, preliminary computations revealed that the estimated coefficients of the other family background measures and of preventive care are only slightly altered by the exclusion of these variables from the equations.

19. Suppose that the periodontal and decay functions are

$$\text{APERI} = a_1 G + a_2 E + u_1 \tag{a}$$

and

$$\text{IDECAY} = b_1 G + b_2 E + u_2, \tag{b}$$

where G is genetic oral health endowment, E is the home environment, u_1 and u_2 are disturbance terms, and intercepts and other independent variables are ignored. Note that a_1, a_2, and b_2 are negative since a more favorable endowment or environment improves oral health. Solve Eq. (a) for G and substitute into Eq. (b) to obtain

$$\text{IDECAY} = b_1 a_1^{-1} \text{APERI} + (b_2 - a_2 b_1 a_1^{-1})E + u_2 - b_1 a_1^{-1} u_1. \tag{c}$$

Clearly, the absolute value of the parameter of E in Eq. (c) is smaller than the absolute value of the corresponding parameter in Eq. (b). Note that APERI is negatively correlated with the composite disturbance term $(u_2 - b_1 a_1^{-1} u_1)$ in Eq. (c). Therefore, if the equation is estimated by ordinary least squares, the regression coefficient of APERI is biased toward zero and that of E away from zero provided E and APERI are negatively related. In the text we make the plausible assumption that this upward bias in the absolute value of the regression coefficient of E is offset by the fundamental difference between the structural parameters of E in Eqs. (b) and (c). That is, we assume that the expected value of the regression coefficient of E understates $|b_2|$ even though it overstates $|b_2 - a_2 b_1 a_1^{-1}|$.

20. Statements concerning statistical significance in the text are based on one-tailed tests except when the direction of the effect is unclear on *a priori* grounds or when the estimated effect has the "wrong sign." In the latter cases two-tailed tests are used.

21. The estimated effects of fluoridation on oral health are not sensitive to the omission of the three region and four size-of-place-of-residence variables from the regressions. This indicates that the fluoridation variable is not simply a proxy for location.

22. *Consumer Reports* (9) cites a report in the *New England Journal of Medicine* which estimates the per capita cost of fluoridation to be about 10 to 40 cents/year (p. 393).

23. The relevant "t" statistics for the three equations in Table 2 are 1.41, 1.79, and 1.48. Note that probable biases in the estimates of the two parents' education coefficients are likely to work toward a finding of no significant difference. The estimate of the direct efficiency effect of father's schooling may be biased away from zero; and the estimate of the direct efficiency effect of mother's schooling may be biased toward zero. The former bias is introduced if father's education serves as a proxy for permanent income (if there is measurement error in current family income). The latter bias is introduced if more educated mothers allocate less time to the production of adolescent oral health because they have a higher opportunity cost of time, and if the opportunity cost of time effect is not fully reflected by the two measures of mother's labor force status. Along similar lines, the estimated father's education effect may be biased upward in the demand curve for preventive care. The mother's education effect is biased downward if oral health is "time-intensive" and if substitution in consumption outweighs substitution in production.

24. These could be thought of as solved "reduced-form" coefficients of the exogenous variables.

25. As is expected on the basis of the education coefficients in Tables 2 and 3, the difference in "total" effects is larger than the difference in direct effects. The test of the significance of the

difference between the "total" effect of mother's schooling and the "total" effect of father's schooling is based on the estimated reduced form—the ordinary-least-squares regression APERI or IDECAY on all the exogenous variables. This procedure is employed because standard errors of solved reduced-form coefficients and standard errors of differences between such coefficients are very difficult to compute. In every case, the estimated reduced-form difference between the schooling coefficient is exactly the same as the solved reduced-form difference. Therefore, the bias introduced by our test is minimal. The test statistics are 1.81, 2.23, and 2.82 for APERI, IDECAY, and IDECAY with APERI, respectively.

26. Manning and Phelps estimate a discriminant function of the probability of obtaining a checkup. They point out that the coefficients in this equation approximate logit coefficients. Since they do not indicate the mean probability of a checkup in their sample, we converted their logit coefficient of the number of dentists into a marginal effect at the mean checkup probability in the HES sample of .7. If m is the marginal effect of a given independent variable, b is its logit coefficient, and p is the probability of a checkup, the conversion formula is

$$m = bp(1 - p).$$

27. The finding that the periodontal index is inversely related to the number of dentists differs from that of Newhouse and Friedlander (43). Using adults in Cycle I of the HES, they report an insignificant positive effect of dentists per capita in the county of residence on the periodontal index. Their result is based on an ordinary-least-squares regression of the periodontal index on the number of dentists and other variables and does not embody the restrictions discussed in Section I.C.

28. There are two "perverse" results that are statistically significant: youths from families in which a foreign language is spoken in the home (FLANG) have better oral health than other youths; and youths whose parents reported a medical difficulty with the youth before the age of 1 year (FYPH) have less decay than other youths. The first of these may be caused by genetic differences in oral health between native Americans and immigrants or the native-born offspring of immigrants. We offer no explanation for the latter finding.

29. For a general discussion of endowment effects in models such as the one employed in this paper, see Tomes (48). A detailed treatment of the role of fluoridation in dental care demand functions appears in Upton and Silverman (49).

30. The reduction in the decay score is taken from the reduced-form coefficient of DENT obtained from the decay function that excludes APERI.

31. In their discriminant estimate of the decision for white children and youths to receive a dental exam, Manning and Phelps specify a price effect that varies with family income. Our extrapolation of their results assumes that (1) family income equals $10,000 (the mean value in the HES); (2) the uninsured price of checkup is $15; and (3) the uninsured probability of a checkup is .7 (the mean in the HES). The reduction in the decay score is obtained from the decay function that excludes APERI.

32. Alternatively, these effects may be attributed to a reduction in per capita income as family size rises with family income held constant. Indeed, the sign of the family size effect is opposite that of the family income effect in all three regressions. Yet something more than a mechanical relationship between family size and per capita income is required to account fully for the contribution of family size to health outcomes. For example, unlike the family size coefficients, the family income coefficients are not always statistically significant. In addition, computations reveal that the impact on physical health of a 1 percent increase in family size is larger in absolute value than that of a 1 percent increase in family income.

33. Some readers may object to the constraint in our recursive model that the direct effect of health manpower on health is zero. For the benefit of these readers, the estimated reduced-form coefficients of the number of pediatricians on obesity, abnormal vision, and anemia are $-.316$ (t = -1.47), $-.010$ (t = -0.07), and $-.168$ (t = -1.57), respectively. The estimated reduced-form coefficients of the number of dentists on the periodontal index and decay are .128 (t = 1.55) and .085 (t = 1.08), respectively. These coefficients give a very different and, in our view, inappropriate

picture of the payoff of a program to expand pediatric manpower compared to a program to expand dental manpower.

REFERENCES

1. American Dental Association, Bureau of Economic Research and Statistics (1969) *Distribution of Dentists in the United States by State, Region, District and County, 1968*, Chicago: American Dental Association.
2. Becker, Gary S. (1965) "A Theory of the Allocation of Time," *Economic Journal* 75 (299).
3. Becker, Gary S. (1974) "A Theory of Social Interactions," *Journal of Political Economy* 82 (6).
4. Becker, Gary S., and H. Lewis (1973) "On the Interaction between the Quantity and Quality of Children," In *New Economic Approaches to Fertility,* (ed.), T. W. Schultz, Proceedings of a conference sponsored by the National Bureau of Economic Research and the Population Council. *Journal of Political Economy,* 81 (2); Part II.
5. Ben-Porath, Yoram, and Finis Welch (1976) "Do Sex Preferences Really Matter?" *The Quarterly Journal of Economics* 90 (2).
6. Birch, Herbert G., and Joan Dye Gussow (1970) *Disadvantaged Children: Health, Nutrition, and School Failure,* New York: Harcourt, Brace and World, Inc.
7. Breslow, L., and B. Klein (1971) "Health and Race in California," *American Journal of Public Health* 61(4).
8. Colle, Ann D., and Michael Grossman (1978) "Determinants of Pediatric Care Utilization," In *NBER Conference on the Economics of Physician and Patient Behavior,* (eds.) Victor R. Fuchs and Joseph P. Newhouse, *Journal of Human Resources* 13, Supplement.
9. *Consumer Reports* (1978) "Fluoridation: The Cancer Scare," 43(7).
10. Daniel, Jr., W. A. (1973) "Hematocrit: Maturity Relationship in Adolescence," *Pediatrics* 52(3).
11. Division of Dental Health of the National Institutes of Health (1970) *Fluoridation Census 1969,* Washington, D.C.: United States Government Printing Office.
12. Division of Dental Health of the National Institutes of Health (1969) *Natural Fluoride Content of Community Water Supplies, 1969,* Washington, D.C.: United States Government Printing Office.
13. Douglas, J. W. B. (1951) "The Health and Survival of Children in Different Social Classes," *Lancet.*
14. Douglas, J. W. B., and J. M. Bloomfield (1958) *Children Under Five,* Allen and Unwin Ltd.
15. Dutton, Diana B. (October 18, 1978) "Hematocrit Levels and Race: An Argument Against the Adoption of Separate Standards in Screening for Anemia," Paper presented at the American Public Health Association Meetings in Los Angeles.
16. Dutton, Diana B., and Ralph S. Silber (1979) "Children's Health Outcomes in Six Different Ambulatory Care Delivery Systems," Mimeograph.
17. Edwards, Linda N., and Michael Grossman (1981) "Children's Health and the Family," In Volume II of the *Annual Series of Research in Health Economics,* (ed.), Richard M. Scheffler, Greenwich, Connecticut: JAI Press.
18. Edwards, Linda N., and Michael Grossman (1979) "The Relationship between Children's Health and Intellectual Development," In *Health: What is it Worth?,* (ed.), Selma Mushkin, Elmsford, New York: Pergamon Press, Inc.
19. Fuchs, Victor R. (1974) "Some Economic Aspects of Mortality in Developed Countries," In *The Economics of Health and Medical Care,* (ed.), Mark Perlman, London: MacMillan.
20. Fuchs, Victor R. (1974) *Who Shall Live?: Health, Economics and Social Choice,* New York: Basic Books, Inc.

21. Goldman, Fred, and Michael Grossman (1978) "The Demand for Pediatric Care: An Hedonic Approach," *Journal of Political Economy* 86(2), Part I.
22. Grossman, Michael (1975) "The Correlation Between Health and Schooling," In *Household Production and Consumption,* (ed.), Nestor E. Terleckyj, New York: Columbia University Press for the National Bureau of Economic Research.
23. Grossman, Michael (1972) *The Demand for Health: A Theoretical and Empirical Investigation,* New York: Columbia University Press for the National Bureau of Economic Research.
24. Haggerty, Robert J., Klaus J. Roghmann, and Ivan B. Pless (1975) *Child Health and the Community,* New York: John Wiley and Sons.
25. Hu, Teh-Wei (1973) "Effectiveness of Child Health and Welfare Programs: A Simultaneous Equations Approach," *Socio-Economic Planning Sciences* 7.
26. Inman, Robert P. (1976) "The Family Provision of Children's Health: An Economic Analysis," In *The Role of Health Insurance in the Health Services Sector,* (ed.), Richard Rosett, New York: Neale Watson Academic Publications for the National Bureau of Economic Research.
27. Kelly, James, E., and Marcus J. Sanchez (1972) *Periodontal Disease and Oral Hygiene Among Children,* National Center for Health Statistics, U.S. Department of Health, Education, and Welfare, Public Health Publication Series 11-No. 117.
28. Kellmer-Pringle, M. L., N. R. Butler, and R. Davie (1966) *11,000 Seven-Year-Olds,* New York: Humanities Press.
29. Kleinman, Joel C., and Ronald W. Wilson (1977). "Are 'Medically Underserved Areas' Medically Underserved?" *Health Services Research* 12(2).
30. Keniston, Kenneth, and the Carnegie Council on Children (1977) *All Our Children: The American Family Under Pressure,* New York: Harcourt Brace Jovanovich.
31. Kessner, David M. (1974) *Assessment of Medical Care for Children.* Contrasts in Health Status, Vol. 3, Washington, D.C.: Institute of Medicine.
32. Lewit, Eugene M. (1977) "Experience with Pregnancy, the Demand for Prenatal Care, and the Production of Surviving Infants," Ph.D. dissertation, City University of New York Graduate School.
33. Manheim, Lawrence (1975) "Health, Health Practices, and Socioeconomic Status: The Role of Education," Ph.D. dissertation, University of California at Berkeley.
34. Manning, Jr., Willard G., and Charles E. Phelps (1978) "Dental Care Demand: Point Estimates and Implications for National Health Insurance," Santa Monica, California: The Rand Corporation, R-2157-HEW.
35. Marmor, Theodore R. (1977) "Rethinking National Health Insurance," *The Public Interest,* No. 46.
36. McClure, F. J., (ed.) (1962) *Fluoridated Drinking Waters,* U.S. Department of Health, Education, and Welfare, Bethesda, Maryland.
37. Michael, Robert T. (1972) *The Effect of Education on Efficiency in Consumption,* New York: Columbia University Press for the National Bureau of Economic Research.
38. National Center for Health Statistics (1972) *Binocular Visual Acuity of Children: Demographic and Socioeconomic Characteristics,* U.S. Department of Health, Education, and Welfare, Public Health Publication, Series 11, No. 112.
39. National Center for Health Statistics (1977) *Blood Pressure of Youths 12-17 Years,* U.S. Department of Health, Education, and Welfare, Public Health Publication, Series 11, No. 163.
40. National Center for Health Statistics (1969) *Plan and Operation of a Health Examination Survey of U.S. Youths 12–17 Years of Age,* U.S. Department of Health, Education, and Welfare, Public Health Publication No. 1000-Series 1, No. 8.
41. National Heart, Lung, Blood Institute's Task Force on Blood Pressure Control in Children (1977) "Report of the Task Force on Blood Pressure Control in Children," *Pediatrics* 59(5), Supplement.

42. Newberger, Eli H., Carolyn Moore Newberger, and Julius B. Richmond (1976), "Child Health in America: Toward a Rational Public Policy," *Milbank Memorial Fund Quarterly* 54(3).
43. Newhouse, Joseph P., and Lindy J. Friedlander (1977) "The Relationship Between Medical Resources and Measures of Health: Some Additional Evidence," Santa Monica, California: The Rand Corporation, R-2066-HEW.
44. Pauly, Mark V. (1980) *Doctors and Their Workshops,* Chicago: University of Chicago Press for the National Bureau of Economic Research.
45. Russell, A. L. (1956) "A System of Classification and Scoring for Prevalence Surveys of Periodontal Disease," *Journal of Dental Research* 35.
46. Tanner, J. M. (1962) *Growth at Adolescence,* Oxford, England: Blackwell Scientific Publications.
47. Theodore, Christ N., and Gerald E. Sutter (1965) *Distribution of Physicians in the U.S., 1965,* Chicago: American Medical Association.
48. Tomes, Nigel (1978) "Intergenerational Transfers of Human and Non-Human Capital in a Model of Quality-Quantity Interaction," Ph.D. dissertation, University of Chicago.
49. Upton, Charles, and William Silverman (1972) "The Demand for Dental Services," *The Journal of Human Resources* 7(2).
50. Welch, Finis (1970) "Education in Production," *Journal of Political Economy* 78(1).
51. Willis, Robert (1973) "A New Approach to the Economic Theory of Fertility Behavior," In *New Economic Approaches to Fertility,* (ed.), T. W. Schultz, Proceedings of a conference sponsored by the National Bureau of Economic Research and the Population Council. *Journal of Political Economy* 81(2); Part II.
52. Zimmer, B. G. (1978) "Impact of Social Status and Size of Family on Child Development," Mimeograph.

AN ECONOMIC ANALYSIS OF THE DIET, GROWTH AND HEALTH OF YOUNG CHILDREN IN THE UNITED STATES

Dov Chernichovsky and Douglas Coate

> One out of every three children under six years of age is living in homes in which incomes are insufficient to meet the costs of procuring many of the essentials of life, particularly food. [Congressional testimony of Charles Upton Lowe, Director of the National Institute of Child Health and Development, 1969 (1).]

I. INTRODUCTION

Interest in the nutritional status of young American children has heightened considerably in the past decade. Much of the concern has resulted from research suggesting varying degrees of undernutrition in low-income American school[1] and preschool children and from evidence indicating a positive association between children's growth and their intellectual development.[2] In this paper we analyze the choice of diet for children 1 to 5 years in the United States and its relation to the children's growth and health. We are particularly interested in the extent to which family income and education may be obstacles to the provision of adequate diets for children in American families. The hypothesis that these obstacles are substantial underlies many government nutrition and income support programs and has led to the congressional mandate of two separate comprehensive national nutrition surveys, the Ten State Nutrition Survey, 1968–70, and the Health and Nutrition Examination Survey, 1971–75.

In a previous paper we used the Ten State Nutrition Survey (TSNS) data to examine the nutritional status of children up to the age of 36 months in poor American families Chernichovsky and Coate (2). The picture that emerged from the analysis of TSNS data was generally contrary to the impressions left by much previous research. The data indicated that low-income parents had pushed the growth of their children through choice of diet nearly as much as possible. Protein, a relatively high-priced nutrient, was consumed in quantities two to three times recommended dietary standards and to an extent where its marginal impact on the growth of children was very small. Family income and mother's education were shown not to be significant (in the statistical sense) barriers to the provision of adequate protein and calorie intakes for children in poor American families. In the TSNS data average protein and calorie consumption was in excess of dietary standards whether the data was stratified by children's age, family income, or ethnic group. Protein intakes in these cross-tabulations were consistently 200 to 300 percent of dietary standards.

In this paper we analyze the Health and Nutrition Examination Survey (HANES) data to provide further evidence on the choice of diet for young American children and its effect on their growth and health. This paper is divided into five sections. In the following section we describe the conceptual framework and specify a general model of children's diet, health, and growth. This is followed by a discussion of the data that includes important descriptive statistics. In Section IV, we present the estimated econometric model. The final section summarizes the research.

II. CONCEPTUAL FRAMEWORK

As a point of departure we postulate that the utility of parents is a positive function of their children's growth. That is, within the bounds of perceived norms, parents desire heavier and taller children.[3] For our analysis it is not necessary that this desire be based on known correlations between current period height and weight of children and their current and future period health status and intellectual development. Rather, we only argue that this desire does exist and

Economic Analysis of Diet, Growth and Health

that parents make sacrifices or forego other pleasures in order to augment the growth of their children.[4]

Although constrained by genetic and physiological factors, parents influence the growth of their children by their choice of diet for the children and by their investment in their children's health (medical care, parental care, sanitary conditions, etc.). The interdependencies among children's growth, children's health and their diet are formalized in the following model.

We begin by relating the parent's choice of the initial diet D_0 for a newborn to birth weight BW, which is a proxy for the infant's demand for food, and initial period socioeconomic influences E_0 that affect the quantity and quality of diet:

$$D_0 = f^o(BW, E_0). \tag{1}$$

In each subsequent period the child's growth G_t is determined by genetic and parental traits Z and by diet D_{t-1} and health status H_{t-1} in the preceding period. Health status can be interpreted as an efficiency parameter that affects the rate at which nutrients are converted into children's growth. Formally,

$$G_t = f(Z, D_{t-1}, H_{t-1}). \tag{2}$$

The diet in each period is a function of the child's growth, which serves as a proxy for appetite or the child's demand for food, and the economic status of the household:

$$D_t = g(G_t, E_t). \tag{3}$$

The child's health status is a function of his diet, growth, and other inputs which produce good health X_t,

$$H_t = h(X_t, G_t, D_t). \tag{4}$$

The levels of X_t are determined by socioeconomic status:

$$X_t = e(E_t). \tag{5}$$

In order to statistically identify certain key relationships and to make the model consistent with available cross-section data, several assumptions are necessary, some of which are explicit in Eqs. (1)–(5). First, birth weight is considered exogenous to our model of children's growth, diet, and health. A more sophisticated model could include birth weight as an endogenous variable and relate it to parental characteristics, diet of the mother, and socioeconomic variables. We also assume that some variables are serially correlated (e.g., diet, household income) or constant (e.g., mother's education, parental traits) over t and that the time increments are infinitesimal.

To isolate the role of diet as a bridge from socioeconomic status to children's growth, we can, given the assumptions detailed above, derive the following simultaneous Eqs. from (2) thru (4):

$$G = g(\hat{D}, \hat{H}, t, Z, BW) \tag{6}$$

$$D = f(\hat{G}, E) \tag{7}$$

$$H = h(\hat{D}, \hat{G}, E) \tag{8}$$

which specify D, G, and H as endogenous variables. Equation (6) is basically a technical relationship, describing how children's growth responds to diet and health levels, given age, birth weight, and parental and genetic characteristics. Equations (7) and (8) are primarily behavioral relationships, explaining the choice of diet in the household for the children, given socioeconomic constraints, and the subsequent influence of diet and growth on health levels.

III. THE DATA

HANES is a national sample of the population of the United States, with oversampling of low income families. The entire HANES sample, which was collected between 1971 and 1975 by the National Center for Health Statistics, contains approximately 28,000 individuals between the ages of 1 and 74. Slightly fewer than 3000 children aged 1 to 5 years were included in the sample. Dietary intake data for the previous 24 hours were collected for children less than 5 years of age by interview of the homemaker. A working sample of 2515 was created by deleting all observations (children) with missing data. The roughly 450 children deleted from the sample did not differ significantly from the working sample in terms of age and sex specific nutrient intakes or height, head, and weight growth. HANES is described in detail by the National Center for Health Statistics (9, 10).

Descriptive statistics for variables collected in HANES relevant to our analysis are presented in Table 1. Endogenous variables in our econometric specifications are selected from the measures of children's diet, health, and growth. Measures of children's growth are height, weight, and head circumference. Measures of children's health are lifetime number of overnight hospitalizations and number of colds in the 6 months prior to the medical history. Children's diet is measured by calorie, protein, calcium, iron, vitamin A, and vitamin C intakes.

Exogenous variables in the growth equations are measures of genetic and parental traits, namely children's age, sex, birth weight, birth order, race, mother's height and weight, and father's height. Exogenous variables in the nutrient intake and health equations are family income, family size, and dummy variables representing education of the household head and whether the head is female.

The mean family income of $9280 is considerably below the 1972 national average of $12,500 and is indicative of the oversampling of low-income families. The mean calorie and protein intakes of 1516 and 56 g are considerably above the protein and calorie standards of roughly 1330 and 26 g for children of

Table 1. Summary Statistics

Variable	Mean	Standard Deviation
Daily calories	1516	584
Daily protein (gm)	55.84	23.06
Daily Vitamin C (mg)	78.45	86.73
Daily iron (mg)	8.01	4.51
Daily calcium (mg)	872	469
Daily Vitamin A (IU)	3576	3743
Weight (kg)	15.48	3.82
Height (cm)	97.86	11.95
Head circumference (cm)	49.23	2.21
Hospitalizations	.30	.45
Colds (last six months)	1.42	1.32
Age (months)	42.63	17.62
Sex (1 = male)	.51	.50
Birth weight (oz.)	115.94	19.80
Birth order	1.45	2.19
Race (1 = non-white)	.23	.42
Mother's height (in.)	64.05	2.69
Father's height (in.)	69.87	3.18
Mother's weight (lbs.)	139.14	29.21
Household income	9280	5563
Household size	5.05	2.03
Years of schooling of household head		
Schooling 1 (1 = less than 12)	.37	.48
Schooling 2 (1 = 12)	.33	.48
Schooling 3 (1 = 13 to 16)	.19	.39
Schooling 4 (1 = more than 16)	.06	.24
Sex of household head (1 = female)	.16	.36

the age and weight corresponding to the sample means.[5] This finding of higher average protein and calorie intakes than dietary standards is not surprising given the similar results from the TSNS, a sample characterized by significantly lower family incomes. Children's intakes of calcium, vitamin A, and vitamin C average two to three times recommended dietary standards in the HANES working sample. In the case of iron, the average intake is two-thirds of dietary standards.

In column 1 of Table 2, levels of growth, health, and nutrient intakes are presented for children in households falling into the upper and lower 30 percentiles of the poverty index (PIR) distribution.[6] There are no significant differences in height, weight, or head growth between these groups, nor in protein, calorie, vitamin A, or iron intakes. In the cases of vitamin C and calcium intakes a statistically significant difference emerges in favor of the higher PIR group. For both groups mean nutrient intake levels consistently exceed dietary standards,

Table 2. Mean Levels of Growth, Nutrient Intakes, and Health for Families in the Upper and Lower 30 Percentiles of the Poverty Index Distribution[a]

	All Families			White Families			Non-White Families		
	Lower Thirty Percentile	Upper Thirty Percentile	t^b	Lower Thirty Percentile	Upper Thirty Percentile	t^b	Lower Thirty Percentile	Upper Thirty Percentile	t^b
Daily calories	1498	1510	.45	1520	1514	−.19	1475	1469	−.10
Daily protein (gm)	54.96	56.14	1.08	56.46	56.26	−.15	53.44	55.00	.55
Daily Vitamin C (mg)	69.12	87.43	5.18	67.64	87.94	4.64	70.62	82.56	1.33
Daily iron (mg)	7.89	8.23	1.56	7.97	8.27	1.01	7.80	7.86	.15
Daily calcium (mg)	819	922	4.82	910	935	.93	72.54	79.67	1.33
Daily Vitamin A (IU)	3577	3694	.42	3621	3595	−.16	3532	4177	1.08
Weight (kg)	15.42	15.51	.53	15.41	15.50	.43	15.42	15.54	.27
Height (cm)	97.63	97.94	.57	97.14	97.99	1.31	98.14	97.51	−.43
Head (cm)	49.17	49.24	.80	49.09	49.24	1.21	49.24	49.30	.24
Hospitalizations	.31	.29	−.72	.32	.30	−.91	.29	.24	−1.03
Colds (last six months)	1.51	1.31	−3.46	1.46	1.29	−2.43	1.57	1.54	−.20

Notes:
[a] The sample sizes for the lower and upper 30 percentiles are, for the entire sample, 1036 and 918; for whites, 521 and 831; for nonwhites, 515 and 87. The imbalance in the white and nonwhite categories results from the use of the entire sample 30 percentile cutoff values for these subsample stratifications.
[b] t values are for significance test of difference between means.

Economic Analysis of Diet, Growth and Health 117

with the exception of iron. There are also no significant differences in hospitalizations, although the lower PIR group had a significantly greater number of colds in the 6 months prior to the medical history. The average family income and household size for the higher PIR group are $14,766 and 4.2. The same figures for the lower PIR group are $3673 and 5.7.

In the remaining portion of Table 2, similar high and low PIR comparisons are made for blacks and for whites in the working sample. The patterns of statistical significance within these stratifications are similar to those in the sample as a whole.

Further information on the nutritional status of young American children can be obtained by examining the diets of children who are underweight for their age and sex. In our HANES working sample, the calorie and protein intakes of children below the 10th percentile in weight for their age and sex are 1440 and 53 g, not significantly different from the working sample means and indicative of more than adequate intakes of these nutrients according to dietary standards. The mean family income for this group of underweight children is $8470. Unless present and past nutrient intakes are not correlated, these numbers imply that influences other than diet may be responsible for producing the condition usually associated with undernutrition. The consideration of the empirical results in the next section will enable us to come to firmer conclusions about the role of socioeconomic variables in the choice of diet by parents for their children and about the subsequent effect of nutrient intakes on children's growth.

IV. EMPIRICAL RESULTS

At the empirical level we have estimated several variations of our model of children's diet, health, and growth. With the exception of calories and protein, the nutrient intake variables did not approach statistical significance on the growth equations, either because of their high correlations with protein and calorie intakes or because they have very small impacts on growth at the margin. The health variables also performed poorly in the growth equations in the statistical sense, apparently because these conditions have minor or very short-term growth effects that are rapidly overcome.

In the presentation of the empirical results, therefore, we emphasize a model with the following endogenous variables: height, weight, head circumference, protein intake, and calorie intake. We also report results for the colds variable and for vitamin C intake.

A. Reduced Forms

The reduced-form relationships derived from Eqs. (6)–(8) relate children's growth, health, and nutrient intakes to genetic and parental traits measured by age, sex, race, parents' heights, mother's weight, birth weight, and birth order;

Table 3. Reduced-Form Estimates[a]

Independent Variables	Weight	Head Circumference	Height	Calories	Protein	Vitamin C	Colds
Constant	4.71 (3.59)	36.21 (34.61)	20.73 (8.45)	−278.48 (−0.79)	6.06 (0.42)	17.76 (0.32)	2.81 (3.36)
Age	0.14 (11.16)	0.13 (14.27)	0.84 (34.20)	18.55 (5.24)	0.16 (1.13)	1.39 (2.47)	−0.01 (−1.88)
Age squared	0.000 (2.38)	−0.000 (−7.90)	−2.49 (−8.84)	−0.09 (−2.45)	0.001 (0.75)	−0.01 (−1.88)	0.000 (0.94)
Sex	0.42 (5.21)	1.12 (17.14)	0.81 (5.32)	149.05 (6.80)	4.17 (4.66)	9.09 (2.61)	−0.02 (−0.43)
Birth weight	0.02 (12.53)	0.01 (9.18)	0.45 (10.12)	0.04 (0.07)	0.01 (0.65)	−0.03 (−0.34)	0.001 (0.90)
Birth order	−0.05 (−1.92)	−0.009 (−0.42)	−0.08 (−1.55)	−1.37 (−0.18)	−0.16 (−0.56)	−0.99 (−0.85)	0.001 (0.06)
Mother's weight	0.01 (5.95)	0.001 (1.13)	0.70 (2.41)	0.64 (1.55)	0.28 (1.67)	0.05 (0.79)	−0.001 (−1.09)
Mother's height	0.07 (4.54)	0.05 (4.32)	0.37 (12.15)	8.38 (1.91)	0.09 (0.51)	−0.37 (−0.53)	−0.001 (−0.16)
Father's height	0.05 (4.37)	0.03 (3.20)	0.24 (9.98)	6.72 (1.89)	0.30 (2.12)	0.39 (0.70)	−0.002 (0.32)
Race	0.40 (3.45)	0.35 (3.82)	1.41 (6.42)	48.31 (1.54)	−2.61 (−2.04)	0.42 (0.85)	0.29 (3.92)
Income	0.000 (2.15)	0.000 (1.77)	0.000 (2.77)	0.000 (0.13)	0.000 (0.60)	0.000 (2.37)	−0.000 (−1.51)
Household size	−0.04 (−1.33)	0.002 (0.11)	−0.10 (−1.93)	18.77 (2.32)	0.80 (2.43)	−0.53 (−4.21)	−0.02 (−1.41)
Schooling 2	−0.01 (−0.11)	0.005 (0.06)	0.15 (0.81)	113.30 (4.25)	3.79 (3.49)	5.19 (1.23)	−0.12 (−1.94)
Schooling 3	−0.01 (−0.08)	0.16 (1.65)	0.34 (1.45)	81.40 (2.39)	3.73 (2.69)	13.92 (2.58)	−0.05 (−0.69)
Schooling 4	−0.63 (−3.26)	0.05 (0.36)	−1.40 (−3.84)	56.97 (1.09)	3.78 (1.78)	9.22 (1.11)	0.08 (0.66)
Sex of head	−0.18 (−0.14)	0.09 (0.95)	−0.30 (−1.28)	125.51 (3.69)	5.55 (4.00)	−1.44 (−2.67)	−0.05 (−0.68)
$\bar{R}^2$	.71	.46	.89	.14	.08	.02	.03
F	278.9	96.0	973.0	18.8	10.5	3.1	5.2
N	2515	2515	2515	2515	2515	2515	2515

Note:
[a] t statistics in parentheses.

and socioeconomic influences measured by household income, household size, and dummy variables indicating the education of the household head, and whether the head is female. The reduced form results are presented in Table 3. We are particularly interested in the children's growth reduced forms because of the information they provide on the significance of the genetic and parental trait variables versus the socioeconomic and behavioral indicators in the determination of children's growth. The results show that the latter set of variables are of limited significance in explaining children's growth. The family income and mother's education coefficients generally have low t values, and the addition of these variables and household size to children's growth regressions that already include the genetic and parental trait variables only slightly reduces the unexplained variance in the dependent variables.[7]

Another interesting aspect of the reduced-form results is the low R^2 values of the nutrient intake equations, none of which exceeds .14. The fact that the exogenous variables in our model explain such a small proportion of the variation in nutrient intakes brings to question the importance of these variables in the diet decision for young children.

B. Children's Growth

The simultaneous-equation estimates of the children's growth equations are presented in Table 4.[8] The protein variable has been excluded from the weight equation because it was statistically insignificant if calories were also included as an explanatory variable. Calories, however, approached statistical significance in these equations even when protein also appeared. The number of calories, then, seems to better explain weight growth than the protein content of the diet. Protein and calories are highly colinear ($r = .82$), so a good portion of the protein influence is captured by the calorie variable. An argument with a similar framework explains why protein appears in the height and head growth equations while calories does not.[9]

In elasticity terms the most important variables in the growth equations are children's age, height of the mother, and height of the father. These results were expected and demonstrate again the importance of variables beyond the influence of the household decision maker in the children's growth process. A result that is surprising is the rather substantial elasticities of children's growth with respect to nutrient intakes. The elasticities (at the means) of height and head circumference with respect to protein are .05 and .25, respectively, and the elasticity of weight with respect to calories is .20. These results imply that an increase in daily protein consumption of 10 percent, or about five and one-half grams, would increase height by an average of one-fifth of one inch and head circumference by an average of one-half of one inch. A 10 percent increase in calorie intake would increase children's weight by an average of seven-tenths of one pound. The protein elasticities in the height and head circumference equations seem particu-

Table 4a. Structural Equation Estimates for Children's Growth and Health: Three-Stage Least Squares[a]

Independent Variables[b]	Dependent Variables			
	Height	Weight	Head Circumference	Colds
Protein	.087		.22	.01
	(4.39)		(11.76)	(1.49)
Calories		.002		
		(5.02)		
Vitamin C				.011
				(.10)
Age	.84	.09	.24	−.17
	(31.52)	(5.71)	(8.57)	(−8.60)
Age squared	−.002	.001	−.002	.001
	(−9.45)	(5.10)	(−7.10)	(6.91)
Sex	.48	.13	.36	−.82
	(2.63)	(1.28)	(1.84)	(−6.11)
Birth weight	.003	.02	.002	
	(8.53)	(11.06)	(.72)	
Birth order	−.19	−.07	.10	
	(−3.82)	(−4.21)	(3.36)	
Mother's height	.51	−.11	.42	
	(20.80)	(−.92)	(23.96)	
Father's height	.40	−.06		
	(18.11)	(6.13)		
Mother's weight	.002	−.008	−.01	
	(1.03)	(5.61)	(−7.30)	
Race	1.24	3.10	−.71	
	(5.19)	(2.53)	(−3.24)	
Income				−.001
				(−6.44)
Household size				.06
				(1.86)
Schooling 2				.72
				(−4.90)
Schooling 3				−1.40
				(−6.83)
Schooling 4				−.79
				(−2.84)

Notes:
[a] $N = 2515$.
[b] Italics indicates endogenous variable.

Table 4b. Elasticities of Selected Variables[a]

Independent Variables[b]	Dependent Variables		
	Height	Weight	Head Circumference
Protein	.05		.25
Calories		.20	
Age	.38	.42	.21
Birth weight	.004	.16	.005
Mother's weight	.004	.07	.01
Mother's height	.33	.46	.54
Father's height	.27	.27	
Race	.003	.001	−.005
Sex	.003	.004	.003

Notes:
[a] Computed at mean values of dependent and independent variables.
[b] Italics indicates endogenous variable.

larly large in light of the fact that protein intake averages more than twice dietary standards. The protein effects on growth seem to be linear throughout the range of intakes characterizing the HANES working sample: that is, we do not appear to be approximating a nonlinear protein effect with a very small impact on growth at the margin.[10]

These substantial elasticities of growth with respect to nutrients that are consumed in excess of dietary standards is consistent with the findings of the evaluation of the Special Supplemental Food Program for Women, Infants, and Children (WIC). This analysis showed that although children in poor American households generally consumed nutrients well in excess of dietary standards, their growth could be accelerated by increasing nutrient intakes. In light of these findings, the WIC evaluators recommended a reassessment of dietary standards and singled out protein in particular (4).

C. Nutrient Intake Equations

The simultaneous equation estimates of the protein, calorie, and vitamin C equations are presented in Table 5. The results are similar for each of the nutrients. Simply stated, they indicate, within the context of our model, that children get the amount of these nutrients that they "ask for." The child's demand for nutrients, represented by weight of the child, is a very important determinant of intakes. The nutrient-weight elasticities are about 1 in each case, and the t values of the weight coefficients are substantial. The family income coefficients approach statistical significance but imply very small elasticities (about .02 in each case).

Table 5a. Structural Equation Estimates for Children's Nutrient Intakes[a]

Independent Variables[b]	Dependent Variables		
	Protein	Calories	Vitamin C
Weight	4.03	78.80	.89
	(12.78)	(10.36)	(.93)
Age	−.18	11.43	2.24
	(−1.06)	(2.80)	(4.27)
Age squared	−.003	−.02	−.002
	(−1.90)	(−5.07)	(−4.89)
Sex	2.25	111.10	9.30
	(2.38)	(4.89)	(2.65)
Income	.001	.003	.001
	(1.92)	(1.46)	(3.03)
Household size	.60	14.27	−.95
	(3.32)	(3.13)	(−1.12)
Schooling 2	2.32	82.10	6.01
	(2.68)	(3.71)	(1.47)
Schooling 3	2.39	52.60	15.35
	(2.28)	(1.94)	(2.93)
Schooling 4	.09	−23.10	9.71
	(.05)	(−.56)	(1.21)
Sex of head	−.02	17.72	−2.87
	(−.03)	(.70)	(−1.16)

Notes:
[a] N = 2515.
[b] Italics indicates endogenous variable.

Table 5b. Elasticities of Selected Variables[a]

Independent Variables[b]	Dependent Variables	
	Protein	Calories
Weight	1.11	.80
Age	.31	.32
Sex	.02	.04
Income	.02	.02
Household Size	.05	.05

Notes:
[a] Computed at mean values of dependent and independent variables.
[b] Italics indicates endogenous variable.

Education of the household head has a positive but nonlinear effect on nutrient intakes. Children in families where the head has 12 years of schooling receive about 5 percent more of these nutrients relative to children in families where the head has less than 12 years of schooling. However, this education differential falls when children in families where the head has college or graduate education are compared to children in families where the head has less than 12 years of schooling.

V. SUMMARY

A primary purpose of this paper was to investigate the extent to which family income and education are obstacles to the provision of adequate diets for young children in the United States. Based on our examination of the HANES data we have found that:

1. Average nutrient intakes of young children are well above recommended dietary standards, with the exception of iron.
2. Average nutrient intakes for children in households of lower economic status are very similar to intakes of children in households of higher economic status. Rates of children's growth are also similar in these households.
3. Family income and education of the household head have statistically significant but very small positive effects on the nutrient intake levels of young children in the model of children's diet, growth, and health estimated in this paper.

These findings are very consistent with those from a similar analysis we performed with the Ten State Nutrition Survey. A most interesting result of the present study is the rather substantial estimated effects of protein intake on children's height and head growth, even though protein is consumed well in excess of dietary standards. This finding and the apparent correlation between children's growth and their intellectual development brings up the question of the adequacy of present protein standards. Could American mothers, who provide very high-protein diets for their children in households at all levels of socioeconomic status, know more about what constitutes an adequate diet for their children than the experts do?

NOTES

1. For examples of research into the problem of undernutrition in American school and preschool children in the United States, see Christakis (3), Owen (12), Sims and Morris (13), and Owen (11).
2. Owen (11), in his review of the effects of nutrition on growth and cognitive development,

concludes that the "evidence, which still should be considered preliminary in nature, . . . [indicates] that bigger is smarter, at least among preschool children."

3. More formally, it could be argued that rates of children's growth enter the utility function in a nonlinear fashion and that excessive rates of growth (e.g., obesity) are negatively related to parent's utility.

4. It is often pointed out that in agricultural societies parents are very concerned about the size of their children because physical strength is an important correlate of individual output. Although a desire for larger children in modern societies may not be based on a similar observation, there is evidence that the height and weight of children at younger ages correlate with their intellectual development and health in later years, and thus with their future earnings.

5. The dietary standards cited in the text are those of the HANES dietary standards committee for children 24–47 months weighing the sample mean of 15.5 kg.

6. As computed in HANES, the poverty index ratio takes account of household income, household size, and household diet requirements as reflected by the age distribution of the household members.

7. Adjusted R^2 values increased by less than .01 when the socioeconomic variables were added to either height, weight, or head size regressions that already contained age, the square of age, sex, parents' height, birth weight, and birth order. It should also be pointed out that the limited significance of the socioeconomic variables does not appear to be due to colinearity with the genetic and parental trait variables. The t values of the socioeconomic variables do not increase markedly even when the genetic and parental trait variables are excluded from the children's growth equations.

8. The results for the colds variable are also presented in Table 4 but are not discussed in the text. Household income and education of the household head are inversely related to the number of children's colds, while protein and vitamin C intakes do not have statistically significant impacts.

9. Because the growth equations formed part of a simultaneous system, traditional F tests could not be employed to test the individual and joint contributions of the protein and calorie variables. Results from ordinary-least-squares regressions indicate that protein makes a significant incremental contribution to explaining the variance in height and head growth when added to regressions containing the other independent variables, while calories does not. When both diet variables are added jointly to height and head growth regressions, the incremental contribution is insignificant. For the weight equation, the incremental contribution to explained variance is significant when the protein and calorie variables are entered individually or jointly to regressions containing the other independent variables.

10. Predicted protein and the square of predicted protein were entered as independent variables in the height and head circumference equations in the final stage of a two-stage least-squares process. No evidence of a nonlinear protein effect was uncovered.

REFERENCES

1. Chase, A. (1977) "Poverty and Low IQ: The Vicious Circle," *New York Times Book Review*, July 17, p. 9.
2. Chernichovsky, Dov, and Douglas Coate (1980) "The Choice of Diet for Young Children and Its Relation to Children's Growth," *Journal of Human Resources*, 15:255–263.
3. Christakis, George, et al. (1968) "Nutritional Epidemiologic Investigation of 642 New York City Children," *American Journal of Clinical Nutrition* 21.
4. Edozien, J. C., B. R. Switzer, and R. B. Bryan (1976) "Medical Evaluation of the Special Supplemental Food Program for Women, Infants and Children (WIC)," University of North Carolina.
5. Health Services and Mental Health Administration DHEW (1972) *Ten State Nutrition Survey*, 1968–1970. DHEW (HSM) 72-8130-8134.

6. Jelliffe, D. B. (1976) "World Trends in Infant Feeding," *The American Journal of Clinical Nutrition* 29.
7. Michael, Robert T. (1972) *The Effect of Education on Efficiency in Consumption*, Columbia University Press, p. 30.
8. National Center for Health Statistics (1976) NCHS Growth Charts, 1976, Monthly Vital Statistics Report, 25, Supp. (HRA) 76-1120.
9. National Center for Health Statistics (1973) *Plan and Operation of the Health and Nutrition Examination Survey*. U.S. Department of Health, Education, and Welfare, Public Health Service Publication No. (HPA) 76-1310, Series 1-No. 10a.
10. National Center for Health Statistics (1977) *Plan and Operation of the Health and Nutrition Examination Survey*. U.S. Department of Health, Education, and Welfare, Public Health Service Publication No. (HRA) 77-1310, Series 1-No. 10b.
11. Owen, George M. (1974) "A Study of Nutritional Status of Preschool Children in the United States, 1968–1970," *Pediatrics* 53, Supplement.
12. Owen, George M., *et al.* (1969) "Nutritional Status of Mississippi Preschool Children," *American Journal of Clinical Nutrition* 22.
13. Sims, Laura S., and Portia M. Morris (1974) "Nutritional Status of Preschoolers," *Journal of American Dietetic Association* 64.

THE DEMAND FOR PRENATAL CARE AND THE PRODUCTION OF HEALTHY INFANTS

Eugene Lewit

I. INTRODUCTION

Despite very high levels of expenditure on health, the United States has lagged significantly behind other developed countries in the reduction of infant mortality. The U.S. infant mortality rate of 16.8 deaths per 1000 live births in 1974 was almost 40 percent above the median rate of 12.1 for 15 other developed countries [Fuchs (11)]. The implications of this level of infant mortality extend beyond the rate itself. In particular, the same factors, prematurity and damaged infants, that are implicated in the high infant mortality rate also importantly influence the health of surviving infants. Less than optimal birth outcomes have been associated with chronic mental, physical, and learning problems which have a deleterious effect on development throughout childhood and during adulthood as well.

Because of the apparent importance of infant and child health in the determination of economic and social well-being, attention has increasingly been paid to expanding the health services available to children and pregnant mothers. In particular, the Maternal and Child Health Care Act, introduced in each of the past several years in the U.S. Congress, is specifically designed to establish a national health care system for this population group in order both to raise health levels and to serve as a model for a comprehensive national health insurance system. This proposed policy to improve health levels by increasing access to and utilization of health care has been advocated despite the lack of conclusive evidence that increased medical care will have a significant impact on infant and child health. To some extent the rationale behind this approach was strengthened by a study from the Institute of Medicine (IOM) (17), *Infant Death: Analysis by Maternal Risk and Health Care,* which concluded that medical care had an important role to play in determining infant mortality. The most significant finding to emerge from the IOM study was the projection that if all pregnant women in New York City in 1968 had received what was defined by the investigators as "adequate" prenatal care, the infant mortality rate could have been reduced by as much as a third. This lower rate would be comparable to those reported contemporarily for other developed countries. Moreover, the study found that there appeared to be a substantial misallocation of care resources among pregnant women when the relative risks associated with individual pregnancies were taken into consideration. Women who were defined to be at high risk were less likely to receive any care, while women who were defined to be at low risk were more likely to receive "adequate" care.

Before plunging ahead with policies designed to increase the utilization of medical services in order to reduce infant mortality and raise child health, several subsidiary topics should be considered. Further investigation is needed of the selection process which results in low levels of utilization of prenatal care by those women who might have benefited most from it. This can be an indication of malfunction of the health care system, but it may also be an indication of the absence of a demand for care by certain individuals. This issue is in a sense highlighted by the additional finding that prenatal medical care had its most substantial impact in reducing infant mortality for those women who were judged to be high socioeconomic risks but much less of an impact on those women who were deemed to be at high medical risk. Since women categorized as being at high socioeconomic risk would be those less likely to receive prenatal care under the present health delivery system, the question as to why these women may not obtain care and the implications of this failure for programs designed to extend care to them become very significant, but they have not been systematically investigated.

A second major thrust currently being advocated to improve pregnancy outcomes is the development of specialized high technology perinatal care centers and the concomitant regionalization of perinatal care in order to use these centers efficaciously. Such care networks, dependent as they are on coordination of care and referral of high-risk pregnancies for expensive high-technology care, depend critically upon appropriate patient as well as physician behavior if they are to realize a reasonable degree of success. In developing these systems much attention has focused on coordination of the activities of providers; less attention has been focused on the behavior of pregnant mothers and their demand for care of this type. Although regional perinatal care networks hold out the promise of delivering high-technology perinatal care in an apparently cost-effective manner, their ultimate success will depend on maternal behavior and the willingness of patients to use the system as developed. The evidence that many "high"-risk mothers may not be using the present, less complex system adequately to insure successful birth outcomes heightens the need for determining the relative importance of the factors that enter into family decisions regarding the current medical care systems so that these factors can receive appropriate consideration in developing and evaluating new systems of care.

The area of household decision making, which has received increased study from economists within the last decade, lends itself to the investigation of the demand for prenatal care services and infant health. Within this area Grossman (14, 15), in particular, has focused on a household production model of the demand for health and the demand for child health and has attempted to determine those variables that are responsible for determining the demand for medical care and the production of health.[1] In addition, a number of economists have turned their attention to examining the determinants of family size and have viewed child services as being a commodity demanded and produced within the household.[2]

It seems, therefore, that a fruitful extension of research on the infant mortality problem with particular attention to the importance of health care services would be to attempt to develop a model of demand for infant health which utilizes the household production perspective. It seems reasonable to presume that the demand for healthy or surviving children can be derived from the demand for child services themselves and moreover, that the demand for medical care in the production of favorable pregnancy outcomes is derived from the demand for surviving children and the expected efficacy of care in producing survival. This is the approach that I take in this paper. I assume that there is a household demand for children which can be satisfied by successful pregnancies and that, in making a determination of the desired number of pregnancies and the expenditure per pregnancy, the household will take into consideration the cost of producing survivors and the postpregnancy cost of children relative to the cost of other goods. In determining the demand for care and in attempting to measure the efficacy of medical care in the production of healthy infants, I view such demand as being derived from the demand for surviving children and I concentrate on the importance of household experience with previous pregnancies in determining both the amount of resources to be expended on a specific pregnancy and the amount of care demanded during a specific pregnancy. Estimates of the consequences of such resource allocations are also presented in order to more clearly define the production process.

In the following section, I develop a model of decision making for expenditures on pregnancies based on prior experience with pregnancy. I attempt to integrate demand for children with the demand for pregnancy and with the demand for expenditures on an individual pregnancy based on prior experience and other exogenous variables.

I then develop and estimate an empirical model to test some of the implications of the theoretical model. Particular emphasis is placed on formulating a system of equations to describe sequential behavior during the pregnancy period. Two demand relationships are formulated. They measure demand for prenatal care by using both the interval to the first prenatal visit and the number of visits as dependent variables. Production functions for birth weight are estimated to assess the results of the pregnancy process, and the implications of the estimated parameters for better understanding the infant health process are discussed.

II. ANALYTICAL FRAMEWORK

The notion that infant and child mortality rates will have a strong influence on family fertility decisions has played an important role in the prediction of population growth patterns by demographers. Observation of family responses to child losses and of high fertility rates in areas with high child mortality experience has led many to conclude that "replacement" is an appropriate model of behavior under these circumstances. By "replacement" is meant the almost automatic attempt to compensate for a dead offspring with a new birth. The basic assumption underlying such hypothesized behavior is that parents have a certain targeted family size and that they will strive to attain this goal even in the face of substantial losses. These models have been used to investigate the high fertility levels associated with high child mortality rates in less developed countries.

Several economists working in the area of fertility determination have pointed out that this simplistic model of replacement does not adequately deal with all the possible relationships between fertility and child mortality. DeTray (6) has pointed out that since pregnancy is not costless, high infant loss rates may, in fact, discourage pregnancies as at the margin the family finds that the utility associated with the probability of the child's surviving is less than the cost of an additional pregnancy.

O'Hara (24) has investigated the cost and return streams associated with investment in children at different stages of their life cycle. He uses a model in which family decisions about children involve the risk of death at each age rather than the overall survival rate. An important aspect of this analysis is that it treats expenditures on children as endogenous and allows parents to adjust expenditures to reflect experience so that changes in the survival rate need not only influence the number of children desired but their average cost ("quality").

Both DeTray and O'Hara stress the importance of the household decision-making process in basing fertility decisions on the allocation of limited resources among competing goals. This framework is shared by much economic research into the economics of fertility and optimal family size which stresses the important distinction between the quantity and "quality" of children [Becker and Lewis (1), Willis (29), DeTray (6), and O'Hara (24)]. However, neither DeTray nor O'Hara have considered the implications of child survival being dependent on the amount of expenditures per child—they and others who have investigated the quantity/quality trade-off have made expenditures conditional on survival or have assumed away infant and child mortality.

In the model developed in the paper, infant survival is conditional on family decisions. In particular, I assume that survival depends on endogenously determined expenditures during the prenatal and perinatal period and that expenditures on children given that they survive the perinatal period are exogenous. The family is assumed to manipulate expenditures during the pregnancy period so as to attain an optimal family size given an exogenously determined reproductive production function and exogenously determined price for surviving children and total resource endowment. Within the context of the model, I examine the affects of income, education, experience, and wantedness of the particular child on the expenditure of productive resources and probable pregnancy outcome.

A. Model of the Derived Demand for Care

Assume a two-good utility space for the individual family with

$$U = U(C, S), \qquad (1)$$

where C = surviving children and S = a composite commodity representing all other goods. Let the price of C equal 1. Further, assume that C is a function of n (number of pregnancies) and p (the probability of survival of an individual pregnancy). p, in turn, is a function of exogenously determined maternal reproductive efficiency RE which varies in the population but is fixed for the individual (although not known with certainty) and EX, expenditures on a particular pregnancy for such items as prenatal care, hospital care, special food, etc. Thus

$$C = np(RE, EX) \qquad (2)$$

where $\partial p/\partial RE > 0$, $\partial p/\partial EX > 0$, $\partial C/\partial p > 0$, $\partial C/\partial n > 0$. Since we are concerned primarily with the determinants of expenditures designed to insure the infant survival, we will assume that K, the cost of a surviving child, is the same for all families but that the fixed cost FX and variable cost EX associated with pregnancy vary in the population. EX is the variable expenditures per pregnancy which the family controls in order to optimize pregnancy outcomes given resource constraints. FX, the fixed cost of a pregnancy, may vary among families but is assumed not to be subject to choice for an individual family. As an example of FX, DeTray (6) has pointed out that pregnancy is not costless because there is always some, perhaps miniscule, risk of maternal mortality. Additional and more substantial costs could be associated with the value of the time lost by the mother either during the confinement surrounding the birth or resulting from her decreased productivity either in the home or market because of the physical stress of pregnancy on her body. As compared with her nonpregnant state, the pregnant woman will probably require more rest, do certain tasks more slowly, and be unable to perform certain very strenuous tasks, particularly during the latter months of the pregnancy period. If such fixed costs result primarily from an effective time loss to the mother, we would expect that the fixed costs of pregnancy would be higher for mothers whose value of time was higher.

Using this notation, we can partition child associated costs as follows:

$$\text{total cost of pregnancies} = n (EX + FX); \tag{3}$$

$$\text{cost of surviving children} = pnK. \tag{4}$$

We can, therefore, write the lifetime income constraint as

$$Y = S + n(EX + FX) + pnK, \tag{5}$$

where all values have been fully discounted. Substituting the expressions for C and S from (2) and (5) into (1) yields

$$U^* = U(p(RE, EX)n, Y - n(EX + FX) - pnK). \tag{6}$$

Taking first partial derivatives of this utility function with respect to EX and n, the two variables that the family can control, yields the following condition for utility maximization:

$$\frac{\partial U^*}{\partial EX} = U_c n \frac{\partial p}{\partial EX} - U_s n \left(1 + \frac{\partial p}{\partial EX} K\right) \tag{7}$$

$$\therefore \frac{\partial p}{\partial EX} = \frac{U_s}{U_c - U_s K}, \tag{8}$$

where $\partial U/\partial C = U_c$ and $\partial U/\partial S = U_s$, and

$$\frac{\partial U^*}{\partial n} = U_c \frac{\partial C}{\partial n} - U_s [(EX + FX) + pK] \tag{9}$$

$$\therefore \frac{U_s}{U_c - U_s K} = \frac{p}{(EX + FX)}. \tag{10}$$

Note that the right-hand sides of both (8) and (10) are identical, being the ratio of the marginal utility foregone from consumption of $S(U_s)$ to the net marginal utility of a surviving infant $(U_c - U_s K)$.

Combining (8) and (10) yields the first-order condition for a maximum,

$$\frac{\partial p}{\partial EX} = \frac{p}{(EX + FX)}. \quad (11)$$

This implies that utility is maximized by equating the marginal product (and hence cost) of the survival function $\partial p/\partial EX$ with the average cost of survival. Expenditures beyond this point will cause the family to allocate too much of S to the production of children; this is particularly true as children may be produced more economically by increasing n, the number of pregnancies.

We can further illustrate the trade-off between increasing expenditures on survival with increasing the number of pregnancies by examining the conditions for producing a given family size (C_0) at minimum cost. As above, let

$$\text{total child cost} = \text{total cost of pregnancies} +$$
$$\text{cost of survivors} = n(EX + FX) + pnK \quad (12)$$

and form the Lagrangian,

$$W = [n(EX + FX) + pnK] + \lambda (C_0 - pn). \quad (13)$$

Now minimize (13) with respect to n and EX, yielding

$$\frac{\partial W}{\partial n} = (EX + FX + pK) - \lambda p \quad (14)$$

$$\therefore \lambda = \frac{(EX + FX) + pK}{p} \quad (15)$$

and

$$\frac{\partial W}{\partial EX} = \left(\left(n + \frac{\partial p}{\partial EX} nK\right) - \lambda n \frac{\partial p}{\partial EX}\right) \quad (16)$$

$$\therefore \lambda = \frac{1 + K\, \partial p/\partial EX}{\partial p/\partial EX}. \quad (17)$$

Combining (15) and (17) and solving for $\partial p/\partial EX$ yields

$$\frac{\partial p}{\partial EX} = \frac{p}{(EX + FX)}, \quad (11a)$$

as above. It is important to note that while the family is assumed free to vary either EX or n in this model to produce its desired number of children, in fact the optimal expenditure per pregnancy is independent of n and is solely a property of the production function p(RE, EX) including the level of RE. While one might want to restrict the p function in several ways, the only requirement for a stable equilibrium is that the average product of EX be greater than its marginal product [i.e., let AP = p/EX and β = EX/(EX + FX) < 1 (the share of variable expenditure in the total cost of a pregnancy), then (11a) requires that $\partial p/\partial EX = \beta \cdot AP$ for cost minimization]. This further implies that the production function has to allow for declining average product at the point of equilibrium. While this doesn't require that marginal product decline, since p has an upper bound of 1 (certainty), marginal product should decline for large enough values of EX.

It should also be noted that the symmetry between cost minimization and utility maximization goes beyond the finding that the marginal conditions (11) and (11a) are identical. The complete generality of C_0 in (13) implies that the resulting minimum cost solution is the same for any desired family size and suggests the following scenario for determining the desired number of children, the desired number of pregnancies, and expenditure per pregnancy. Using the marginal condition (11a) and knowledge of p(RE, EX) and RE, one could solve for optimal p (p*) and associated optimal expenditure per pregnancy EX*. Together these parameters determine the total price of a surviving child,

$$\text{Total price per survivor} = \frac{(EX^* + FX) + p^*nK}{p^*}. \quad (18)$$

Using this price, the family's initial endowment and their utility function one could determine the desired number of children C* and the desired number of pregnancies, $n^* = C^*/p^*$. Of particular interest would be how these equilibrium values would be affected by changes in the exogenous variables Y, FX, RE and changes in the production function p(EX, RE). Note, for example, that if RE is a neutral shift parameter, e.g., $p = RE \cdot f(EX)$, then the marginal condition (11a) is independent of the level of RE and so therefore is EX*. Thus for any given level of RE, p* would be determined and so would C* and n*. While under these conditions varying RE will not effect EX*, different endowments of RE may result in different levels of n* and C*. In particular, since by definition $\partial p^*/\partial RE > 0$, $\partial c^*/\partial RE$ should also be positive because the cost of a surviving child declines as RE increases. The effect of different levels of RE on n* is, however, undetermined and depends on the trade-off between the increase in C* and the fact that with the increase in p* fewer pregnancies per survivor are required.

B. The Production Relationship

Having demonstrated the importance of the marginal condition (11a) in determining expenditures on pregnancy and pregnancy outcomes and the symmetry between the cost minimization and utility maximization approach, I shall now concentrate on a number of exercises in comparative statics in an effort to determine the effect of changes in the exogenous variables on pregnancy expenditures.

In Figure 1, the equilibrium condition implied in (11a) is shown graphically. In the figure, expenditures EX are plotted on the X axis and average product AP,

Figure 1. Equilibrium conditions, constantly falling average product.

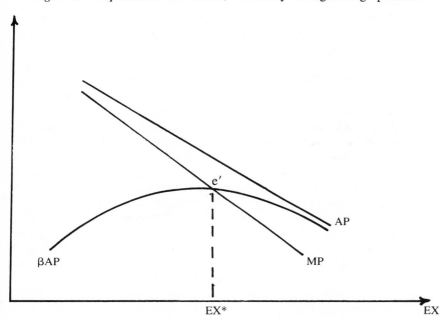

marginal product MP, and average cost βAP are plotted on the Y axis; ∂p/∂EX = MP > 0 but declining ($\partial^2 p/\partial EX^2 < 0$)—consistent with the notion that since p approaches 1 as an upper bound, marginal product is likely to fall.

To determine the equilibrium point e', we only use the necessary condition that AP be falling at the equilibrium point and the implication that AP > MP. As shown on the figure, AP > MP for all values of EX so that both AP and MP fall as EX increases. This is not true for βAP (average cost), however. Note that

$$\frac{\partial AP}{\partial EX} = \frac{1}{EX + FX}\left(\frac{\partial p}{\partial EX} - \frac{p}{EX + FX}\right) \qquad (19)$$

and that (19) taken in conjunction with the marginal condition (11a) implies that βAP < MP for small values of EX, rises to its maximum at βAP = MP, point e', and then falls as MP > βAP.[3] As EX → ∞, βAP → AP so that beyond e' βAP lies between MP and AP. The point e', the point of intersection between MP and βAP, defines EX* and its associated parameters p*, C*, and n* under *ceteris paribus* conditions.

In the discussion that follows, variations of the production relationships as graphed in Figure 1 are used to derive relationships between expenditures and changes in other exogenous variables.

C. Income Effect

Consider first the effect of a change in income Y on EX*. If children are a normal good ($\partial C/\partial Y > 0$), we would expect the desired number of children C* to increase as income rose. Unless Y enters the calculation of EX*, p* will not change and the only way that higher-income families can increase family size is to increase the desired number of pregnancies n* until a new equilibrium between S and C* is reached.[4]

However, within the confines of our model a reason for expecting EX* to be positively related to income can be demonstrated through consideration of the relationship between income and fixed cost. It was argued above that a primary component of fixed cost is the loss of mother's time due to pregnancy. In addition to the actual time lost during the birth and maternal recovery period, even healthy mothers will find, particularly late in the prenatal period, that they tire easily, may require more sleep, and may find certain tasks, particularly those requiring strength or agility, difficult if not impossible to perform. Such a reduction in the effective productivity of maternal time, as well as time actually lost, may be regarded as the fixed time cost of a pregnancy. It is important to distinguish this time cost from the variable time cost which may be associated with visits for prenatal care or time spent in classes in preparation for childbirth and child care. Such variable time expenditures are endogenous and should properly be included in the expenditures (EX) category of pregnancy costs. Even if the strictly defined fixed time loss is the *same* for women of different income levels, the cost associated with this time loss may not be. Standard results from household production theory imply that the value of time is positively correlated with income, and, hence one would expect that fixed costs FX, *ceteris paribus*, would be positively correlated with income and that β (the share of variable cost in total cost) would be less at each level of EX as income rises. The implications of this line of reasoning are demonstrated in Figure 2. AP and MP are as before, and the time loss associated with pregnancy is assumed the same for females at high (Y_H) and low (Y_L) income levels; however, fixed cost for higher-income mothers is higher than for lower-income mothers and correspondingly $\beta_H < \beta_L$ (where β_H is β for high-income mothers, and β_L is β for low-income mothers). Therefore, $\beta_L AP$ lies above $\beta_H AP$ and intersects MP at e_L to the left of e_H, the intersection point for $\beta_H AP$. The implication is that $EX^*_L < EX^*_H$ and that expenditures on pregnancy will be positively correlated with income.

This not unexpected prediction is a testable hypothesis generated by the model rather than being an assumption as in other studies. Moreover, the model implies that marginal and average product will be negatively correlated with income.

D. Production Function and Efficiency Effects

Having examined the expected effect of changes in income on pregnancy expenditures, I would now like to examine the effect of differences in RE

Figure 2. Effect of income on the demand for inputs.

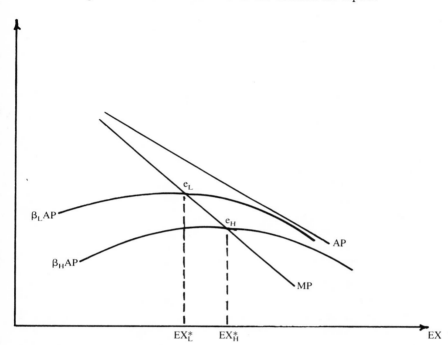

(exogenously determined reproductive efficiency) and differences in the production function itself on expenditures. Thus far, we have characterized the production function as having a positive but falling marginal product and falling average product around the equilibrium point and as being asymptotic to $p = 1$ as EX becomes large. Such a function is shown in Figure 3 and labeled P_0. Note the positive intercept a implies that even if EX were zero there would still be some possibility of survival. This seems quite reasonable given the positive albeit low infant survival rates observed in extremely poor countries and among primitive populations.

The probability of output with no endogenously chosen input poses an interesting problem in defining differences in efficiency among different producers. Efficiency strictly defined refers to the ratio of total output to total input and is said to increase as that ratio increases. It is not unusual to observe different firms in the same industry producing at different levels of efficiency, with different factor proportions and different levels of output. These observed differences in efficiency may be attributed among other things to differences in the technologies employed, economies or diseconomies of scale, or differences in the level of

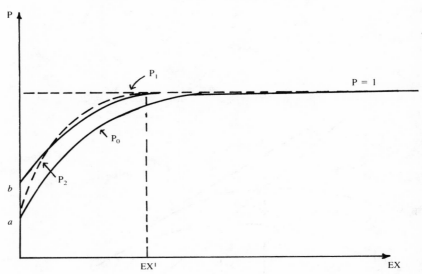

Figure 3. Survival production functions.

some otherwise undefined input usually called entrepreneurial ability. Recently, this notion of differences in efficiency in production has been extended to the production of consumption commodities within the household. So-called environmental variables, of which education is the most widely studied, have been thought to affect the production of commodities in the home by changing the marginal product of inputs in the production of household commodities. Michael (19) has tested the hypothesis that such an increase in efficiency in the household acts as though it augmented family income, increasing the consumption of luxuries and decreasing the consumption of necessities among those families with more education at a given income level. Grossman (14) and Inman (16) have extended this notion of increased efficiency resulting from higher levels of education to the production of health.

In the production of healthy infants, however, increases in the marginal product of variable inputs are not the only source of increased efficiency. For example, in Figure 3, a mother whose reproductive production function is represented by P_1, derived from P_0 by merely shifting the function upward by the amount ($b - a$), the difference in the level of output with zero variable inputs, is a more efficient producer of healthy infants then a mother on P_0 for most levels of output despite the fact that the marginal product of EX is required to be the same for all levels of EX below the level where p approaches 1 asymptotically.

Referring again to Figure 3, consider function P_2 with the same intercept as P_0 but a higher marginal product. In a sense, the advantage of P_2 relative to P_0 represents the advantage accruing to the usually examined environmental vari-

Prenatal Care and Infant Health

ables like entrepreneurial ability or education. Moreover, it is possible to think of alternative production relationships entailing lower intercepts but higher marginal products (P_1 vs. P_2, for example). Individuals with these kinds of production functions may be less efficient for some lower levels of EX but more efficient at higher levels. Production functions such as those graphed in Figure 3 are not widely used in economics since situations involving positive output with no endogenous inputs or with limitations on output are rarely encountered. They do have relevance to the production of human characteristics since both genetic endowment (a possible variant of the intercept) and asymptotically limited output may be features in the production of survival or of particular skills and talents. For example, Grossman (14) uses an asymptotically contained production relationship to define the production of healthy time in a given time period (e.g., one can be healthy no more than 7 days a week or 365 days a year).

I now would like to consider the probable effects of changes in the conditions of production on the endogenously determined household decision variables EX* and p*. Consider the function

$$p = 1 - Ae^{-BEX}, \qquad (20)$$

where p is probability of survival, EX expenditures, and A, (A < 1) and B positive coefficients which will reflect the different notions of efficiency discussed above. Note that (20) has all the desirable attributes we have discussed earlier,

$$\frac{\partial p}{\partial EX} = ABe^{-BEX} > 0 \qquad (21)$$

$$\frac{\partial^2 p}{\partial EX^2} = -AB^2 e^{-BEX} < 0 \qquad (22)$$

and

$$\lim_{EX \to \infty} p = 1 \qquad (23)$$

Note further that $(1 - A)$ is the level of p when EX = 0. Moreover,

$$AP = \frac{1 - Ae^{-BEX}}{EX} \qquad (24)$$

may either rise and fall with increasing EX or fall for all levels of EX depending on the values of the parameters A and B.

Referring to Figure 4, consider the effect of a change in A on EX*. The curves MP_0 and βAP_0 are presented as the initial condition. They intersect at e_0 and the resulting optimal expenditure is EX_0^*. Now suppose that A decreases [note that a decrease in A increases the intercept $(1 - A)$ and shifts the p function upward]. A decrease in A will cause the βAP curve

Figure 4. Effect of a change in A on EX*.

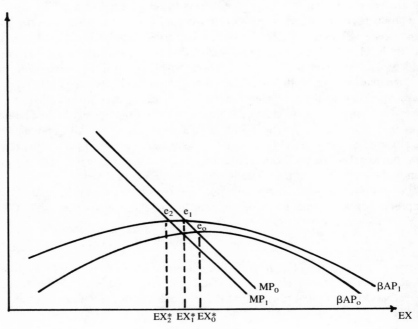

$$\beta AP = \frac{1 - Ae^{-BEX}}{EX + FX} \qquad (25)$$

to shift upward to βAP_1. If MP_0 is not affected, the new point of intersection will be at e_1 and the optimal $EX^*_1 < EX^*_0$. In other words, an upward shift in the production function which doesn't change marginal product over the relevant range will cause expenditures to decline because the same optimal level of p* can be had more cheaply.

With the production function given in (20), however, marginal product (21) is also a function of A; in fact,

$$\frac{\partial MP}{\partial A} = Be^{-BEX} > 0 \qquad (26)$$

This implies that with a decrease in A, MP will fall to a new level MP_1 and the new intersection e_2, where $MP_1 = BAP_1$, will be at an even lower level of EX, EX^*_2.

We now turn to an analysis of the effect of changes in the slope of p, that is, changes in B on EX*.

Recalling the optimal condition (11), utility is maximized when

$$\frac{\partial p}{\partial EX} = \frac{p}{(EX + FX)} = \beta AP. \qquad (11)$$

Substituting for $\partial p/\partial EX$ from (21) and for βAP from (25) yields

$$ABe^{-BEX} = \frac{1 - Ae^{-BEX}}{EX + FX} \qquad (27)$$

or

$$1 = Ae^{-BEX}(1 + BEX + BFX). \qquad (28)$$

If we implicitly differentiate (28) with respect to B, remembering that EX will change as B changes to maintain the optimal point, we obtain

$$\frac{dEX}{dB} = \frac{FX - FX \cdot BEX - BEX^2}{B^2(EX + FX)}. \qquad (29)$$

Since the denominator of (29) is always positive, the sign of dEX/dB will depend on the sign of the numerator and will be positive, negative, or zero, depending on whether

$$FX \lesseqgtr BEX(FX + EX), \qquad (30)$$

respectively. Thus, the effect of changes in the slope of p on EX* is indeterminate *a priori*. Moreover, with FX predetermined, the sign depends on the optimal value of EX that corresponds to a specific B and hence on differences in the actual parameters of the production function itself.

How may we interpret these results in light of the earlier discussion of efficiency and particularly our concern about shifts in reproduction efficiency? It is tempting to interpret shifts in the intercept as being analogous to shifts in reproductive efficiency. This is particularly true because changes in A may affect p independently of the level of expenditure. Thus, A has the features of a pure endowment effect that might be related to the genetic or biological production state of the mother. With the functional form used as an example above, a shift in A which raises the intercept will cause marginal product to decline—perhaps a result not consistent with everyone's notion of reproductive efficiency; however, given the asymptotic nature of any production function for p it would seem almost inevitable that an increase in the intercept would lead to a decrease in the slope as p approached 1.

It is tempting to view B as a measure of an efficiency effect associated with an environmental variable such as education. This is because changes in B affect only the marginal product of the inputs rather than the intercept. Also note that

$$\frac{\partial p}{\partial A} = -e^{-BEX} < 0, \qquad (31)$$

so that the impact of changes in A on p depends on the value of B. If A is regarded loosely as RE, an exogenously determined set of biological factors, then the size of the parameter B will determine the marginal product and mix of inputs in a manner analogous to household environmental variables discussed elsewhere in the literature.

In summary, the model yields the following predictions:

1. A change in RE, which is essentially incorporated in shifts in the entire production function, will tend to be negatively correlated with EX* (note, however, that since the total price per survivor is also negatively correlated with RE, C* should be positively correlated with RE as children become cheaper to produce);
2. Changes in environmental variables such as education, which change only the marginal product of inputs, will have an ambiguous effect on EX*. Changes associated with decline in MP should cause expenditures to fall; however, if MP rises, the sign of the resulting change in expenditures is ambiguous. (Note, however, that if upward shifts in MP are to be regarded as representing increased efficiency, the total price per survivor should fall under these circumstances, although EX* may increase; this should also lead to an increase in C*.)

E. Experience Effects

Thus far in our discussion we have treated the household decision-making process as though it occurred in a certain world with complete information. In fact, p is a probability and outcomes are uncertain so that a natural extension of the model might be to include expected utility analysis and a discussion of decision making under uncertainty. Using a model similar to the one developed in this paper, however, Ben-Porath (2) has shown that such added complications are not likely to lead to significantly more fruitful models in the absence of more *a priori* restrictions on the utility function of the individual family. In addition, such an extension is not likely to shed much light on the significant question as to how expectations are formed and modified during the family formation process. In fact, while I have developed above a one-period model with decisions on critical variables being made at the beginning of the family formation period, family formation takes place in a substantial time continuum with one pregnancy following another in sequence. Accordingly, I shall concentrate on the sequential nature of this decision-making process in the remainder of this section. During this sequential process many parameters may change including expectations about income, attitudes toward and the demand for children and expectations about the production function for children p(RE, EX). In particular, in the terminology of the previous section, families' expectations about both RE and $\partial p/\partial EX$ might well change, leading them to revise their decisions about EX*, p*, C*, and n*.

Prenatal Care and Infant Health

Substantial work on the effect of exogenous child attributes on fertility decisions has been reported in a series of papers by Welch and Ben-Porath, [Ben-Porath and Welch (4); Welch (25); Ben-Porath and Welch (3); and Ben-Porath (2)]. Although they have been primarily concerned with the effect of the sex mix of surviving children on family size, Ben-Porath and Welch have indicated that an extension of their model to account for other exogenously determined attributes such as mortality is possible. The extension of such a model to measure empirically the effect of pregnancy losses on the decision to have an additional child has been investigated by Williams (28).

Welch (26) provides a mathematically detailed sequential decision model of the effect of prior experience with the sex of offspring on the decision to have additional children. His conclusion based on what he feels are reasonable estimates of the relevant parameters is that the effects of learning (i.e., changing expectations about the sex ratio) are not dominant. Williams (28) applies Welch's derivation to the case of completely exogenously determined mortality but without definite predictions about measurable parameters. Neither considers the situation which has been our concern so far, that is, the effect of prior experience with pregnancy on endogenously determined expenditures during current and future pregnancies.

Because Welch does provide an elegant argument about the expected effects of experience on expectations, it is worth calling attention to differences in the determination of the sex ratio and infant loss rate. For one thing, I have argued that the infant loss rate is partially endogenously determined by family expenditures on pregnancy—historically the sex of the unborn child has not been subject to parental control. Secondly, the expected values of the sex ratio and infant loss rate within the population are of a different order of magnitude, particularly in developed countries: Welch estimates the sex ratio at about .51 male while recent U.S. data indicate that survival of offspring during late pregnancy (over 20 weeks gestation) and infancy (less than 1 year) exceeds .95. Thus, if the family plans three pregnancies, the probability that all three will be boys is .125 while the probability that all three will survive is .857, seven times greater. Moreover, if there is some desire for balance, as Welch assumes, the probability in three pregnancies of a mix of sexes is .75 while the probability of a single loss only is .135.[5] If we adopt the information theory approach to the value of information (i.e., that the more unlikely an event is the more information its occurrence contains), then there is more potential information to be gained from family experience about the potential survival of progeny than there is to be learned about potential differences in the sex ratio.

Let us now consider the implications of gaining experience about child survival patterns as children are born. Recall that the original model we considered was a one-period decision model where decisions were made about EX^*, p^*, C^*, and n^* at the beginning of the childbearing period and were held throughout the period of family formation. Since p^* is less than 1, at the end of n^* pregnancies there will be some distribution of surviving children among families whose initial goals were all the same. Some families will have achieved C* children; these we will call the lucky ones. Other families will have fewer surviving children, with the most unlucky families having no surviving children. Following Welch, we shall discuss the impact on a family of each of these different outcomes as encompassing an income and a learning (price) effect.

Consider the unlucky family for whom after n* pregnancies, C = 0; they have experienced an income loss equal to n*(EX* + FX). If there is no learning (reformulation of expectations) or disutility associated with pregnancy losses, they will be in the same situation as a family starting at that lower income level. In this rather extreme example, they may optimize again and continue attempting to have children with a new C*' and EX*' appropriate to their new lower level of total income. Their loss in the production of children will not be absorbed totally in the demand for children but rather spread throughout all consumption. Note, however, that if C is a normal good (i..e, $\partial C^*/\partial Y > 0$), then they will desire a smaller family after they have experienced these losses than initially (C*' < C*). In addition, if the income effect on EX* is as we derived above, they will spend less on these subsequent pregnancies (EX*' < EX*). It does not seem unreasonable to expect similar behavior from all families whose attained family size is less than they initially targeted—this, in fact, is a form of the replacement behavior we discussed earlier.[6]

Consider now the learning or informational effect of having experienced n* pregnancies with something other than C* successes. We are dealing with a second type of uncertainty here—that is, we have assumed up to this point that the production function p = p(RE, EX) and the level of RE were known with certainty and that the only uncertainty of outcome resulted from p* being less than 1. In the real world, it is very unlikely that either RE or p = p(RE, EX) will be known with certainty and, at best, the household can only form estimates as to their values. The parameters in our model of interest to the household are RE and $\partial p/\partial EX$, and we shall denote their estimate of these values by *RE* and *MP*. Fortunately our earlier discussion of the symmetry between utility maximization and cost minimization as well as our exercises in comparative statistics regarding these variables will be of use in analyzing the effect of learning about these parameters on behavior.

First, we should distinguish between losses which convey information about the production process and losses which do not. Pregnancy losses due to apparently random events outside the pregnancy process itself such as accidents or deaths by violence presumably convey little information about expected outcomes resulting from subsequent pregnancies. Their influence on subsequent pregnancy related decisions will be felt only as the result of a pure income loss. On the other hand, during a pregnancy specific information about the pregnancy process and the value of inputs and expected outcomes may be gained whether or not the resulting infant survives (e.g., information about potential Rh sensitization resulting from pregnancy is usually obtained as a result of specific blood typing of parents during the first pregnancy). Let us consider the effect of such information on subsequent pregnancy decisions whether or not an unexpected loss has occurred.

Consider again the case of the household with no success after n* pregnancies. If they do not reevaluate *RE* and *MP*, then they will reformulate a new decision plan using these same parameters and only the income effect analyzed above will cause a change in EX*. However, if p* was initially thought to be high, they may be tempted to revise their estimate of p* which depended on their estimate of p = p(EX, RE). They may revise their estimate of RE which we have loosely defined as being associated with the height of the function or their estimate of MP.[7]

Let us briefly recall the results of the previous section where we examined the effects of changes in RE and MP using the function $p = 1 - Ae^{-BEX}$ as an example of the type of functional relationship which might well relate RE and EX to p. Recall that a change in A which we felt could be interpreted as being related to biological efficiency (RE) was seen to be negatively correlated with EX* and that this effect was only reinforced because of related induced shifts with this function in MP. Recall also that shifts in B which we interpreted as only affecting MP and as being associated with efficiency in production and with environmental variables were negatively correlated with EX*.

Thus, we hypothesize the following affects of experience on pregnancy behavior:

1. Couples who experience a level of C < C* with expenditure EX* will experience an income loss effect which will tend to reduce EX*' and C*', the new targets during the second decision-making period. Depending on the size of the income effect they may decide to continue (i.e., n*' > n*) or cease having additional children.
2. If couples further experience a learning effect which they interpret as reflecting a biological reproductive efficiency effect similar to the change in A discussed above, they will tend to increase their target EX* to EX*' to compensate for their less productive position. Notice that such an increase in EX is contrary to the income effect and the question of which effect dominates becomes largely a matter of empirical determination. In addition, an increase in EX* associated with learning from bad experiences will raise the price of surviving children and may reduce desired family size (i.e., C*' < C*) in a manner reinforcing the income effect.

If we relax one more element of the initial model, we can move toward a more realistic formulation of the fertility-related decision-making process. Initially, we assumed a one-period model where households chose the utility maximizing values of EX, C, n, and p and stuck to that decision throughout the period of family formation. In this section we recognize that since p* is likely to be less than 1, certain families will not attain C* after n* pregnancies even though they expend EX* per pregnancy. Regardless of any revisions in their expectations concerning p*, it is logical to regard these families as being in a situation similar to other families with a reduced income level and allow them the possibility of trying again (i.e., have pregnancies beyond n*). Moreover, if they also revise their expectations regarding p = p(RE, EX) this will also affect their decision regarding C*, n*, EX*, and p*.

It is easy to see that we can further relax our assumptions about when decisions are made and consider a complete sequential decision-making procedure of a special kind. For example, consider the situation cited earlier where $p^* = .9$ and $C^* = 3$ so that n* = 3.33 or 3 to 4. After the first pregnancy most couples will experience a success but some will have a failure. In particular, they now know that they will require at least four pregnancies to have three children. In addition, if they regard the loss as indicating a lower than expected child production potential, they may revise their decisions about future expenditures on pregnancy and desired family size. Within the context of our one-period model we may regard such couples as modifying their behavior to include possibly new values of EX*, C*, n*, and p* as though these new values would apply to *all future* pregnancies. Moreover, it is not unlikely that if p* is considerably lower than in our example or C* considerably higher, couples who experience success may reevaluate their position in light of favorable results after each pregnancy. So long as households follow the procedure of minimizing the cost of producing the remaining children they desire, we need only be concerned with how their experience alters their view of the forces influencing the marginal condition (11a) in order to make predictions about EX*. The stock of children already attained as well as the sequential path of pregnancies which lead to the current levels of C and S will influence decisions as to whether to have additional pregnancies but will not influence our predictions about the effect on expenditures per birth.

III. THE PREGNANCY PROCESS: AN EMPIRICAL FORMULATION

A. Data

The ideal data with which to investigate the implications of the preceding model of interaction between the demand for children, pregnancy experience, expenditures on pregnancy, and pregnancy outcomes would be a form of panel data which would include for a sample of women information on expenditures and outcomes over a series of pregnancies. With such a data set, it would be possible to estimate the effect of pregnancy input levels and outcomes experienced on expenditures during any indexed pregnancy and also to investigate the effect of the experienced efficacy of pregnancy inputs and experienced pregnancy outcomes on the decision to have additional children.

Unfortunately, such data sets are not yet available and investigators have to compromise between examining survey data that contain information on the reproductive histories of individual women but little about expenditures on individual pregnancies and cross-sectional birth certificate data which contain information on prenatal care utilization and pregnancy experience for a cohort of pregnancies in a given year. Williams (28) has used data from the 1965 National Fertility Survey to investigate the effect of experience with prior pregnancies on the decision to have additional children. In addition, she has investigated completed family size and mortality rates as determined by prior pregnancy experience and socioeconomic variables as they have been measured in the National Fertility Survey and variations in birth weight using data from the National Natality Survey (1964–66). Although the model developed in this paper has touched on issues similar to those investigated by Williams, its thrust is toward a number of interesting and potentially expensive public policy options which revolve around the efficacy of medical care in producing favorable pregnancy outcomes. For this reason, I decided to utilize a data set containing information on the amount of prenatal care received during a specific pregnancy, pregnancy outcome, and pregnancy history.

The data set used is a sample of birth certificates from the January–June 1970 New York City birth cohort as recorded by the New York City Department of Health. There are several advantages to using this data source. The number of observations is large, over 67,000 births, and New York City birth certificates contain information not only about the birth, but about parents' ages, their place of birth, race, parents' education, and the reproductive history of the mother. Information is also available about the amount of medical care utilized during pregnancy, about the mother's health, and finally, about the status of the infant at birth.

Unedited data tape listings of all births registered in New York City in 1970 and 1970 fetal deaths were provided by the New York City Department of Health. Although the original intent of the study was to include information on mothers who experienced a fetal death in the empirical estimates, it was soon determined that comprehensive and accurate information on fetal deaths was not available. In addition to the well-known problem of incomplete fetal death certificates, the data for 1970 were affected by the legalization of abortion in New York State as of July 1 of that year. The city's Health Department was not equipped to differentiate in collecting vital statistics for that year between elective abortions and spontaneous fetal deaths. Both were recorded on the same forms and maintained on the same data tape files. Moreover, there seemed to be good reason to believe that the number of "suspicious" spontaneous abortions increased early in 1970 in anticipation of the legalization of elective abortion later that year. Accordingly, data on fetal deaths were not used in estimating either demand or outcome functions.

The data set used was restricted to births occurring between January and June 1970 because it appeared that decisions concerning births occurring after this date could have been affected in perhaps a transient way by the pending legalization of abortion after June 1970. Because of the very large New York City birth cohort, the data set even after this restriction was still very large. In addition, the data were subjected to an editing procedure to accommodate missing data items. Individual records were edited as follows: (1) if values for the dependent variables were not recorded, the observation was eliminated; (2) if only a small number of independent variables were not recorded, relevant mean values of the missing data item were used in place of the missing value; (3) if a large number of independent variables were not recorded, the observation was eliminated.

The edited data set contained 54,280 observations, approximately 13,000 observations having been eliminated due to insufficient recording of data items to justify the observations' inclusion.

In addition to the information systematically recorded on the birth certificate, New York City is divided geographically by the Health Department into Health Areas and Health Districts. Using information on the Health District of residence of the parents which is available on the birth certificate, individual observations were linked with measures of the geographic availability of prenatal care providers including gynecologists/obstetricians, prenatal clinics in hospitals and health stations, and special-emphasis maternal and infant/child care centers. Using these augmented data, one can estimate the effect of the availability of facilities on their utilization by individuals, a topic of significant policy concern. Moreover, the use of individual observations tends to minimize simultaneous determination problems which might arise in making such estimates because of the relationship in more aggregative units of observation between the availability of facilities and the potential demand for services.

B. Determinants of Prenatal Care Utilization: Empirical Specification

In Section II, I examined the demand for different levels of pregnancy inputs in very general terms. The empirical model, estimated using New York City birth certificate data, focuses on the demand for and use of prenatal care in the pregnancy process. This is not because other endogenous inputs such as maternal nutrition or cigarette smoking are not considered important, but rather because information on these other inputs and behaviors is not available. The restriction of the investigation to prenatal care need not be viewed in a negative light. Of all the potential inputs, prenatal care is one of the few that produces health (both maternal and infant) primarily during the pregnancy period. Life style factors such as proper diet, adequate sanitary conditions, and cigarette smoking affect individual health whether pregnant or not. Moreover, to the extent that the prenatal care process educates mothers and fosters modifications in their life styles favorable to health, its beneficial effect on outcome may operate through changes in these other inputs. More extensive and intensive prenatal medical care and its extension to prepregnancy, postpartum, and family planning services have been the primary policy instruments advocated in the attempt to reduce infant mortality levels in the United States. The Institute of Medicine study, the establishment of Maternal and Infant Care projects, the establishment of neonatal intensive care units, and the recently voiced concern about the adequacy of existing health insurance plans to encourage adequate prenatal care (Muller *et al.*, 21) all seem to indicate that the emphasis is and will probably continue to be on increasing health services utilization in order to reduce infant mortality. Accordingly, it would appear that an empirical investigation of the determinants of the demand for prenatal care would be of substantial interest exclusive of the theoretical framework discussed earlier.

1. Dependent Variables

In the data, the demand for prenatal care may be measured by the number of prenatal visits (NUMVISIT) and the interval from the last menstrual period to the first prenatal visit (INTERVAL). While not strictly a traditional demand measure, INTERVAL has several important attributes which make its investigation of interest. First, it is accepted medical wisdom that early care is preferred—that definite benefits can be derived from monitoring pregnancies early in their course and that specific health risks can be identified early and as a consequence treated more effectively. Second, it is likely that the earlier prenatal care is initiated the more visits will be consumed. This is because there is an established protocol for scheduling prenatal visits during the course of even a normal pregnancy. Of course, individual mothers may violate this protocol and physicians may vary in the degree to which they adhere to the protocol, so the relationship between INTERVAL and NUMVISIT is not a strict identity.[8]

An aspect of this proxy demand measure that makes it of particular interest is that the interval is largely determined by the mother based on her consideration of her needs and wants. Once she begins prenatal care, decisions on the amount and type of care to be consumed will be jointly determined by the mother and her physician. The physician's interpretation of the care required by the maternal condition as well as the mother's evaluation of the information supplied by the physician will be important factors in determining NUMVISIT. In fact, attempts by the physician to encourage the mother to follow a protocol will only reinforce this interrelationship.

Several problems arose in utilizing INTERVAL and NUMVISIT as coded on New York City birth certificates as dependent variables in these demand estimates. In the case of the INTERVAL variable, a problem arose because it was difficult to assign a meaningful value to the INTERVAL measure for mothers who had no prenatal care. Although it may appear natural to define INTERVAL as zero for these pregnancies, this procedure will lead to inappropriate parameter estimates because the interval to the first visit is an inverse measure of the demand for care (i.e., the shorter the interval the greater the demand). For this reason, attempts to estimate demand equations over the entire range of possible values for INTERVAL, including INTERVAL defined as zero when no care is sought, may bias parameter estimates toward zero.

Two different procedures were followed to accommodate this discontinuity in the dependent variable. First, for the entire sample, a function with a dependent variable (NOVISIT) corresponding to the visits/no visits dichotomy was estimated and for a sample restricted to those pregnancies with at least one prenatal visit, a function was fit using the continuous variable INTERVAL as a dependent variable. There are, however, several limitations to this approach. One is that functions with dichotomous dependent variables present special problems of estimation. A second objection is that it may be difficult to interpret the results of this two-step estimation procedure and that in estimating the continuous function the information contained in the observations where there were no visits has been discarded.

An alternative strategy was to set an outer limit on the probable duration of a pregnancy and transform INTERVAL into a new variable as a function of the hypothesized potential pregnancy period. A scan of the data indicated that 350 days was a reasonable outside limit on the duration of a pregnancy (measured from the date of the last menstrual period) and the new dependent variable for visit interval was defined as

$$\text{INTER}(350) = (350 - \text{INTERVAL}) \quad \text{if} \quad \text{INTERVAL} \neq 0$$
$$\text{INTER}(350) = 0 \quad \text{if} \quad \text{INTERVAL} = 0. \tag{34}$$

This technique has the benefit of transforming the interval measure into a continuous variable that will be positively correlated with the demand for care. That is, the earlier prenatal care is initiated, the larger will be INTER(350) and in the absence of any prenatal care INTER(350) will be zero. The primary drawback to using this transformation is that the size of the estimated coefficients could depend on the scaling of the variable. The constant selected, 350 days, is somewhat arbitrary, and a smaller constant would have moved the scatter of points with INTER(350) > 0 closer to those observations with INTER(350) = 0, while a larger constant would have spread the two groups further apart. The sign and extent of the possible bias in the estimated coefficients that is dependent on the scaling factor appears *a priori* to be indeterminate.

A problem arose in using the number of prenatal visits (NUMVISIT) as either an independent or dependent variable. This resulted from the fact that in transferring information on NUMVISIT from the actual birth certificates to the EDP files which were utilized in this study only one column was allocated to this variable. Hence, although NUMVISIT could theoretically take any value on the actual birth certificate, it only could take on the values 0–9 on the data tapes. As a substantial number of observations were coded with 9 visits (45 percent), this raised the possibility of introducing a bias into equations where this variable appeared.

The distribution of actual prenatal visits from a hand-coded one-day sample of birth registrations was available. From this sample, I determined the distribution

Table 1. Definitions of Dependent Demand Variables

Variable Name	Definition
INTERVAL	Interval measured in days between date of mother's last menstrual period and the date of her first prenatal care visit
INTER(350)	A transformation of the INTERVAL variable: (1) Equals (350-INTERVAL) if INTERVAL $\neq$ 0 (2) Equals 0 if INTERVAL = 0
NUMVISIT	Number of prenatal care visits
NOVISIT	Dummy variable that equals one if mother had no prenatal care visits, zero otherwise

of observations with 9 or more visits by race/nativity groupings (white, black, foreign-born, Puerto Rican). The actual number of visits of mothers with 9 or more visits were tightly grouped in the range from 9 to 12 visits but sharply skewed toward larger numbers, with a maximum at 22 prenatal visits for a single pregnancy. The mean number of visits in this range was 11.04 visits, and this mean did not vary significantly among the race/nativity groupings. Accordingly, this mean, 11.04, was substituted for 9 in the actual data whenever 9 was coded on the data tape, and NUMVISIT, when it was utilized as either a dependent or independent variable, was allowed the values 0 through 8 and 11.04.[9] Dependent variable names and definitions are summarized in Table 1.

2. Independent Variables

Following the rationale developed in the theoretical model, prenatal care is considered an input in the production of healthy infants and the demand for care is accordingly viewed as derived from the demand for the ultimate product. Generally, the demand for a factor of production depends on its price, its value in production, and the desired level of output. In the household production context, the desired level of output depends, among other things, on income, tastes, and the parameters of the household production function.

Using the 1970 New York City birth certificate data set, demand for prenatal care relationships was estimated using a number of available proxy measures of the idealized potential independent variables. These proxy variables are named and defined in Table 2. The same set of independent variables was used in estimating relationships for the various measures of prenatal care utilization. In addition, as discussed below, appropriately scaled measures of the length of the interval to the first prenatal visit were included as independent variables in the NUMVISIT equations to account for the effect of medical visit scheduling protocols on the number of visits demanded. Summary statistics for the dependent and independent variables are presented in Appendix Table A-1.

Table 2. Definitions of Independent Variables

Variable Name	Definition
INTERVAL	Interval measured in days between date of mother's last menstrual period and the date of her first prenatal care visit
INTER1	Dummy variable that equals one if INTERVAL is greater than zero (i.e., if there is at least one visit)
INTER2	Dummy variable that equals one if INTERVAL is greater than 90 days
INTER3	Dummy variable that equals one if INTERVAL is greater than 180 days
INTER(350)	A transformation of the INTERVAL variable: (1) Equals (350-INTERVAL) if INTERVAL > 0 (2) Equals 0 if interval equals zero
INTER(350)2	Dummy variable that *equals* one if INTER(350) is less than 210 days
NUMVISIT	Number of prenatal care visits
AGEMOTH	Mother's age in years
AGEMSQ	Mother's age in years squared
EDMOTH	Years of formal schooling completed by mother
EDFATH	Years of formal schooling completed by father
LEGIT	Dummy variable that equals one if the child was born in wedlock
RACEN	Dummy variable that equals one if the child is of Negro race (if either parent is Negro the child is coded as Negro)
FORMOTH	Dummy variable that equals one if the mother was born outside the U.S. and its possessions
PRMOTH	Dummy variable that equals one if the mother was born in Puerto Rico
CHILDLIV	Number of previous children born alive to mother still living at time of indexed birth
TOTLOSS	Sum of number of previous children born alive, now dead and previous fetal deaths at all gestation ages
TBO	Total pregnancy order not including current birth (i.e., TBO = CHILDLIV + TOTLOSS)
%LOSS	Percent of previous pregnancies not surviving to date of current pregnancy (i.e., %LOSS = TOTLOSS/TBO)
FIRST	Dummy variable that equals one if this is first pregnancy
LAST	Dummy variable equals one if previous pregnancy ended in a fetal death
MIC	Dummy variable equals one if mother resides in geographically defined health district which had an active Maternal and Infant Care Project health facility in 1970
CLINIC	Number of hours per week per hundred pregnancies of prenatal clinic time in all (municipal, voluntary, and MIC projects) facilities in health district where mother resides
OB/GYN	Number of obstetrician/gynecologists (in private practice) per hundred pregnancies in the health district where the mother resides

The rationale behind the inclusion of each variable and its expected effect on demand are as follows:

1. AGEMOTH, AGEMSQ—A "J-shaped" statistical relationship has been consistently observed between maternal age and reproductive loss [Nortman (23)]. This relationship has been publicized with emphasis on the increased risk to very

young and older (over 35 years) mothers. If mothers perceive this relationship as evidence of a biological relationship between age and reproductive efficiency, maternal age would be a consideration in their demand for care. In particular, to the extent to which age-related variations in reproductive efficiency reflect a pure endowment effect (a shift in the production function), the model would predict a positive correlation between the increased risk associated with age and the demand for prenatal care.

2. EDFATH—Unfortunately, the New York City data do not provide information on family income. As a proxy for permanent family income, economists have frequently used father's education. This procedure is based on past experience of a consistently observed relationship between permanent family income and father's education. Regarding family income (Y), I have argued that Y affects the demand for inputs during pregnancy primarily because of its positive correlation with the fixed cost of pregnancy. Y will tend to increase the demand for inputs. Whether or not Y will ultimately be positively correlated with the demand for medical care depends not only on this result but on the shadow price and productive value of medical care and other pregnancy inputs. It seems reasonable, however, to expect a positive relationship between care and Y. An additional function of prenatal care not explicitly discussed above is to maintain the health of the mother during and after pregnancy. To the extent to which the demand for maternal health is a function of family income (Grossman, 14), there is an additional reason to expect a positive income effect.

Although EDFATH is reported frequently in the New York City birth certificate data set, it is not always available. This was particularly true for illegitimate births. Moreover, for illegitimate births, it is unlikely that father's education will be as good a proxy for income as for legitimate births. A sufficiently large sample of illegitimate births did contain information on EDFATH so that a mean value of EDFATH for illegitimate births could be calculated and used to replace mising data. In addition, a dummy variable to indicate illegitimate births was included in all regressions in part to capture the potential bias ascribable to the inaccurate measurement of EDFATH for these pregnancies.

3. LEGIT—A dummy variable set equal to 1 if the child was born in wedlock. In the context of the model developed earlier, no account is specifically made of the effect of legitimacy status on the demand for care. It would appear, however, that, other things equal, illegitimate children may be less "wanted" and that holding other parental characteristics constant, family income might be considered lower in single-parent families as total household resources (primarily time which could be sold in the market or used in household production) would be lower. If we would expect a "wantedness" or income effect, then we expect that the demand for care for legitimate births will be greater than for births out of wedlock.

4. EDMOTH—Education of mother has generally been interpreted as a general environmental variable in other household, health, and children's health production models. In particular, more educated mothers may be more aware of recom-

mended prenatal care procedures, may be more oriented toward "preventive" care, and may be more able to use pregnancy inputs effectively. For these reasons, we would expect more educated mothers to demand healthier infants and perhaps demand more care. On the other hand, to the extent to which more educated mothers have a larger stock of health to begin with and, accordingly, a higher level of reproductive efficiency, we may observe an endowment effect. Such an effect, associated with a pure shift in the production function, should lead to a reduction in the demand for pregnancy inputs, particularly if it is associated with a decline in the marginal product of a pregnancy input. Thus, it is difficult to predict the expected relationship between mother's education and the demand for medical care.

5. CHILDLIV—The number of children of the indexed mother living at the time of the indexed birth captures many different hypothetical demand effects. In particular within the context of a semisequential decision model, CHILDLIV is a measure of experienced reproductive success (holding total births constant) and also a measure of the stock of children a household has already acquired. Both these factors would tend to indicate that high levels of CHILDLIV should tend to depress the demand for care. In particular, other things equal, the experience of successful pregnancies may be taken as information of a high level of reproductive efficiency which may be negatively correlated with the demand for care. Moreover, in a sequential context, families may attach a declining value to additional children, or alternatively some families with larger targeted family sizes may optimize their demand for child services by producing larger amounts of "lower"-quality children; in either case, the demand for care should decrease as CHILDLIV increases.

Family size as measured by CHILDLIV may also be related to the shadow prices of time and other inputs in the reproductive process. Hence, there may be economies of scale associated with the preparation of nourishing meals as family size increases; older children's time may substitute for the mother's in the production of other household commodities during the pregnancy period; and the presence of young children may increase the cost of prenatal visits if babysitters are required. The expected effect of these shifts, while frequently predictable for individual situations, cannot usually be determined in the aggregate for most families and is not identifiable in the data set used to estimate the model.

If we hold the number of pregnancies constant in a cross section, then CHILDLIV will reflect past experience with pregnancy and possibly information about individual reproductive efficiency. As discussed previously, this may have both an income and a price effect. The net effect is somewhat ambiguous. *Ceteris paribus,* increasing levels of C may have an income effect which encourages increases in expenditures on subsequent pregnancies and a reproductive efficiency effect which discourages such increases—only if the reproductive efficiency effect is interpreted largely as representing increases in the marginal product of factors will the two effects reinforce each other.

6. TBO, %LOSS, FIRST, LAST—A group of experience variables measuring not only the extent of experience with pregnancy (TBO), but also the nature of that experience. In particular, %LOSS and LAST are measures of previous pregnancy failures (child and fetal deaths). To the extent that these failures are interpreted as indicating a reduction in reproductive efficiency, they should be positively correlated with the demand for care. FIRST, a dummy variable for first pregnancies, is entered with no *a priori* assumption as to its effect on the demand for care but as a measure of the effect of the absence of previous pregnancy experience. Similarly, TBO is a cumulative measure of the extent of all pregnancy experience.

7. CLINIC, OB/GYN, MIC—Measures of the relative availability of care. Within this data set, there are no measures of prices actually paid or charged for prenatal care. It seems reasonable to assume that since all births occurred within New York City during a 6-month period, all mothers faced the same distribution of nominal market prices for prenatal care. Net physician prices might have reflected differences in physician characteristics and the extent of individual health insurance coverage, but theoretically "any" woman could have purchased prenatal care from any provider at his/her going fee. Within such a market, a significant determinant of the net total cost of physician visits is the availability of services within a local geographic area. Measurements of obstetrician/gynecologists per capita (OB/GYN), obstetrical clinic hours per capita (CLINIC), and the existence of local Maternal and Infant Care special project care centers (MIC) are all measures of the relative accessibility of prenatal care providers and accordingly may be viewed as being negatively correlated with the price of care. We would expect the demand for care to be a negative function of price and hence positively correlated with these measures of the availability of prenatal care providers.

8. RACEN, FORMOTH, PRMOTH—These particular dummy variables reflect differences in the population not elsewhere captured in the set of independent variables discussed in the model or available in the data set. As such, these attributes may have environmental, genetic, socioeconomic, or cultural aspects. Of particular interest is the consistently reported finding that blacks have higher infant death rates and lower weight infants on average than whites. Within the multivariate model estimated in this study, these variables are entered to distinguish between differences in experience which may ultimately affect demand and outcomes and biologically determined differences in reproductive efficiency which are reflected in population-based infant death statistics.

9. INTERVAL, INTER2, INTER3—These measures of the length of the interval between the mother's last menstrual period and her first prenatal visit are included to measure and control for the effect of the recommended prenatal care protocol on the number of visits demanded. The nonlinear terms (INTER2, INTER3) are introduced because the protocol is nonlinear; i.e., during the first and second trimesters (90-day periods) of pregnancy one visit a month is recommended; in the third trimester, a visit every 2 weeks is recommended during the first 2 months and a weekly visit thereafter until delivery.

The inclusion of the INTERVAL variables in the NUMVISIT function takes account of the sequential nature of decisions during the pregnancy period. Because of the specific sequential nature of this process, one can ask the question, "What is the nature of the determination of the amount of prenatal care demanded as a result of the patient-physician interaction given the patient's initial decision on when to seek care?" Moreover, because of the presumed sequential causality running from INTERVAL to NUMVISIT, this part of the system may be regarded as recursive and, therefore, ordinary-least-squares estimation techniques may be employed.[10]

C. Determinants of Prenatal Care Utilization: Empirical Results

1. Decision to Seek Prenatal Care

In Table 3 are reported the results of classical least-squares regressions of the dichotomous dependent variable NOVISIT (a dummy variable that equals 1 if the mother had no prenatal care visits and 0 if she had at least one visit) on an increasingly inclusive group of independent variables. The coefficients in the table may be interpreted as the change in the probability that the mother had no visits per unit of the independent variable or if a characteristic is represented by a dummy variable, as the change in the probability associated with the presence of the characteristic.

Coefficients of maternal age reported in Eqs. (b) and (c) indicate that maternal age has a nonlinear effect on the probability of seeking prenatal care. The probability of seeking care increases, although at a falling rate, with increasing maternal age, reaching a peak at approximately 38.9 years and falling afterward. This is somewhat ironic in that many studies indicate that *very* young mothers are potentially at high risk of an adverse pregnancy outcome. However, this finding may also indicate that to the extent that care produces favorable outcomes the apparent high risk of very young mothers is due to their lower utilization of care. On the other hand, the evidence indicates that older mothers (over 35), another high-risk group, are among the most likely to seek care—perhaps because of the significant amount of publicity that increased risk associated with the more advanced childbearing years has attracted.

Both mother's and father's education were found to be positively correlated with an increased propensity to seek care. The coefficient on father's education, while much larger in (a), which I would tend to interpret as an income effect, is equal to the coefficient on mother's education in (c). The relative impact of mother's education, which can be interpreted as an efficiency variable and as such may theoretically have both positive and negative effects on the demand for care, is positive but is reduced in size by its relationship to family size (CHILDLIV) and maternal age (AGEMSQ) in (b).

Not surprisingly, legitimate births are more likely to have prenatal care, proba-

Table 3. NOVISIT: Dichotomous Dependent Variable that Equals 1 if Mother Had No Prenatal Visits

Coefficient of OLS Regression Estimates: Entire Sample
N = 54,280
(t-statistic in parentheses)

Variables in the Equation	(a)	(b)	(c)
AGEMOTH	−0.0009	−0.0068	−0.0068
	(6.04)**	(6.34)**	(6.28)**
EDMOTH	−0.0011	−0.0007	−0.0006
	(4.23)**	(2.33)**	(2.20)*
EDFATH	−0.0021	−0.0014	−0.0006
	(5.83)**	(3.98)**	(4.01)**
LEGIT	−0.0436	−0.0403	−0.0399
	(20.97)**	(19.04)**	(18.69)**
RACEN	0.0114	0.0077	0.0052
	(6.12)**	(4.07)**	(2.57)*
FORMOTH	−0.0029	0.0009	0.0007
	(1.50)	(0.45)	(0.34)
PRMOTH	0.0211	0.0191	0.0176
	(9.09)**	(8.19)**	(7.29)**
AGEMSQ		0.0001	0.0001
		(4.52)**	(4.50)**
TBO		−0.0028	−0.0028
		(1.47)	(1.44)
CHILDLIV		0.0108	0.0107
		(4.78)**	(4.71)**
FIRST		−.0019	−0.0020
		(0.90)	(0.92)
%LOSS		0.0003	0.0078
		(0.05)	(1.14)
LAST			−0.0097
			(2.25)*
MIC			0.0092
			(4.62)**
CLINIC			0.0100
			(3.99)**
OB/GYN			0.0006
			(0.66)
Constant	0.1190	0.1876	0.1872
$\bar{R}^2$	0.0239	0.0275	0.0281

Notes:
*Indicates coefficient significant at 5% level.
**Indicates coefficient significant at 1% level.

bly reflecting a hypothesized "wantedness" effect and perhaps an additional income effect. The dummy variables for black (RACEN), Puerto Rican (PRMOTH), and foreign-born mothers (FORMOTH) are, in a sense, measures of ignorance inasmuch as their coefficients measure differences among mothers with different racial, ethnic, or nativity characteristics which are not otherwise explained. Both Negro and Puerto Rican mothers are less likely to receive care, while being foreign-born has no statistically significant effect. Interestingly, the size of the coefficient on RACEN (Black mother) is reduced by over one-half by the introduction of experience and availability-of-care variables in regressions (b) and (c). This would tend to indicate that there are less "unexplained" factors affecting the utilization of prenatal care facilities by blacks than might be attributed to the racial characteristic itself.

The coefficients of the experience variables TBO, CHILDLIV, FIRST, %LOSS, and LAST introduced in regressions (b) and (c) lend limited support to the hypothesized effects of experience. In particular, the probability of not seeking care increases with each live child already in the family, a finding consistent with either an experience or a family size effect. In addition, the probability of seeking care is higher for those mothers whose most recent previous pregnancy ended in a fetal death—probably an experience effect.

The variables associated with the availability of prenatal care providers had different effects on the care/no care decision. The availability of clinic time (CLINIC) appeared to increase the probability of seeking care; however, the availability of obstetrician/gynecologists (OB/GYN) had no significant effect. Surprisingly, the availability of a federally funded Maternal and Infant Care (MIC) project facility in a health district was associated with an increased probability of failing to seek care. One possible explanation for this finding is that MIC projects are established primarily in areas where the utilization of prenatal care is already low so that project centers serve as markers of population groups with a low demand for care. This explanation would be convincing if these MIC projects had been established in the period immediately prior to 1970. However, the MIC program has been underway since 1962, so that adequate time had elapsed within the individual health districts to encourage increased use of prenatal care facilities by 1970. While it could be argued that these projects were established primarily in areas where the availability of alternative providers of care was so limited that even after the MIC facilities had been in existence for 8 years they could not, at the level at which they were operating, compensate for the lack of other facilities in a given geographic area, this is not completely borne out in the data. The simple correlation coefficients between MIC and CLINIC and between MIC and OB/GYN are .49 and −.17, respectively, indicating that MIC areas did tend to have a somewhat reduced availability of private physicians, but this may have been well compensated for by the presence of clinic facilities. Of course, neither CLINIC nor OB/GYN completely adequately measure the amount of provider capacity in a health district.

Of note is the considerable reduction in the coefficient of father's education which accompanied the introduction of the three availability of care measures in regression (c). At first blush, it would appear that increased accessibility to care providers can redress differences in the demand for care which are related to differences in permanent income (as captured in EDFATH). This conclusion has to be tempered, however, by the finding of a positive coefficient on the MIC dummy which would appear to require further investigation.

2. Length of the Interval to the First Prenatal Visit

Tables 4 and 5 report estimates of the demand for care, with two different measures of the interval between the date of the mother's last menstrual period and the date of her first visit used as dependent variables. The regressions reported in Table 4 were run only on that subsample of 52,552 mothers who had at least one visit, and in Table 5 INTER(350) is the dependent variable. As the regression coefficients reported in these tables are remarkably similar, only overall effects will be discussed and exceptions to the general similarity noted.

As was found with regard to the visit/no visit dichotomy, the relationship between maternal age and the length of the interval to the first visit appears nonlinear. The interval is found to be shortest for mothers of 34 to 35 years despite the fact that the risk of an adverse outcome rises steeply as maternal age increases beyond this age. As in the case of the visit/no visit equation, both mother's and father's education are negatively correlated with the length of the interval to the first visit, but the effect of father's education is much stronger, particularly after taking account of the experience variables. Thus in regressions (b) and (c) in Tables 4 and 5, the coefficient of father's education is approximately 3.6 times as great as the coefficient of mother's education. This result is consistent with a hypothesized strong income effect as captured in father's education and a weaker productive efficiency effect as captured in mother's education.

Illegitimate births have their first prenatal visit, if any, delayed about a month as compared with legitimate births—although the measured delay is smallest for that subsample of mothers with at least one visit. Black, Puerto Rican, and other foreign-born mothers also delayed first visits. The coefficients of RACEN and PRMOTH are comparable in size to the coefficient of LEGIT, although they are not as amenable to explanation.

The coefficients of the experience variables indicate no specific birth order effect *per se*, although the coefficient of FIRST reported in Eq. (c) of Table 5 would tend to indicate a slight shortening of the interval associated with first births. The coefficients of CHILDLIV in both tables tend to indicate that the presence of live children or alternatively previously successful pregnancies tend to substantially extend first visit intervals from 5 to 7 days per live child. On the other hand, the effect of negative experience seems to be concentrated in only the outcome of the previous pregnancy. Although a history of bad experience as measured by %LOSS tends to reduce the first visit interval, the coefficient on

Table 4. INTERVAL: Interval Measured in Days between Date of Mother's Last Menstrual Period and the Date of Her First Prenatal Visit

Coefficients of OLS Regression Estimates for the Sample of Mothers Making at Least One Prenatal Visit, N = 52,552

(t-statistic in parentheses)

Variables in the Equation	(a)	(b)	(c)
AGEMOTH	−0.9389	−9.0350	−8.9093
	(20.86)**	(26.61)**	(26.25)**
EDMOTH	−1.0116	−0.5745	−0.6114
	(11.46)**	(6.53)**	(6.95)**
EDFATH	−2.7976	−2.2317	−2.2019
	(24.31)**	(19.47)**	(19.23)**
LEGIT	−27.9751	−24.1184	−23.2206
	(41.49)**	(35.53)**	(33.98)**
RACEN	28.8665	25.9259	24.1485
	(48.04)**	(43.24)**	(38.15)**
FORMOTH	10.3381	13.1904	12.7555
	(16.98)**	(21.75)**	(21.01)**
PRMOTH	25.2312	24.1240	22.0326
	(38.85)**	(32.67)**	(28.87)**
AGEMSQ		0.1331	0.1312
		(21.75)**	(21.46)**
TBO		0.6227	0.5849
		(1.04)	(.97)
CHILDLIV		4.9878	4.9310
		(6.95)**	(6.87)**
FIRST		−1.0475	−1.1454
		(1.55)	(1.69)
%LOSS		−8.2524	−3.8456
		(4.39)**	(1.79)
LAST			−6.1099
			(4.50)**
MIC			4.7311
			(7.64)**
CLINIC			0.0160
			(3.09)**
OB/GYN			0.0768
			(.24)
Constant	198.9807	294.9729	290.9597
R̄²	0.1946	0.2156	0.2176

Notes:
*Indicates coefficient significant at the 5% level.
**Indicates coefficient significant at the 1% level.

Table 5. INTER(350): 350—Interval Measured in Days between Date of Mother's Last Menstrual Period and the Date of her First Prenatal Visit; Equals Zero if Mother had No Prenatal Visits

Coefficients of OLS Regression Estimates: Entire Sample
N = 54,280
(t-statistic in parentheses)

Independent Variable	(a)	(b)	(c)
AGEMOTH	1.0979	10.1455	10.0024
	(20.99)**	(25.78)**	(25.42)**
EDMOTH	1.2372	0.7012	0.7314
	(12.09)**	(6.88)**	(7.16)**
EDFATH	3.2645	2.5799	2.5523
	(24.42)**	(19.38)**	(19.02)**
LEGIT	35.7779	31.3818	30.4144
	(46.32)**	(40.39)**	(38.89)**
RACEN	−30.7679	−27.1017	−24.8127
	(44.18)**	(38.99)**	(33.80)**
FORMOTH	−9.6078	−13.1955	−12.7261
	(13.48)**	(18.59)**	(17.91)**
PRMOTH	−28.9406	−27.4190	−25.0155
	(33.60)**	(32.15)**	(28.37)**
AGEMSQ		−0.1468	−0.1447
		(20.67)**	(20.40)**
TBO		−0.2356	−0.1991
		(.34)	(.29)
CHILDLIV		−6.7611	−6.6779
		(8.15)**	(8.06)**
FIRST		1.4600	1.5506
		(1.86)	(1.97)*
%LOSS		8.4228	2.6557
		(3.84)**	(1.06)
LAST			7.8569
			(4.96)**
MIC			−6.5521
			(8.97)**
CLINIC			−0.3300
			(.54)
OB/GYN			−0.1000
			(.28)
Constant	127.35	20.67	24.91
R̄²	.194	.216	.218

Notes:
*Indicates coefficient significant at 5% level.
**Indicates coefficient significant at 1% level.

%LOSS loses significance when LAST is introduced into the regressions. This is probably due to the high correlation between %LOSS and LAST (simple correlation coefficient equals .67) and the fact that recently experienced losses as captured by LAST may have a more substantial impact on behavior than those experienced earlier in the childbearing life cycle which are reflected in %LOSS. Moreover, LAST measures a specific pregnancy related loss (a fetal death), while %LOSS includes a small number of child deaths as well as fetal losses. The coefficient of LAST is not insubstantial in any set of regressions and, in fact, is larger than the coefficient of CHILDLIV in all regressions in which they both occur. The significant negative coefficient on LAST is consistent with interpreting previous failures as indicating a reduction in reproductive efficiency of the production function shift type rather than representing a change in the value of the marginal product of prenatal care inputs.

Turning our attention to the availability of care variables (MIC, CLINIC, OB/GYN), we note that, as in the case of the visit/no visit regressions reported in Table 3, being located in a designated MIC project area tends to be associated with a decreased demand for care (as measured by a lengthened first visit interval). The availability of obstetrician/gynecologists has no significant effect on the interval and the effect of increased availability of clinic time is paradoxical. The coefficient of CLINIC is only significant in the interval regressions restricted to those mothers who had at least one visit [Table 4, regression (c)]. In this equation, it has a positive sign, indicating that an increase in clinic hours per pregnancy in a health district is associated with an increase in the length of the first visit interval. This finding is somewhat contrary to expectations, particularly as this variable was previously found to be positively associated with initiating care in the first place (Table 3). To some extent the opposite effects of this variable as reported in Tables 3 and 4 may account for its failure to achieve significance when INTER(350) is used as the dependent variable since these regressions are run over the entire sample. Although an increase in CLINIC may be associated with a decrease in the price of prenatal care (i.e., the more clinic hours available per pregnant mother, the lower the *net* cost of a visit), CLINIC is at best an imperfect measure of either the time or information costs of prenatal care. The numerator, scheduled clinic hours, will not usually completely capture differences in individual clinic production capacities and the denominator, pregnancies per health district, is only a measure of potential demand for and not the actual use of a facility. The ratio of district-specific clinic capacity to actual use would be a better surrogate index of the availability (cost) of clinic care to an individual pregnant mother, but such a measure is not available.

3. *Number of Prenatal Visits*

In considering the demand for and utilization of prenatal care as measured by the number of prenatal visits (NUMVISIT), we shall primarily be concerned with contrasts between demand functions estimated solely on predetermined variables and those which take the length of the interval to the first visit into account. The latter group of relationships will enable us to assess the impact of physician input on the utilization of care. These two sets of regressions are presented in Table 6.

Equations (a) through (c) in Table 6 contain only independent variables which measure the socioeconomic characteristics of households, pregnancy experience, and the availability of care facilities, and as such these regressions can be said to represent reduced-form demand equations. The signs and size of the coefficients reported in these equations are consistent with the effect reported for the estimated INTERVAL equations. Thus, the number of visits is a nonlinear function of mother's age reaching a maximum at about 36 years—an age just below the average woman's final years of fecundity, the period when the risk of abnormal pregnancy outcomes is greatest. Mother's and father's education both tend to increase the number of visits, but the importance of EDFATH relative to EDMOTH is reduced when compared with the results of the interval equations—the elasticity at the means of EDFATH is .11 and of EDMOTH .06 [based on regression (c)]. The signs of the coefficients on LEGIT, RACEN, FORMOTH, and PRMOTH are consistent with previously reported effects and indicate a substantial amount of variation in utilization associated with these characteristics which is not otherwise well explained. The experience variables (CHILDLIV, FIRST, %LOSS, TBO, LAST) behave as expected. There appears to be a reduction in the number of visits of approximately .3 visit per live child in the home, while first pregnancies receive an added amount of care. As in the previously discussed estimates, %LOSS, the cumulative experience variable, while it has a positive effect when entered alone, is dominated by LAST (last pregnancy ended in fetal death = 1) when both are entered into the equation.

The coefficients of the variables representing the availability of care providers are consistent with the results of the interval regressions. Mothers in MIC-designated areas have .4 fewer visits, while the availability of clinic hours tends to increase the number of visits, although not substantially [the elasticity of CLINIC in regression (c) is .01]. The coefficient of OB/GYN has a negative sign, although it is not statistically significant.

In regressions (d) through (f) in Table 6, I examine the effect of including the information on interval to the first visit in the NUMVISIT estimates. In regression (d), I have used incremental dummy variables to measure the effect of the length of the interval and to take account of the nonlinearity of the suggested obstetrical protocol for scheduling expectant mothers for prenatal care.[11] In regression (e), I use the transformed variable INTER(350) and a dummy variable INTER(350)2 to account for nonlinearity. In regression (f), the sample is restricted to only those mothers who had at least one prenatal visit and the definition of INTERVAL as the actual interval to the first visit is followed. INTER2 and INTER3 once again are introduced to account for nonlinearities.

The effect on the coefficients of the predetermined variables of including these

Table 6. NUMVISIT: Number of Prenatal Care Visits

Coefficients of OLS Regression Estimates: Entire Sample, N = 54,280
(t-statistic in parentheses)

Independent Variables	(a)	(b)	(c)	(d)	(e)	(f)[a]
AGEMOTH	0.271	0.360	0.355	0.126	0.067	0.097
	(14.62)**	(19.07)**	(18.85)**	(8.21)**	(4.46)**	(6.23)**
EDMOTH	0.060	0.041	0.041	0.026	0.019	0.022
	(12.23)**	(8.41)**	(8.28)**	(6.53)**	(4.92)**	(5.53)**
EDFATH	0.096	0.073	0.073	0.014	−0.0006	0.008
	(15.04)**	(11.37)**	(11.33)**	(2.68)**	(0.11)	(1.47)
LEGIT	1.328	1.255	1.224	0.450	0.346	0.381
	(35.42)**	(33.69)**	(32.65)**	(14.69)**	(11.41)**	(12.14)**
RACEN	−1.116	−0.973	−0.848	−0.312	−0.135	−0.244
	(33.51)**	(29.18)**	(24.10)**	(10.81)**	(4.77)**	(8.34)**
FORMOTH	−0.121	−0.268	−0.254	0.032	0.112	0.063
	(3.56)**	(7.88)**	(7.46)**	(1.14)	(4.09)**	(2.27)**
PRMOTH	−0.976	−0.875	−0.785	−0.221	−0.066	−0.158
	(23.71)**	(21.39)**	(18.58)**	(6.40)**	(1.94)	(4.51)**
AGEMSQ	−0.004	−0.005	−0.005	−0.002	−0.0008	−0.001
	(12.73)**	(14.63)**	(14.49)**	(6.07)**	(2.80)**	(4.40)**
CHLDLIV		−0.299	−0.286	−0.101	−0.093	−0.098
		(22.88)**	(7.19)**	(3.15)**	(2.95)**	(3.01)**
FIRST		0.225	0.229	0.189	0.183	0.192
		(6.05)**	(6.08)**	(6.21)**	(6.11)**	(6.24)**
%LOSS		0.250	0.030	0.008	−0.048	−0.014
		(3.64)**	(0.25)	(0.08)	(0.50)	(0.14)

	(1)	(2)	(3)	(4)	(5)	(6)[a]
TBO			−0.006	−0.015	0.001	−0.007
			(0.18)	(0.57)	(0.01)	(0.24)
LAST			0.302	0.103	0.074	0.086
			(3.98)**	(1.67)	(1.23)	(1.39)
MIC			−0.417	−0.251	−0.229	−0.247
			(11.91)**	(8.88)**	(8.16)**	(8.61)**
CLINIC			0.230	0.230	0.230	0.230
			(7.52)**	(8.48)**	(9.81)**	(8.91)**
OB/GYN			−0.024	−0.018	−0.021	−0.020
			(1.39)	(1.35)	(1.54)	(1.12)
INTER1				9.099		
				(141.66)**		
INTER2				−1.248		−0.678
				(48.24)**		(13.38)**
INTER3				−2.770		0.282
				(92.89)**		(7.13)**
INTER(350)					0.030	
					(110.21)**	
INTER(350)2					0.119	
					(3.21)**	
INTERVAL						−0.024
						(50.57)**
Constant	1.690	0.657	0.748	−2.245	−0.175	8.963
$\bar{R}^2$	0.133	0.149	0.152	0.447	0.460	0.347
N	54280	54280	54280	54280	54280	52552

Notes:

*Coefficient statistically significant at 5% level.
**Coefficient statistically significant at 1% level.
[a] Regression (f) restricted to these observations for which NUMVISIT > 0.

various measures of the interval to the first visit is substantial. In the case of mother's age, there is a tendency for the maximum number of visits to shift to older ages. This is particularly pronounced in regression (f) where the maximum number of visits is reached at age 48—an age at which most women are no longer fecund. The tendency of the number of visits to continue to rise among older pregnant women, particularly those who have made an initial physician contact, would indicate that physicians appear to encourage higher-risk mothers, as measured by age, to obtain more prenatal care.

The effect of mother's and father's education is also altered in these specifications. Father's education, which was seen to be more important in determining the interval to the first visit, is only significant in specification (d). Moreover, in all three specifications, (d) through (f), mother's education is significant and has a larger positive effect. This finding tends to support the notion that there is a productivity effect associated with mother's education which would tend to increase the productivity of prenatal visits and encourage their purchase. The absence of a strong father's education effect, which we have viewed as a proxy for an income effect, may partially result from the way in which prenatal care and obstetrical delivery is priced by many private physicians. Most private physicians charge a single fee for prenatal care and obstetrical delivery—there may typically be no individual charge for each prenatal visit. For this reason, income may affect the decision as to whether to purchase an entire package of care but not individual visits as they cannot frequently be purchased separately.[12] Pricing in clinics and MIC centers is varied. There is no charge for care in municipal and MIC facilities, while clinics in voluntary hospitals charge on a per-visit basis. Many users of "voluntary" facilities are subsidized by Medicaid and so are unaffected by the method of charging for prenatal care.

The introduction of interval measures in the number-of-visits regressions substantially reduces the size of the previously largely unexplained differentials associated with legitimacy and the race/ethnicity/nativity dummies. In fact, the significant positive coefficients on FORMOTH in regressions (e) and (f) indicate that foreign-born mothers tend to have slightly more visits than native-born mothers once one takes account of the longer interval they wait before initiating care. In general, it would appear that as much as 80 percent of the apparent reduction in the number of prenatal care visits associated with illegitimacy or being black or Puerto Rican can be attributed to the longer intervals these mothers wait before initiating care and not to substantially reduced levels of care once they make contact with the medical care system.

The variables indicating a history of pregnancy losses, %LOSS and LAST, as well as TBO are not significant determinants of the number of visits, holding the interval constant. This would tend to indicate that physicians are not likely to modify their care protocol for a specific pregnancy to take account of a history of previous pregnancy losses. On the other hand, both CHILDLIV and FIRST remain significant in regressions (d) through (f), although the size of the estimated coefficients is reduced. These findings may reflect a "wantedness," maternal time, or "portfolio" effect, particularly when contrasted with the lack of statistical significance of the other experience measures. Thus, if demand is highest for the first child and declines with each successive birth, one would expect the demand for care to be as reflected in the coefficients of CHILDLIV and FIRST in regressions (d) through (f). It is equally plausible to argue that the presence of children in the home effectively changes the value of the mother's time in such a way so as to discourage the mother from leaving home to seek prenatal care as frequently as she might do if there were no children present. Lastly, it is possible that these findings represent a "portfolio" effect (although this might better be captured in the coefficient of TBO); that is, people who plan on having large families, other things equal, may spend less per child and less per pregnancy and have fewer visits. In any random cross section, completed family size (CHILDLIV) is likely to be correlated with desired family size, particularly when one has controlled for maternal age and other family characteristics.

The estimated coefficients of MIC, CLINIC, and OB/GYN in regressions (d) through (f) are consistent with those in regressions (c).

The coefficients of the various measures of the interval to the first visit [INTER1, INTER2, INTER3, INTER(350), INTER(350)2, INTERVAL] all demonstrate the very strong effect of the interval to the first visit and obstetrical protocols on the actual number of visits. For example, the coefficients of the three dummy variables INTER1, INTER2, INTER3 in regression (d) would indicate that, other things equal, women who begin care during their first trimester will have 9.1 visits, those who begin during their second trimester 7.9 visits, and those in the third 5.1 visits. The number of visits corresponding to each trimester is similar to the number of visits which would result from following the protocol for prenatal care as recommended by obstetricians. Following this protocol, a mother would have approximately 12 visits if she began during the first trimester, 10 if she began in the second, and 7 if she began in the third. Of course, the recommended visit schedule is for an uneventful pregnancy; pregnant mothers at high risk of obstetrical problems or actually suffering complications of pregnancy would be expected to have more visits. The persistent difference of between two and three visits between the recommended protocol and the estimated coefficients may have been the result of the upper bound of 11.04 placed on the dependent variable NUMVISIT. This upper bound was necessitated by the manner in which the data was recorded (see Section III. B. 1).

D. Determinants of Pregnancy Outcome: Birth Weight—Empirical Specification

Having investigated empirically and theoretically the factors associated with the demand for and utilization of prenatal care, it would appear important to investigate empirically the effect of these factors and the utilization of prenatal care on pregnancy outcomes. In this section, I focus on estimating demand/ production functions for birth weight as an indicator of pregnancy outcomes.

1. Dependent Variable

There are a number of good reasons for choosing birth weight, as measured in grams (or pounds), as a measure of the success of a pregnancy. Birth weight is universally and fairly accurately measured and recorded. It is a good objective indicator of the condition of the child at birth, particularly as it is the best single predictor of subsequent infant mortality. Low birth weight has been found to be associated, among surviving infants, with mental and physical growth retardation during subsequent childhood years. It is a continuous variable which doesn't present special problems of estimation in a regression format. It would appear amenable to medical intervention, particularly through the control of the diet of pregnant women.

Low birth weight is the most important proximate cause of infant mortality. Generally, 2500 g is considered the demarcation point between infants of low and adequate birth weight. In 1960, for instance, the U.S. death rate for infants weighing between 2001 and 2500 g was 2.3 times the rate for all births, and the probability of death rose rapidly as birth weight fell below 2000 g. The risk of neonatal death (death during the first month of life) was over 30 times greater for infants weighing under 2500 g as compared with infants weighing more than 2500 g. In addition, in comparing the unfavorable infant mortality rates in the United States with other developed countries, Geijerstam (12) has suggested that they may be due to a considerable extent to differences in birth weight distributions.

Birth weight captures at least two aspects of the pregnancy process in one measure. It is an index of maturity. During the gestation period, the fetus grows and develops from a single fertilized egg into a complex organism designed to cope independently with the external environment. This development and maturation takes time. Infants born too early will not have developed the necessary body systems to deal effectively with their new environment and may thus be unable to survive or only able to survive with their impaired autonomous functions artifically supported. During the gestation period, the infant is also growing in size. There is a strong, stable relationship between the length of gestation period, gestation age, and birth weight. So-called intrauterine growth curves have been developed to represent the functional relationship between birth weight and gestation age [Williams (27)].

Recent evidence has accrued to indicate, however, that birth weight is more than just a surrogate for gestation age. Using a large sample of California births (1.5 million), Williams (27) has demonstrated that when either birth weight or gestation age is included in a sophisticated logit function to estimate their adjusted effects on survival, birth weight is by far the stronger variable, explaining

93 percent of the variation in survival as compared with 54 percent for gestation age. Infants who are born prematurely, by gestation age criteria, but who are of adequate or above average birth weight are much more likely to survive than infants born at term but at low birth weights.[13]

Although prematurity, whether measured by shortened gestation ages or low birth weight, is associated with substantially increased risk of infant mortality, follow-up studies have also indicated an association between prematurity and reduced physical development and intellectual achievement among school-aged children. Recent research has pointed out an association between being a "low birth weight for gestation age" infant and having learning deficiencies measured many years later in elementary and even junior high schools. These findings have lead many researchers to speak of such infants' development as being seriously compromised by prenatal "undernutrition," "malnutrition," or "fetal deprivation." However, there does not seem to be any conclusive evidence that increasing the nutritional intake of mothers of such infants during pregnancy would have significantly corrected this situation.[14]

2. Independent Variables

Although there have been many studies of medical and socioeconomic correlates of birth weight, none has used a demand/production format such as I have suggested in this paper. Hence, it is difficult to specify a traditional production function for birth weight. This is particularly true as fetal growth *in utero* is a biological process which cannot be observed and except for animal experiments, it is socially unacceptable to experiment with outcomes by varying inputs except by specific supplementation.

It appears to be possible to identify three endogenously determined inputs: prenatal care, maternal nutrition, and several other maternal habits during pregnancy. Of these other maternal habits, smoking during pregnancy has been identified as resulting in reduced birth weight. Excessive alcohol use has also been associated with reduced birth weight. In addition, certain drugs may have a deleterious effect on pregnancy outcomes, although no drug in general use has specifically been associated with low birth weight.

The mechanism by which prenatal care may affect birth weight is not very clear. Part of the prenatal care process should be to provide the mother with information on the hazards of smoking and the value of proper nutrition—not infrequently a supplemental prenatal vitamin pill may be prescribed. To the extent to which this information has value and the mother follows instructions, this aspect of the prenatal care process should beneficially affect birth weight. A second mechanism by which prenatal care may increase birth weight would be by increasing gestation age. Specific therapeutic procedures may be effective in certain cases in prolonging the gestation period and thus give the fetus more time to mature and grow. Despite the fact that the mechanisms by which prenatal care may affect birth weight are not too clear, there are several very good reasons for including measures of the extent of prenatal care in a production function for birth weight. First, data are available to measure the aggregate effects of prenatal care on birth weight in a large sample and perhaps shed some light on the value of such care as typically practiced. Second, this measurement is important as prenatal and perinatal care are two areas of major policy concern. And lastly, prenatal care is the only measure of endogenously determined pregnancy specific inputs present in the data set available for this study and in most other large data sets. As input measures, prenatal care is quantifiable by both the number of visits (NUMVISIT) and the interval to the first visit (INTERVAL). Both measure aspects of care that have been stressed in the medical literature as being important components of healthy pregnancies.

Unfortunately, neither measures of maternal nutrition nor health-related habits such as smoking are reported in the data. Several proxies are available, primarily the education of both parents. As suggested previously, EDFATH may be viewed as a proxy for permanent family income. It seems reasonable to expect that if nutrition is a normal good, it should be positively correlated with income. Two aspects of nutrition are important and both should be correlated with income: the nutrition of the mother during pregnancy and the nutritional state of the mother's body immediately prior to pregnancy. EDMOTH may serve as an index of the mother's health and nutritional state prior to pregnancy as individual health and education levels have been shown to be correlated consistently. More educated mothers may be less likely to smoke if education increases their awareness of the hazards of smoking and enables them to better translate such information into behavior favorable to health. Education may also increase the productivity with which a given budget is used to produce adequate nutrition and the productivity of the prenatal care input itself.

There are a number of variables in the data which may reflect either exogenous biological or genetic factors, exogenous environmental factors, or endogenous behavioral factors. Such variables include maternal age (AGEMOTH, AGEMSQ), race of the child (primarily RACEN),[15] and nativity of mother (FORMOTH, PRMOTH). In the case of maternal age, females at either extreme end of the fecund age spectrum may be biologically less able to produce adequately developed infants. On the other hand, there may be different environmental and behavioral factors operant in pregnancies at different ages which may affect birth weight.

As previously discussed, race (particularly the black/white dichotomy) has been demonstrated to be associated with differences in birth weight. However, this finding is open to any or a combination of biological, environmental, or behavioral interpretations. Although neither Puerto Rican or other immigrant groups represent distinct genetic or ethnic groups, they may represent different genetic types than the "average" native-born white mother. On the other hand, immigrant status may be associated with socioeconomic factors, such as different levels of income for a measured level of education, language barriers which make the optimal utilization of medical care difficult, or different attitudes toward pregnancy and childbearing.

Several of the experience variables (CHILDLIV, TOTLOSS, FIRST) have been shown to be predictors of birth weight in subsequent pregnancies [Williams (28)]. To the extent to which these experience variables measure reproductive efficiency, this is to be expected and their inclusion in the current study justified. Moreover, to the extent to which they may represent a pattern of repeated behavior deleterious to fetal development they should be included in an exhaustive list of possible independent variables to determine the extent to which their value as predictors is mitigated or enhanced by subsequent behavior.

In addition, LEGIT, the legitimacy dummy variable, is included in the estimated function for birth weight. Although I find it difficult to make a strong case for including legitimacy as an input into a production process, one could argue that if marriage is productive of particular efficiencies accruing to the household, then some net benefit may be observed in the case of pregnancy also. On the other hand, one could argue that for illegitimate pregnancies, father's education is a poor proxy for income and that as a first step in adjusting for this difference, a dummy variable to measure legitimacy should be included in any production specification. Overall, it would appear that so little is known about the pregnancy production process that the exclusion of a potentially important variable such as legitimacy *a priori* is unwarranted.

Lastly, I have included gestation age (GESTN) in the birth weight equation. In a more fully developed system gestation age would be regarded as only an intermediate endogenous variable to be determined simultaneously with birth weight; however, in the system I am estimating it is difficult to identify separate functions for both birth weight and gestation age. Accordingly, for reasons enumerated above, it was decided to investigate birth weight. Gestation age is an important predictor of birth weight but not necessarily a productive element *per se;* however, the inclusion of gestation age in the birth weight equation may serve a useful purpose by allowing us to ask, "What is the true contribution of a particular factor when gestation age is held constant?" The "adjustment" value of gestation age is particularly important in accurately estimating the effect of prenatal care on birth weight. Because of the protocol implicit in the prenatal care process most women, once they begin care, will have regularly scheduled visits until they deliver. If we include a measure of the date of the first visit as one measure of the prenatal care input, then to a large extent a measure of the number of visits becomes a fairly good proxy for the duration of the pregnancy and therefore of gestation age. If we do not include gestation age as an independent variable in estimating the effect of the amount of prenatal visits on birth weight, we will substantially overestimate the productive value of such visits as there is, for most women, a biologically determined relationship between gestation age and birth weight.

3. Determinants of Birth Weight: Empirical Results

The results of regressing birth weight on a group of predetermined socioeconomic and pregnancy experience variables as well as measures of prenatal care inputs are reported in Table 7. A simple additive function was specified. Regressions (a) through (c) may be interpreted alternatively as demand, produc-

Table 7. WGHT: Birth Weight (g)

Coefficients of OLS Regression Estimates: Entire Sample N = 54,280
(t-statistic in parentheses)

Independent Variables	(a)	(b)	(c)	(d)
AGEMOTH	26.369	21.544	14.638	7.253
	(7.80)**	(6.22)**	(4.55)**	(2.25)*
EDMOTH	−0.297	0.768	0.546	−0.190
	(0.33)	(0.86)	(0.65)	(0.23)
EDFATH	−0.616	0.732	0.863	−0.977
	(0.53)	(0.62)	(0.79)	(0.90)
LEGIT	99.285	102.593	82.145	56.826
	(14.54)**	(14.99)**	(12.90)**	(8.86)**
RACEN	−116.110	−123.632	−91.234	−71.682
	(19.13)**	(20.21)**	(16.04)**	(12.39)**
FORMOTH	46.184	54.949	53.499	60.414
	(7.44)**	(8.79)**	(9.22)**	(10.42)**
PRMOTH	−31.148	−36.702	−28.397	−8.115
	(4.15)**	(4.89)**	(4.07)**	(1.16)
AGEMSQ	−0.371	−0.329	−0.204	−0.101
	(6.03)**	(5.26)**	(3.51)**	(1.74)
CHILDLIV		21.848	21.168	27.449
		(9.34)**	(9.75)**	(12.62)**
TOTLOSS		−36.672	−31.313	−32.912
		(9.15)**	(8.42)**	(8.90)**
FIRST		−8.526	−21.242	−23.762
		(1.31)	(3.52)**	(3.96)**
GESTN			8.674	8.375
			(27.12)**	(86.96)**
INTER1				140.207
				(8.83)**
INTER2				−30.800
				(5.54)**
INTER3				1.320
				(0.19)
NUMVISIT				13.637
				(14.67)**
Constant	2720.299	2771.136	462.781	470.766
$\bar{R}^2$	0.031	0.035	0.169	0.178

Notes:
*Statistically significant at 5% level.
**Statistically significant at 1% level.

tion, or outcome equations, while equation (d) can be interpreted as a production or outcome equation. As in previous regressions, the effect of mother's age is nonlinear—birth weight rises with maternal age up to age 35.5 and falls subsequently. This finding is consistent with the generally accepted notion that pregnancies are at greater risk as maternal age increases beyond 35 years. The coefficients on the group of characteristic dummy variables LEGIT, RACEN, FORMOTH, and PRMOTH are all significant and consistent with the demand-for-care equations. However, unlike the case of the previously reported demand-for-care equations, these significant coefficients need not be regarded as indicators of ignorance. Several of them, RACEN, PRMOTH, and perhaps FORMOTH, may be regarded as indicating potential genetic or biological differences in physiological pregnancy production processes associated with racial or ethnic differences. Surprisingly, the coefficients on mother's and father's education are negative and insignificant. *A priori,* I would have expected positive significant coefficients on these variables since (1) mother's education may be regarded as a measure of reproductive efficiency which should result in increased birth weights; and (2) father's education may be regarded as a proxy for permanent income and has been shown to be related to the demand for and use of prenatal care and presumably other pregnancy inputs.

In regression (b), the experience variables CHILDLIV, TOTLOSS, and FIRST are entered. Coefficients on these variables are significant or nearly so and the signs are as expected. The indication is that previous good experience as measured in CHILDLIV tends to repeat itself in the form of higher birth weights on subsequent pregnancies and that a history of pregnancy losses is associated with lower birth weights. Other things equal, therefore, it would appear that families are justified in modifying their demand for care during a particular pregnancy based on their experience with prior pregnancies.

In regression (c), gestation age is entered into the equation. In a more complete empirical model of the pregnancy production process gestation age would be viewed as an intermediate outcome variable similar to birth weight. In the system estimated in this paper, gestation age has been included primarily to more accurately measure the independent effect of other variables on birth weights. As reported in Eq. (c), there is a strong gestation period growth effect on birth weight—approximately 8.7 g for each day of gestation.

In regression (d), the variables measuring prenatal care inputs are entered, INTER1, INTER2, INTER3, and NUMVISIT. We note that holding gestation age constant there is a 140-g increase in birth weight associated with starting care during the first trimester. The gain associated with starting care during the second trimester falls to 110 g and is also 110 g for pregnancies where the initiation of care is delayed to the third trimester. Thus, the minimal increase in birth weight associated with any amount of prenatal care is 110 g—a not inconsiderable amount when one considers that infants of 2500 g or less are at markedly increased risk.

The coefficient of NUMVISIT indicates that, holding gestation age constant, weight increases approximately 13.6 g with each visit. Thus, the total gain associated with a full complement of prenatal care comprising 12 visits and starting in the first trimester would be 303 g as compared to a pregnancy without any care inputs. This increment in birth weight is equal to about 12 percent of the 2500 low-birth-weight–high-risk classification marker and about 10 percent of the mean birth weight for the entire sample.

It is worth comparing the size of the coefficients of the four maternal characteristic dummies (LEGIT, RACEN, FORMOTH, PRMOTH) in Eq. (a) with their value in regression (d) to determine whether the introduction of the experience, gestation age, and prenatal care variables has an effect in reducing the otherwise apparently largely unexplained differences between these groups. The coefficient on LEGIT, while still positive and significant, is reduced by 44 percent primarily due to the importance of legitimacy status in determining the demand for care. The coefficient on RACEN is reduced by about 40 percent, with half of that decrease attributable to the fact that blacks apparently have shorter gestation periods and the other half to the fact that they use fewer prenatal care inputs. Foreign-born mothers apparently bear even heavier children than expected, particularly when account is taken of their lower utilization of prenatal care inputs. The most substantial effect is on the coefficient for Puerto Rican mothers. It drops from a highly significant -31 g to an insignificant -8 g. This would indicate that practically all the birth weight differential that can be attributed to this ethnicity characteristic can be explained by differences in the utilization of prenatal care. Overall, it would appear that a significant proportion of birth weight differentials that have been traditionally attributed to maternal characteristics such as race or to legitimacy status can be explained by variations in the demand for care among these different groups. That there are significant differentials remaining even after taking account of care inputs and the level of individual reproductive efficiency argues that there are still some important variations to be explained and that biological differences should be seriously considered as a possible explanation for certain of these differences.

IV. CONCLUSIONS AND POLICY IMPLICATIONS

Perhaps the most significant finding reported in this paper is the important contribution to birth weight associated with a full complement of prenatal care. Because of the very critical importance of birth weight as a determinant of infant survival and of the quality of the child during the developmental years, this finding is of significance in developing policies to increase infant and child health. The finding of a positive association between prenatal care and birth weight takes on added importance because the effect is measured in a multiple regression context wherein other variables such as gestation age, race, and maternal age are simultaneously controlled.

Also of interest is the finding that differences in the level of prenatal care input account for a substantial portion of the differences in birth weight otherwise attributed to race, ethnic, and legitimacy status. Taken as a whole, the system of demand and production equations implies a substantial absence of knowledge as to why race, ethnicity, or legitimacy status should affect the demand for care but does offer differences in care inputs as an avenue by which they affect birth weight. A similar relationship between maternal age, birth weight, and care is also observed. The coefficients of maternal age and age squared are substantially reduced in size and statistical significance [Table 7, column (d)] when measures of prenatal care utilization are included in birth weight equations.

Despite the finding of an important positive effect of care on birth weight, these results have to be tempered by the realization that a specific avenue by which care might affect birth weight has not been identified. Neither the model nor the data has allowed us to adequately separate the effect of care from the effect of other pregnancy-specific inputs which may significantly affect birth weight and are probably correlated with the amount of care demanded and consumed. These other factors could include maternal nutrition, smoking, alcohol consumption, and general concern for a successful outcome. Although all these factors may be modified as a consequence of prenatal care inputs, there is reason to believe that traditional prenatal care may not have always been appropriate. In particular, the Committee on Maternal Nutrition of the National Research Council observed in 1970 that "current obstetric practice in the United States tends to restrict weight gain during pregnancy. . . . one may raise the question of whether this practice is in effect contributing to the large number of low-birth weight infants and to high perinatal and infant-mortality rates" [National Academy of Sciences (22)]. This committee further suggested that a target gain of 20 to 25 lb was reasonable for a full-term pregnancy and cautioned against weight reduction programs that distort normal gains during pregnancy. In the context of the favorable effect of care on pregnancy, it is important to realize that these guidelines were issued in 1970 in response to a concern about prenatal care protocols which tended to stress restricting maternal weight gain. To the extent that these obstetrical practices were successful in restricting weight gain, prenatal care might have actually contributed to lower birth weights in the cohort we are studying.

On the other hand, the finding of no significant effect on birth weight of either mother's or father's education both when care is either excluded from or included in outcome regressions tends to indicate that other variables such as nutrition and smoking, which are probably related to education, may not be operating systematically through these variables.

Given the apparent importance of care as an input in the production of healthy infants, attention naturally focuses on the determinants of the demand for care. In this paper, I have stressed the importance of family decisions in determining the utilization of prenatal care both through determining the interval to the first visit and the number of visits. In particular, the decision whether to seek care at all and the decision as to when during the pregnancy period to initiate care are both determined largely within the households and are important ultimate determinants of the amount of care actually consumed and of birth weight and pregnancy outcomes. In this context, I have to note that unfortunately both the theoretical and empirical models were unable to adequately explain the large differences in the demand for care that were attributable to legitimacy status, race, and ethnicity. The differences in the initial demand for care which are identified with these factors in the regressions reported in Tables 3–5 are not reduced by consideration of pregnancy experience, family size, or the availability of sources of care, and these differences have been found to be important determinants of the number of visits and birth weight.

These negative points aside, there are several areas of consistency between the empirical estimates and the theoretical formulation which are worthy of emphasis. In particular, the finding that the demand for care is responsive to experience with previous pregnancies gives credence to the notion that where information is available regarding expected outcomes households will react so as to make informed decisions. The finding of a much greater responsiveness of the demand for care to differences in father's education than mother's education is consistent with the prediction from the model of a strong income effect and an ambiguous education/efficiency effect. In fact, to the extent to which the categorical variables LEGIT, RACEN, FORMOTH, and PRMOTH reflect differences in family income levels not adequately measured by EDFATH the responsiveness of the demand for care to differences in income levels may be greater than indicated.

The finding of a consistent negative relationship between the presence of Maternal and Infant Care projects and the demand for and utilization of care is certainly worthy of further study. As is pointed out in the body of the paper, these projects were established in areas of high pregnancy wastage, low care utilization, and low levels of availability of other care providers. After almost 8 years of operation, one would have expected that one measure of their success would be the finding of a positive impact of their presence on the utilization of prenatal care services by individuals in their designated service areas. Careful evaluation of the services offered by MIC projects and the manner in which they are marketed to their designated low-utilization populations would appear warranted at this point in time. In particular, consideration should be given to the possibility of not only a malfunction in this program but also to the possibility that these findings are suggestive of a level of operation and funding by current MIC projects which is too low to adequately compensate for the minimal level of availability of more traditional providers of care in MIC districts.

Two other avenues of investigation would appear to be worth pursuing in light of the findings presented here. First, it would appear extremely important to estimate a model of the pregnancy demand/production process with better measures of care inputs, income, and other defined inputs such as maternal diet, smoking, and alcohol consumption. Since these three factors have been identified as being important determinants of birth outcomes, it would be important in evaluating the potential contribution of prenatal care to improved birth outcomes to separate out the extent to which the degree of utilization of prenatal care has served as a proxy measure for these other behaviors which may be more causally related to low birth weight and other negative birth outcomes. In addition, it would be important to identify the extent to which prenatal care contributes to favorable outcomes by promoting appropriate behavior with regard to diet, smoking, and alcohol consumption in expectant mothers.

Secondly, in the years since the data set used in this study was assembled, cheap, legal abortion and more recently amniocentesis have become widely available. While not readily acceptable to many mothers for religious, moral, or other reasons, cheap legal abortion does present a reasonable substitute for a low-input suboptimal pregnancy particularly as the postabortion potential exists for most women to have a child under what may be more favorable circumstances. Secondly, amniocentesis coupled with legal abortion present to certain high-risk mothers an identified prenatal procedure which can be performed early in pregnancy and may materially affect outcome. Even though in the vast majority of cases amniocentesis results will prove negative and hence will not materially affect outcomes, the procedure may increase the demand for care on the part of high-risk mothers. This increased demand for care could potentially affect outcomes beyond the identification of damaged fetuses. Thus, it would appear that any additional research in this area should try to take account not only of previously not identified maternal behaviors but also of these recent developments which may materially affect both the demand for care and birth outcomes.

ACKNOWLEDGMENTS

This paper is adapted from my Ph.D. dissertation, "Experience with Pregnancy, The Demand for Prenatal Care and the Production of Surviving Infants" (C.U.N.Y. Graduate Center, 1977). Helpful comments on my dissertation research were provided by Michael Grossman, Victor Fuchs, Elliot Zupnick, Harold Hochman, Barry Chiswick, Fred Goldman, Robert Willis, Melvin Reder, Finis Welch, and Ann Williams. Very able research assistance was provided by John Wolfe, Christy Wilson, Don Wright, Terry Eschavez, Rajneish Ghei and Barbara Chamberlain. My initial research on infant health was begun at the National Bureau of Economic Research as part of the health economics project funded by grants from the Robert Wood Johnson Foundation and the National Center for Health Services Research.

APPENDIX A

Table A-1. Means and Standard Deviations of Dependent and Independent Variables for the Entire Sample (N = 54,280)

Variable	Mean	Standard Deviation
MOTHAGE	25.16	5.51
EDMOTH	10.94	3.37

(continued)

Table A-1—(Continued)

Variable	Mean	Standard Deviation
EDFATH	11.90	2.60
LEGIT	0.79	0.41
RACEN	0.29	0.46
FORMOTH	0.22	0.42
PRMOTH	0.17	0.38
TBO	1.31	1.74
CHILDLIV	1.11	1.48
TLOSS	0.19	0.63
FIRST	0.41	0.49
%LOSS	0.07	0.20
AGEMSQ	663.38	300.21
MIC	0.37	0.48
CLINIC	0.53	0.52
OB/GYN	0.50	0.83
LAST	0.06	0.23
INTER(350)	219.46	71.71
INTER1	0.97	0.18
INTER2	0.61	0.48
INTER3	0.18	0.39
INTER(350)2	0.36	0.48
NUMVISIT	7.99	3.30
NOVISIT	0.03	0.18
WGHT	3176.14	569.27
WGHTSQ	10.41	3.44
GESTN	277.89	24.15

NOTES

1. See in addition, Friedman and Leibowitz (9), Goldman and Grossman (13), Inman (16), Colle and Grossman (5) and Edwards and Grossman (7, 8).

2. See for example, Willis (29), DeTray (6), Michael (20), Ross (25), Ben-Porath and Welch (3), and Becker and Lewis (1), to mention only a representative sample.

3. Since $(\partial p/\partial EX - p/EX) < (\partial p/\partial EX - p/(EX + FX))$, (19) may be positive. This would be particularly likely where EX was small relative to FX.

4. We could also assume that because p is a measure of child health and because child health as an aspect of child quality might be related to income as is the quality of other goods, $\partial p/\partial Y > 0$ (Grossman, 14). This argument, while reasonable, is an unnecessary assumption in our relatively simple model and reduces the power of the symmetrical production/utility maximization relationship which can greatly simplify the model.

5. Moreover, it is unlikely that there is any desire for "balance" between dead and surviving children as there might be between boys and girls.

6. Notice, however, that because pregnancies and children occur in discrete rather than continuous bundles, this effect may be muted as C attained approaches C*. For example, if $p^* = .9$ and $C^* = 3$, then 3.33 pregnancies will be required on average to attain C*. The family may estimate n* at 3–4 pregnancies judging itself more lucky if it takes only 3 and less lucky but still not unfortunate if it

takes 4. After completing 3 pregnancies with 2 survivors, the family may well continue on to the fourth pregnancy under the original scenario without revising any of its initial targets.

7. Since p = f(EX, RE) is asymptotic to 1, the height and the slope of p will probably be related because for some sufficiently large value of EX, p will approach the asymptote. We shall, however, continue to analyze these effects separately.

8. One possible pattern is to seek early confirmation from a physician that the mother is pregnant and then delay subsequent visits until late in the pregnancy if the pregnancy is otherwise uneventful.

9. An alternative estimation procedure would have been to use a "Tobit" estimation technique to estimate parameters of the demand relationship. This technique, which is a maximum-likelihood hybrid of probit and multiple regression analysis, is designed to provide consistent, asymptotically efficient estimates when dependent variables tend to cluster at an upper or lower bound. Using this technique, one estimates an index from which the probability of observing a nonlimit value for the dependent variable and the dependent variable's expected value, conditional on a set of exogenous variables, can be determined in a single step. As the concentration of values at the limits decrease, "Tobit" estimates approach those of ordinary least squares. Although perhaps preferred to the actual estimation technique employed, this approach was not followed because (1) there was no computer program to perform this estimation supported, debugged, or available at the computer installation at which the computations were performed; (2) this estimation technique requires that the parameter values be estimated using an iterative procedure which can be prohibitively expensive with a data set the size of the one available; and (3) the results with a dummy variable MAXVISIT (equal to 1 if the limit value 9 was coded for NUMVISIT) used as an independent variable were not sensitive to this limitation.

10. In a fully recursive system, a sequence of endogenous variables might be represented by a causal chain without any feedback, e.g., a → b → c. Knowledge of the sequence may be used to identify equations for a, b, and c as, for example, a will appear as a predetermined variable in the equations for b and c, but neither b nor c in the equation for a. Moreover, if the disturbances in the equation for b are uncorrelated with a and the disturbances for c uncorrelated with a and b, then it can be shown that the application of ordinary-least-squares estimation to the equation for each variable yields maximum-likelihood estimates [Johnston (18)].

11. The protocol is nonlinear, i.e., during the first and second trimester (90-day period) one visit a month is recommended; in the third trimester, a visit every two weeks is recommended during the first two months and a weekly visit thereafter until delivery. In terms of the marginal dummy variables INTER1, INTER2, and INTER3, we would expect the coefficient of INTER1 (equals 1 if INTERVAL is greater than zero) to be greater than either INTER2 or INTER3 but that INTER3 should be greater than INTER2 since it is primarily during the third trimester that the nonlinearity becomes most acute and the period between visits progressively shorter.

12. Perhaps the only way to purchase these services separately would be to start care very late and persuade the physician to reduce the fee accordingly as fewer prenatal visits would be expected. If this was a widely followed practice, then we might expect income proxy variables, such as father's education, to affect the interval but not the number of visits with the interval held constant.

13. As compared with gestation age as an index of infant health at birth, birth weight is much more accurately measured. Birth weight is determined simply by weighing the infant at birth. Gestation age is typically determined by the time interval between the date of the mother's last menstrual period and the date of delivery. If the date of the last menstrual period is only recorded at birth or at a prenatal visit near the birth date, there is considerable room for recall error. Only the month may be recalled, or a recording error may enter the reported data if the calendar year has changed between conception and birth. In editing the New York City birth certificate data for this research, substantial recording errors of several years or even decades were encountered in all calculated time intervals. These errors resulted from inaccurate reporting of the date of the last menstrual period and other important dates. Another error is introduced in calculating gestation age because a substantial proportion of conceptions occur prior to marriage. Since mothers are frequently

reluctant to report this fact, they will tend to report the date for their last menstrual period as occurring after their marriage. This problem is so pervasive that it has become standard procedure in calculating intrauterine growth curves to disregard births at high weights for very short gestation periods as resulting from this sort of reporting error. At the other end of the gestation age spectrum are the very high gestation ages that are reported for women who presumably missed several menstrual periods prior to actually conceiving. Although much smaller in number than the number of births at low gestation age, they do present a problem in interpretation.

14. It is possible that many mothers of such infants are unable physiologically to deliver adequate nutrients to their fetuses *in utero* regardless of the maternal intake of nutrients. Such mothers might be regarded as having a reduced level of reproductive efficiency which may act to diminish the marginal product associated with the pregnancy input food.

15. It is conceivable that birth weight may be partially genetically determined. It has long been noted that black intrauterine growth curves are lower than those for whites. While one might be tempted to associate this finding with environmental deprivation during pregnancy, there is also substantial evidence that black infants at low birth weight have substantially lower infant mortality rates than comparable white infants. Hence, one is tempted to speculate that part of the apparent difference in intrauterine growth between the races may be genetically determined and that black infants of moderately low birth weight may be more "mature" than whites of similar birth weight.

REFERENCES

1. Becker, G. S., and H. F. Lewis (1972) "Interaction Between Quantity and Quality of Children," in *Economics of the Family*, (ed.), T. W. Schultz, Chicago: University of Chicago Press.
2. Ben-Porath, Y. (1973) *More On Child Traits and the Choice of Family Size*, M. Falls Institute, Discussion Paper No. 731.
3. Ben-Porath, Y., and F. Welch (1973) *Chance, Child Traits and the Choice of Family Size*, R-1117-NIH/FR, Santa Monica: Rand Corporation.
4. Ben-Porath, Y. and F. Welch (1976) "Do sex preferences really matter?" *Quar. J. Econ.* XC:285–307.
5. Colle, A. and M. Grossman (1978) "Determinants of Pediatric Care Utilization," *Jr. of Human Resources* 13:115–158, Supplement.
6. DeTray, D. (1972) "Child Quality and the Demand for Children," In *Economics of the Family*, (ed.), T. W. Schultz, Chicago: University of Chicago Press.
7. Edwards, L. N. and M. Grossman (1977a) *An Economic Analysis of Children's Health & Intellectual Development*, NBER Working Paper No. 180.
8. Edwards, L. N. and M. Grossman (1977b) *The Relationship Between Children's Health & Intellectual Development*, NBER Working Paper No. 213.
9. Friedman, B. and A. Leibowitz (1975) *The Bequest Motive in Human Capital & the Health Care of Children*, Unpublished paper.
10. Fuchs, V. R. (1974) "Some Economic Aspects of Mortality in Developed Countries," in *The Economics of Health and Medical Care*, (ed.), M. Perlman, New York: Wiley.
11. Fuchs, V. R. (1978) *The Great Infant Mortality Mystery or What Caused the Slump?* Unpublished paper.
12. Geijerstam, G. (1969) "Low Birth Weight & Perinatal Mortality," *Public Health Reports* 84:939–948.
13. Goldman, F. and M. Grossman (1978) "The Demand for Pediatric Care: An Hedonic Approach," *Journal of Political Economy* 86.
14. Grossman, M. (1972) *The Demand for Health: A Theoretical and Empirical Investigation*, New York: Columbia University Press.
15. Grossman, M. (1974) "The Correlation Between Health and Schooling," in *Household Production and Consumption*, (ed.), N. E. Terleckyj, New York: Columbia University Press.

16. Inman, R. P. (1976) "The Family Provision of Children's Health: An Economic Analysis," in *The Role of Health Insurance in the Health Services Sector,* (ed.), R. N. Rosett, New York: National Bureau of Economic Research.
17. Institute of Medicine. (1973) *Infant Death: Analysis by Maternal Risk and Health Care.* Washington, D. C.: National Academy of Sciences.
18. Johnston, J. (1972) *Econometric Methods,* 2nd ed. New York: McGraw-Hill.
19. Michael, R. T. (1972) *The Effect of Education on Efficiency in Consumption,* New York: National Bureau of Economic Research.
20. Michael, R. T. (1973) "Education and the derived demand for children," *J.P.E. Supplement* (March/April).
21. Muller, C., M. Krasner, and F. S. Jaffe (1975) "An index of insurance adequacy for fertility-related health care," *Medical Care* XIII:25–36.
22. National Academy of Sciences (1970) *Maternal Nutrition and the Course of Pregnancy.* Report of the Committee on Maternal Nutrition, National Research Council. Washington, D. C.
23. Nortman, D. (1974) "Parental Age as a Factor in Pregnancy Outcome and Child Development," *Reports on Population/Family Planning,* no. 16.
24. O'Hara, D. J. (1972) *Change in Mortality Levels and Family Decisions Regarding Children.* R-914-RF. Santa Monica: Rand Corporation.
25. Ross, S. (1974) *The Effect of Economic Variables on the Timing and Spacing of Births,* Ph.D. dissertation, Columbia University.
26. Welch, F. (1974) *Sex of Children: Prior Uncertainty and Subsequent Behavior.* R-1510-RF. Santa Monica: Rand Corporation.
27. Williams, R. L. (1974) *Outcome-Based Measurements of Medical Care Output: The Case of Maternal and Infant Health,* Ph.D. dissertation, University of California, Santa Barbara.
28. Williams, A. D. (1976) *Fertility and Reproductive Loss,* Ph.D. dissertation, University of Chicago.
29. Willis, R. J. (1972) *The Economic Determinants of Fertility Behavior,* Ph.D. dissertation, University of Washington.

EARLY LIFE ENVIRONMENTS AND ADULT HEALTH:
A POLICY PERSPECTIVE

Anthony E. Boardman and Robert P. Inman

I. INTRODUCTION

Adult health, like the weather, receives a great deal of attention but seems largely immune to significant change.[1] The present system of adult health care is directed primarily at curative care; our few adventures into preventive adult care seem to have offered only mixed blessings.[2] While curative advances have added small gains to our expected length of life and have no doubt reduced morbidity, curative care promises only marginal relief for many adult illnesses.

Yet improvements in adult health yield significant social gains. Rice's (31) original study on the cost of illness estimated that the present value of income lost due to premature (before 65) adult deaths was about 16 percent of annual wage income.[3] Further, Vaupel (36) provides convincing evidence that these losses from early adult death are particularly biased toward the economically disadvantaged. The economic consequences of adult morbidity parallel those for premature death. Bartel and Taubman (1) estimate the annual lost earnings from such adult illness as (nonfatal) heart diseases is $1.1 billion, from mental illness as $1.7 billion, from respiratory illness (bronchitis, emphysema, asthma) as $0.8 billion, from arthritis as $1.2 billion, and from ulcers as $0.8 billion. Luft (20), working with a larger sample of individuals using the very broad distinction "sick or not sick," had estimated aggregate annual loss in earnings from nonfatal illness as $23 billion. As interesting is the distribution of those losses. They are borne disproportionally by individuals in semiskilled and physically demanding occupations which require regular hours—for example, assembly line work. For Luft's sample, black females when sick suffered an average decline of 38 percent in their incomes, white females 33 percent, black males 45 percent, and white males 36 percent. Poor adult health strikes our economy as a significant and a regressive tax on wage income. Additional "second-round" economic losses—such as a child's altered schooling plans in response to a parent's illness—and the surely important psychic losses that accompany major sickness only increase the social burden of adult ill health.

The design of public policy to reduce the social losses from poor adult health requires a careful understanding of the likely causes of adult illness—in particular, those causes that can be altered by public policy. Yet even knowledge of the extent which predetermined causes of adult illness (e.g., a genetic predisposition toward a particular disease) or simply bad luck play in an individual's disease history can be important for policy design. For if serious illness is truly random, then social insurance for the disabled which compensates for more than simply medical expenses becomes desirable.[4] The research below marks a first effort at detailing the pattern of likely causes of adult illness with an eye toward this pattern's implications for the design of national health care policies.

The analysis is admittedly aggregative and, from the point of view of medical science, assuredly crude. We are not, however, recommending individual treatment strategies. Our goal is more modest, yet more speculative. Can we connect loosely defined attributes of adult health behavior to broadly categorized adult health outcomes and, from this connection, draw preliminary inferences as to an appropriate strategy for the design of a national health policy? We shall focus particularly on the potential role of early life environments and health behaviors on subsequent adult health.

The sample for our empirical analysis is drawn from a twins' panel maintained by the National Academy of Science–National Research Council (NAS-NRC); a detailed description is available in Jablon (16) or Bartel and Taubman (1). The sample is drawn from the population of white male twins born in the United States between 1917 and 1927. All individuals in our study are military veterans and were alive in 1974, the date of the last questionnaire submission. Thus, our sample is likely to be healthier than a randomly drawn sample of white, U.S.

males born during the same period. For each individual the date at which that person was first diagnosed as ill with a particular disease is recorded. Diseases have been classified by the International Classification of Diseases (ICDA) code. A maximum of 54 diseases may be recorded over each individual's life (birth to 1974). Unfortunately, the accuracy of the diagnoses cannot be checked, but the NAS has indicated that disease information from VA and military records is generally reliable. Other disease histories are self-reported. It is some consolation that for the four adult diseases examined here—heart disease (ICDA codes 400–404, 410–429), psychoses and neuroses (codes 290–302, 305–309), arthritis (codes 710–718), and bronchitis, emphysema, asthma (466, 490–493)—approximately 70 percent of the reported diagnoses come from the pre-1954 period when the VA and military health records were the major source of diagnostic information [see Bartel and Taubman (1), also Tables 1 and 2].[5] While far from ideal, this individual data base with longitudinal health histories does provide a unique opportunity to examine likely causes of adult health from the health policy perspective.

The results are suggestive, though hardly decisive. We find that many of the hypothesized behavioral and environmental determinants of adult health prove to be statistically significant and quantitatively important. Our work is generally consistent with the conclusions of more carefully structured experimental work on the determinants of adult health, but we control here for a wide range of other intervening variables not usually examined. While our empirical analysis includes an impressive list of possible determinants, the results also make clear that these variables do not adequately account for all variations in individual health prospects. Indeed, the percent of variation in illness states explained in our analysis never exceeds .2 and is generally less than .1. We conclude from such results that either better models are needed or that there simply exists a permanent and large random (i.e., nonbehavioral, nonenvironmental) component to adult healthiness. No doubt the truth lies somewhere in between. Finally, our results allow us to *tentatively* identify the effects of improved environments on adult health. The impacts of improved early life environments on an individual's adult health prospects look particularly significant. Favorable changes in early life health behavior can reduce the chances of major adult illness by as much as one-half or more.

II. TOWARD A BEHAVIORAL MODEL OF ADULT HEALTH

While numerous biological, social, and economic factors have been examined as potential causes of adult illness, and in many instances found to be correlated with the presence of disease, as yet no clearly decisive causal agents have been discovered which fully explain the incidence of major adult illnesses. While animal experiments have isolated possible biological causes of disease, Neyman

(25) has rightly criticized these studies as incomplete for their neglect of social, environmental, and economic causes of human illness. He raises the important question of whether such laboratory studies will ever be able to correctly test a fully general model of the human disease process. Recent work with human populations by physicians and biostatisticians has attempted to establish the correlation of illness with socioeconomic background variables, controlling whenever possible for genetic and early life determinants with samples of twins. Because of small sample sizes, however, these studies are able to examine only one or two background variables at a time. Bias due to omitted (uncontrolled) variables may result, or potentially important interactions between social and economic variables (e.g., stress and poverty) must be left unexplored. Studies employing larger sample sizes to permit the introduction of multiple variables as potential determinants of human disease have generally used aggregated data for collections of populations—for example, average health statistics for residents residing within cities or within states. The danger here is that what is true for an average of a population need not be true for any single individual within the sample cities. In particular, a factor which has important health effects for a small subsample of residents in each city may go unnoticed when populations are aggregated. Still, broad-based health consequences can be successfully identified using these macrodata bases; see, for example, the recent work on air pollution and health by Lave and Seskin (19). The most productive research strategy, however, is likely to employ a combination of laboratory research on the biological processes of disease with controlled and "natural" experiments yielding individual rather than aggregate data on behavior, environment, and health. The NAS-NRC sample is one such individual-based data set.[6]

What *causes* illness? For the dominant diseases affecting adults, we generally do not know. While biological models of disease processes have been proposed and tested, they have not been well integrated with alternative social and economic hypotheses relating behavior, environment, and health. Nor is the current research record much better on the determinants of behaviors (presumably) conducive to better health. While economic, sociological, and psychological models of health habits and behavior have been proposed, the supporting evidence is only now forthcoming.[7] As of now, we know of no compelling integrative theory of health and behavior which yields well-defined, testable hypotheses and which can be addressed with data bases such as ours. It is too ambitious a task for us to suggest such a theory here. Nonetheless, exploratory empirical analyses can be performed with the idea that systematic patterns within the data will motivate the search for such a theory. That is our task here.

We shall impose a broad structure on our analysis which we feel is consistent with much of the recent work on behavior and health, but for the most part detailed structural hypotheses regarding variable effects are not offered. Our empirical analysis is best viewed as a reduced form exercise in which "net" health and health behavior effects of key variables are to be estimated. The framework is simple. Parents "endow" their children with a promise of a future wealth position (X_1, a bequest), a level of health (X_2, a childhood disease history and health behavior), a level of human capital skills (X_3, education), and attitudes toward work and expectations toward life's prospects (X_4, articulated through reasons for first job choice). These skill, attitude, wealth, and health endowments in turn help to determine an individual's adult behavior and environment and, finally, that individual's adult health. We measure an individual's adult behavior and environment generally along five dimensions: work success (Y_1, measured as a series of agree to disagree responses to questions about respondent's current job); the existence of a social support network to cope with stress and/or illness (Y_2, measured as current marital status and number of residences to reflect dislocation); adult smoking behavior (Y_3, measured as years smoked prior to illness); adult drinking behavior (Y_4, measured by total alcoholic consumption in a year and the extent of heavy drinking); and adult preventive care behavior (Y_5, measured as the number of years prior to illness in which respondent had a preventive checkup).[8] The analytic framework can be summarized as a six-equation system:

$$Y_i = \psi_i (X_1, X_2, X_3, X_4; u_i) \quad (i = 1 \ldots 5) \qquad (1)-(5)$$

$$H = \mu(X_1, X_2, X_3, X_4; Y_1, Y_2, Y_3, Y_4, Y_5, H_{-t}; u_6), \qquad (6)$$

where H is a measure of current adult health, H_{-t} is an index of previous adult health, and u_1 - u_5 and u_6 are the unobserved determinants of adult health behavior and current adult health, respectively. Within this recursive specification, early life environments play a dual role on adult health, first through their impact on $Y_1, \ldots, Y_5$ which then impact on health (the indirect effect of $X_1, \ldots, X_4$) and second, through their effect on health directly (the direct effect). The total (indirect plus direct) effect of $X_1, \ldots, X_4$ can be estimated as well from the reduced form equation:

$$H = \rho(X_1, X_2, X_3, X_4, u_7), \qquad (6')$$

upon the substitution of Eqs. (1)–(5) into (6)[9]. Table 1 summarizes our measures of X_1 through X_4 and Y_1 through Y_5 and provides our prior "guess" as to the variable's likely direct effect on adult health. The table is largely self-explanatory.[10]

Of central importance to the estimations of (1)–(6) and of (6′) is how we choose to measure childhood and adult health. For our analysis, we shall measure health status by the presence or absence of specific illnesses indicating the presence of illness by 1 and the absence by 0. We prefer this (0,1) disease specific approach to the measurement of adult health for two reasons. First, for an analysis of health determinants to provide "value-free" information on the provision of healthiness, we should ideally use only cardinal monitors of biological function—e.g., body temperature, blood pressure, physical strength, white

Table 1. Determinants of Adult Illness

Variables	Mean[a] (s.d.)	Expected Direct Illness Effect[b]	Estimated Direct Illness Effect[c]			
			Arth.	Bron.	Heart	Psy./Neu.
Early Life Environments						
FATHERS EDUCATION (X_1)	9.30	−	−	+*	+	−*
(In years)	(3.50)					
FATHER SES	34.83	−	+	−	+	+
(Blau-Duncan measure of occupational status)	(22.63)					
90–96 = doctors, lawyers						
30–34 = skilled laborers						
0–14 = farmers, unskilled labor						
CH CROUP (X_2)	.31	+	+*	−	+	+
(Croup as a child = 1, 0 otherwise)	(.46)					
CH ASTHMA	.01	+	−	+	−*	−
(Asthma as a child = 1, 0 otherwise)	(.12)					
CH HAYFEVER	.21	+	+	+*	+	+*
(Hayfever as a child = 1, 0 otherwise)	(.41)					
CH ECZEMA	.03	+	−	+*	+	−
(Eczema as a child = 1, 0 otherwise)	(.17)					
CH RHEUMATIC FEVER	.04	+	+	+	+*	−
(Rheumatic fever as a child = 1, 0 otherwise)	(.19)					
CH ST. VITUS	.01	+	−	−	−*	−
(Chorea as a child = 1, 0 otherwise)	(.08)					
WEIGHT 25	155.00	+	−	−	+	+
(Wt. at age 25 in lbs.)	(34.08)					

HEIGHT	69.74	—	+	+*	+	—*
(In inches)	(2.57)					
SMOKE < 25	.62	+	—	+*	+*	+
(Before age 25 = 1, 0 otherwise)	(.48)					
CH CHECKUP	.02	—	+	+	—*	—
(As a child = 1, 0 otherwise)	(.13)					
(X_3)						
EDUCATION	13.26	—	—*	+	—	—*
(Years of schooling)	(3.11)					
VOCED	.33	—	—	—	+	+
(Have received vocational training = 1, 0 otherwise)	(.47)					
EARLY POP.	356.9	—	+	—	—	+
(Population of childhood town, 000's)	(620.4)					
NO. OF BROT & SIS	3.01	+	—*	—	+	+
(Number of brothers and sisters)	(2.23)					
OLDER BROT & SIS	1.86	+	+	—	—	—*
(Number of older brothers and sisters)	(1.94)					

Adult Life Environments

(Y_1)						
(1) = Agree completely to						
(9) = Disagree completely						
POSITION ASPIRED	4.96	+	+	+	—	—
(Has reached that position)	(2.71)					
USE TRAINING	4.66	+	—*	+	+	—
(In full)	(2.73)					
RESPONSIBLE POSITION	7.52	—	—	—	+	+
("Too much" responsibility)	(2.22)					
NO TIME	6.66	—	+	—	—*	+
(to complete work)	(2.66)					
UNEASY	7.79	—	—	+	+	—
(At work)	(7.15)					

(*continued*)

Table 1 (Continued)

Variables	Mean[1] (s.d.)	Expected Direct Illness Effect[2]	Estimated Direct Illness Effect[3]			
			Arth.	Bron.	Heart	Psy./Neu.
GET ALONG	1.84	−	−	−*	−	−
(With co-workers)	(1.53)					
FINANCE ACH	4.53	−	−*	+	+*	−*
("*Not* achieved what I had hoped for")	(2.89)					
TRAIN ADEQUATE	7.15	−	+	−	−	−
(Training not adequate)	(7.36)					
CHANGE EMPLY	2.31	+	+	+*	+	+*
(Number of times of employment change)	(2.56)					
MARITAL (Y_2)	.15	+	−	−	+	+
(Single = 1, 0 otherwise)	(.36)					
NO. RESIDENCES	3.01	+	+*	+	+	−
(Number of residences)	(2.14)					
Years smoked before illness: (Y_3)						
YEARS SMOKE	9.88	+		No results		
(Arth.)	(12.62)					
YEARS SMOKE	10.29	+				
(Bron.)	(12.89)					
YEARS SMOKE	9.83	+		due to multicollinearity		
(Psy./Neu.)	(12.59)					
YEARS SMOKE	10.25	+		with SMOKE < 25		
(Heart)	(12.88)					
PUBLIC	.84	n.p.	−*	+	−	+
(Public school = 1, 0 otherwise)	(.36)					

PRIVATE	.003				—	—
(Private school = 1, 0 otherwise)	(.06)					
PAROCHIAL	.10	n.p.		+*	—	—
(Parochial school = 1, 0 otherwise)	(.30)				—*	
			(X_4)			
Entered occupation for:						
PAY	.51	+	+	+	+	—*
(= 1, 0 otherwise)	(.47)					
FINSUCCESS	.55	+	+	—	+	—*
(= 1, 0 otherwise)	(.46)					
INDWORK	.51	—	+	—	—	+
(= 1, 0 otherwise)	(.47)					
LIKEWORK	.83	—	—*	+*	—	—*
(= 1, 0 otherwise)	(.34)					
JOB SECURITY	.70	—	—*	—	—	—
(= 1, 0 otherwise)	(.43)					
NO OPTION	.23	+	—	—	—*	—
(= 1, 0 otherwise)	(.33)					
(Other Variables)						
MOTH EDUCATION	9.84	—	+	—	—	+*
(Mother's education in years)	(2.97)					
BOTH PARENTS	.91	—	+*	—	+	+
(Reside with = 1, 0 otherwise)	(.28)					
BLOOD DISEASES (ICDA Code, 280-289)	.01	n.p.	+	—	+*	—
	(.10)					
MENTAL ILLNESS (ICDA Code, 290-315)	.10	n.p.	+*	+*	+	+*
	(.30)					
DISEASES OF NERVOUS SYSTEM (ICDA Code 320-389)	.06	n.p.	—*	—	—*	—*
	(.24)					
DISEASES OF CIRCULATORY SYSTEM (ICDA Code, 390-458)	.03	n.p.	—*		—	—
	(.17)					

(continued)

Table 1 (Continued)

Variables	Mean[1] (s.d.)	Expected Direct Illness Effect[2]	Estimated Direct Illness Effect[3]			
			Arth.	Bron.	Heart	Psy./Neu.
DISEASES OF RESPIRATORY SYSTEM (ICDA Code, 460-519)	.07 (.26)	n.p.	−*	−*	−*	−*
DISEASES OF DIGESTIVE SYSTEM (ICDA Code, 520-577)	.13 (.33)	n.p.	−*	−*	−*	−*
DISEASES OF GENITOURINARY SYSTEM (ICDA Code, 580-629)	.05 (.22)	n.p.	−*	−	+	−
DISEASES OF SKIN (ICDA Code, 680-709)	.08 (.27)	n.p.	+	−	−	+*
DISEASES OF MUSCULOSKELETAL SYSTEM (ICDA Code 710-738) (Y_4)	.12 (.32)	n.p.	+*	−	−*	−
TOT CONS (Total alcoholic consumption)	579.6 (976.4)	+	−	+	−*	+
HEAVY CON (Number of times per year you are a ''heavy consumer'') (Y_5)	5.9 (26.45)	+	+	+	+*	−

Variable	Mean (SD)					
CHECK-UP (Years of regular preventive check-up as an adult) (Y_6)	13.68 (6.57)	−	+*	−	+	+
Sick = 1, Not Sick = 0 with:						
INFECTIONS & PARASITIC DISEASES (ICDA Code, 0-136)	.07 (.25)	n.p.	−*	−*	−*	−*
NEOPLASMS (CANCERS) (ICDA Code, 140-239)	.01 (.10)	n.p.	−	−	−*	−
ENDOCRINE, NUTRITIONAL, METABOLIC (ICDA Code, 240-279)	.01 (.10)	n.p.	−*	−	−	+
CONGENTIAL ANAMOLIES (ICDA Code, 740-759)	.01 (.10)	n.p.	+	+*	+	+
ILL-DEFINED CONDITIONS (ICDA Code, 780-796)	.01 (.10)	n.p.	+	−	−	+
ACCIDENTS, POISONS, VIOLENCE (ICDA Code, 800)	.01 (.10)	n.p.	+*	−	−	−*
AGE	50.1 (2.9)	+	+*	+*	+*	

Notes:

[a] Mean and standard deviation of variable.

[b] A (+) indicates the variable is expected to increase the likelihood of adult illness; a (−) indicates the variable is expected to reduce the likelihood of adult illness. The notation (n.p.) represents "no prediction."

[c] A (+) indicates the variable had an estimated positive effect on the likelihood of adult illness; a (−) indicates a negative effect. A (*) represents significance at the .1 level or higher employing a one-tailed t test.

cell counts. Given this vector of biological facts, a specification for each attribute could then be estimated. Yet the large number of biological facts required for such analysis to be of public policy interest makes this approach impractical. Some aggregation will be needed. However, the commonly used health measures such as activity restriction or self-assessed health status move us too far in the other direction. Such measures mix individual preferences and biological facts so we can no longer be sure whether we are measuring health or current preferences. We can learn to adapt to illness, and in such cases some "biologically true" illnesses (i.e., malfunction) may not be reported as an activity restriction. This seems to be a particularly serious problem for chronic diseases. Our use of a disease state measure (sick = 1, not sick = 0) of health seems a useful compromise. The measure incorporates a number of biological facts, called symptoms, but the aggregation is done consistently through reasonably well-defined standards of medical and social opinion of what malfunctions constitute an illness. Policy bias will be mininized. Second, our use of the sick/not sick dichotomy is exactly what is at issue for preventive health care policy. The (1, 0) specification for adult diseases allows us to estimate the probability that an individual with a given vector of X and Y characteristics will contract disease i. Reducing that probability through the manipulation of X and Y is the aim of any preventive health care policy. The diseases chosen for our analysis—on the basis of sample size and policy importance—are adult arthritis, bronchitis-emphysema-asthma, heart and circulatory illnesses, and psychoses and neuroses. Table 2 summarizes the prevalence of these illnesses within our sample; the mean value reflects the percent of our sample which has contracted the illness since discharge from the service ($\simeq$ age 25).

Finally, we have adopted a logistic specification of the disease process which specifies the probability of illness contraction as a logistic function of the early (X) and adult (Y) behaviors and environments and prior health. Specifically, if p is the probability of illness, then (6) becomes

$$p = \frac{1}{1 + e^{-\{\beta'X + \delta'Y + \epsilon H_{-t} + u_6\}}}.$$

The logistic specification of the disease process has been used in previous empirical work on health prospects by Truett et al. (35) and Inman (14). The specification is not completely general, however. The marginal effects of the X and Y variables on the probability of illness [= $\beta_i p(1 - p)$ and $\delta_i p (1 - p)$ for X_i and Y_i, respectively] are *a priori* constrained to interact with prior health in a particular way. Specifically, if prior illnesses increase the probability of contracting a current illness ($\epsilon > 0$), then the marginal effects of early and adult behaviors and environments will be larger in absolute value for those with poor prior health.[11] Thus if smoking is bad for your health (β, $\delta > 0$, increasing p) it will be even more damaging if you have had previous illness. Conversely, if preventive checkups are good for health (β, $\delta < 0$, reducing p) then the checkups are

Table 2. Illnesses

Variable Name	Mean (s.d.)	
ARTHRITIS	0.0933 (0.291)	Arthritis = 1 (ICDA Code 711-718, 724, 727, 731, 734); otherwise = 0.
BRONCHITIS	0.0233 (0.151)	Bronchitis = 1 (ICDA Code 490-493); otherwise = 0.
HEART	0.0517 (0.222)	Heart = 1 (ICDA Code 410-429); otherwise = 0.
PSY./NEU.	0.0786 (0.269)	Neuroses and psychoses = 1 (ICDA Code 290-302, 305-309); otherwise = 0.

even more valuable for those with a history of bad health. For the adult diseases studied here, this interactive pattern does not seem an unreasonable prior constraint.

III. EMPIRICAL RESULTS

Tables 1 and 3 summarize our results. Parameter estimates and their standard errors are presented in Table 3; Table 1 summarizes the qualitative results for the direct effects of the X and Y variables on illness for comparison to their *a priori* expected effects. The results presented in Table 3 are for ordinary-least-squares (OLS) estimation of Eqs. (6) and (6') using as our dependent variable, 1 if sick, 0 if not sick for the appropriate illness. The use of OLS to estimate a logistic relationship has well-known problems,[12] but maximum-likelihood procedures for our sample size and number of explanatory variables proved prohibitively expensive for all four diseases. However, experiments with subsets of X and Y variables did not show any serious differences between OLS and maximum-likelihood estimates of marginal effects. For completeness, then, we have chosen to present the OLS estimates. The coefficient values can be interpreted as the marginal effect on the probability of illness of a small change in the associated X or Y variable. Columns 1, 3, 5, 7 measure the total (direct plus indirect) effects of the early life environment variables, and columns 2, 4, 6, 8 capture the direct effect of each X and Y variable.

The qualitative results reported in Table 1 show that in most instances the early and later life environmental variables affect health as initially expected. When significant (denoted by *), most behavioral and environmental variables had their anticipated positive (health-reducing) or negative (health-increasing) effects, though of course there are some surprises. For example, father's education was a statistically significant cause of increased risk to lung diseases, but when the quantitative effects are calculated the impact is small. This is generally the case for other environmental variables of unexpected sign (mother's education in psychoses/neuroses, total alcoholic consumption in heart, adult preventive checkups in arthritis). The one important exception is the effect of the job choice

Table 3. Adult Health Equations

	Arth.		Bron.		Heart		Psychoses/Neuroses	
	1	2	3	4	5	6	7	8
X_1								
FATHER'S EDUCATION	-.00185	-.00145	.0025	.00221	.00289	.00177	-.0068	-.0064
	(.00251)	(.00247)	(.0013)	(.00130)*	(.00191)*	(.00189)	(.0023)*	(.0021)*
FATHER'S SES	.00022	.000185	-.00022	-.00022	-.000017	.000021	.000064	.000085
	(.00034)	(.000337)	(.00017)	(.00017)	(.00026)	(.00026)	(.000315)	(.00029)
X_2								
CH GROUP	.0259	.0239	-.0050	-.0029	-.0035	.00021	.0099	.00262
	(.0145)*	(.0143)*	(.0074)	(.0075)	(.0110)	(.0011)	(.0133)	(.0124)
CH ASTHMA	-.0298	-.0096	.0227	.0018	-.0691	-.0733	-.0369	-.0366
	(.0552)	(.0546)	(.0285)	(.0028)	(.0422)*	(.0415)*	(.0508)	(.0471)
CH HAYFEVER	.0114	.0134	.0486	.0519	.0073	.0155	.0355	.0353
	(.0164)	(.0163)	(.0085)*	(.0085)*	(.0125)	(.0125)	(.0151)*	(.0140)*
CH ECZEMA	-.0146	-.0196	.0358	.0331	.0293	.0237	-.0012	-.0050
	(.0376)	(.0371)	(.0193)*	(.0195)*	(.0287)	(.0284)	(.0346)	(.0321)
CH RHEUMFEV	.0301	.0449	.0219	.0187	.0501	.0447	-.0025	-.0174
	(.0353)	(.0348)	(.0182)	(.0182)	(.0269)*	(.0266)*	(.0032)	(.0301)
CH ST. VITUS	-.0364	-.0661	-.0185	-.0216	-.0751	-.0803	-.0268	-.0123
	(.0700)	(.0689)	(.0361)	(.0361)	(.0535)*	(.0502)*	(.0645)	(.0595)
WEIGHT 25	-.0001	-.000106	-.000037	-.000053	.000074	.000058	.00020	.000162
	(.0002)	(.000194)	(.000101)	(.000102)	(.00015)	(.00015)	(.00018)	(.000167)
HEIGHT	.00047	.00114	.00022	.0196	.00055	.00072	-.0045	-.00389
	(.0025)	(.00267)	(.00014)*	(.00140)*	(.0021)	(.0021)	(.0025)*	(.00231)
SMOKE < 25	-.0085	-.0104	.0147	.0157	.0155	.0249	.0106	.0078
	(.0139)	(.0139)	(.0021)*	(.0073)*	(.0105)*	(.0106)*	(.0128)	(.0121)
CH CHECKUP	.0263	.0202	-.00048	.0011	-.0694	-.0727	.0205	-.0083
	(.0479)	(.0480)	(.0024)	(.0252)	(.0366)*	(.0367)*	(.0442)	(.0414)

X_3								
EDUCATION	-.0032	-.0052	.00053	.000122	.00097	-.00018	-.0063	-.00435
	(.0025)	(.0025)*	(.00127)	(.00134)	(.0018)	(.00196)	(.0023)*	(.00221)
VOCED	.0037	-.0034	-.00335	-.00462	.0077	.0046	.0072	.0077
	(.0145)	(.0143)	(.00747)	(.00754)	(.0110)	(.0110)	(.0133)	(.0124)
PUBLIC	-.0729	-.0706	.0144	.0130	-.0205	-.0217	.0170	.0004
	(.0276)*	(.0272)*	(.0142)	(.0143)	(.0210)	(.0208)	(.0754)	(.0235)
PRIVATE	.0764	.0443	-.020	-.0269	-.0655	-.0755	-.0499	-.0643
	(.0049)	(.0499)	(.0077)	(.0109)*	(.0210)	(.0208)	(.0308)	(.0287)
PAROCHIAL	-.0143	-.0059	.0382	.0334	-.0578	-.0546	.0112	-.0073
	(.0335)	(.0334)	(.0172)*	(.0174)*	(.0255)*	(.0255)*	(.0308)	(.0287)
X_4								
PAY	.0095	.0042	.00048	.0099	.0111	.0093	-.0245	-.0199
	(.0153)	(.0151)	(.00788)	(.00795)	(.0116)	(.0116)	(.0141)*	(.0131)*
FINSUCCESS	.0033	.0073	-.00739	-.00802	.0108	.0079	-.0298	-.0248
	(.0151)	(.0149)	(.00777)	(.00783)	(.0115)	(.0114)	(.0139)*	(.0128)*
INDWORK	.0065	.0046	-.0053	-.00804	-.0029	-.0091	.0141	.0039
	(.0150)	(.0148)	(.0077)	(.00781)	(.0114)	(.0114)	(.0138)	(.0128)
LIKE WORK	-.0398	-.0457	.0223	.01901	-.0145	-.0191	-.0318	-.0445
	(.0209)*	(.0209)*	(.0108)*	(.0109)*	(.0160)	(.0160)	(.0193)*	(.0181)*
JOBSECURITY	-.0249	-.0238	-.0107	-.0102	-.0084	-.0055	.0093	-.0101
	(.0163)*	(.0162)*	(.0083)	(.0085)	(.0124)	(.0124)	(.0189)	(.0140)
NO OPTION	-.0233	-.0122	-.0159	-.0168	-.0276	-.0279	.0094	-.00043
	(.0210)	(.0203)	(.0105)*	(.0106)*	(.0156)*	(.0155)*	(.0189)	(.0175)
OTHER X VARIABLES								
MOTH EDUCATION	.0036	.0031	-.0010	-.000876	-.00125	-.00061	.0045	.0052
	(.0028)	(.0028)	(.0014)	(.00148)	(.0022)	(.0022)	(.0026)*	(.0024)*
BOTH PARENTS	.0353	.0395	-.0052	-.00627	.0186	.0194	-.0014	.00048
	(.0232)*	(.0231)*	(.0119)	(.0120)	(.0177)	(.0176)	(.0214)	(.0199)
EARLY POP	.000008	.0000093	.0000036	-.0000029	-.0000069	-.0000046	.000012	.000007
	(.000013)	(.0000131)	(.0000067)	(.0000068)	(.000010)	(.00001)	(.000012)	(.000011)

(continued)

Table 3—(Continued)

	Arth.		Bron.		Heart		Psychoses/Neuroses	
	1	2	3	4	5	6	7	8
NO. OF BROT & SIS	-.0044	-.0062	.00052	-.00034	.00143	.0000043	.0034	.0033
	(.0047)	(.0047)*	(.00245)	(.00246)	(.00369)	(.0035)	(.0043)	(.0044)
OLDER BROT & SIS	.0050	.00504	-.00163	-.00083	-.00604	-.00466	-.0089	-.0084
	(.0054)	(.00535)	(.00279)	(.00281)	(.00414)*	(.00410)	(.0049)*	(.0046)*
Y_1								
POSITION ASPIRED		.00108		.0018		-.00059		-.00083
		(.00305)		(.0016)		(.0023)		(.0026)
USE TRAINING		-.00488		.00087		.00059		-.0019
		(.00297)*		(.00156)		(.0023)		(.0026)
RESPONSIBLE POS		-.00134		-.000099		.0031		.0021
		(.00327)		(.0017)		(.0025)		(.0028)
NO TIME		.00096		-.00099		-.0045		.00137
		(.00268)		(.0014)		(.0021)*		(.00231)
UNEASY		-.000712		.00069		.0022		-.00021
		(.00216)		(.00131)		(.00165)		(.00186)
GETALONG		-.0054		-.0038		-.0037		-.00016
		(.0045)		(.0023)*		(.0034)		(.00389)
FINANCE ACH		-.0051		.00133		.0024		-.0054
		(.0024)*		(.00124)		(.0018)*		(.0020)
TRAIN ADEQ		.0011		-.0010		-.0014		-.00004
		(.0021)		(.0011)		(.0015)		(.00178)
CHANGE EMPLY		.0015		.0025		.0027		.0074
		(.0026)		(.0014)*		(.0021)		(.0023)
Y_2								
MARITAL		-.0115		-.0036		.00155		.0147
		(.0181)		(.0095)		(.00138)		(.0156)

198

NO. OF RESIDENCE	.0119 (.0033)*	.00076 (.00175)	.0014 (.0025)	−.00157 (.00287)
Y_3 YEARS SMOKE	n.i.	n.i.	n.i.	n.i.
Y_4 TOT CONS.	−.0000036 (.0000074)	.0000045 (.000038)	−.0000089 (.0000057)*	.00000 (.00000)
HEAVY CONS	.00005 (.00025)	.000141 (.00013)	.00038 (.00019)*	−.00002 (.00021)
Y_5 CHECK-UP	.00213 (.00100)	−.000075 (.00053)	.00095 (.00077)	.00085 (.00086)
Y_6 0–136 (INFECT)	−.027 (.015)*	−.0137 (.0075)*	−.0201 (.0110)*	−.0262 (.0124)*
140–239 (CANCERS)	−.0102 (.0271)	−.0095 (.0141)	−.0454 (.0205)*	−.0096 (.0233)
240–279 (ENDOCRINE)	−.0667 (.0381)*	−.0249 (.0194)	−.0145 (.0285)	.00069 (.0322)
280–289 (BLOOD)	.0323 (.0562)	−.0141 (.0310)	.0982 (.0462)*	−.0263 (.0514)
290–315 (MENTAL)	.0867 (.0599)*	−.0142 (.0311)	.0141 (.0449)	.0931 (.0053)*
320–389 (NERVOUS)	−.0174 (.0130)*	−.0315 (.0068)*	−.0408 (.0099)*	−.0405 (.0112)*
390–458 (CIRCULATION)	−.0347 (.0157)*	−.00063 (.00811)	−.0093 (.0120)	−.0131 (.0135)
460–519 (RESPIRATORY)	−.0240 (.0132)*	−.0114 (.0069)*	−.0424 (.0101)*	−.0255 (.0113)*
520–577 (DIGESTION)	−.0403 (.0135)*	−.0215 (.0071)*	−.0423 (.0103)*	−.0469 (.0117)*

(*continued*)

Table 3—(Continued)

	Arth.		Bron.		Heart		Psychoses/Neuroses	
	1	2	3	4	5	6	7	8
580-629 (GUT)		-.0251		-.0093		.00147		-.0054
		(.0191)*		(.0098)		(.0145)		(.0164)
680-709 (SKIN)		.0104		-.0095		-.0137		-.0283
		(.0159)		(.0083)		(.0122)		(.0138)*
710-738 (MUSCLE)		.108		-.0022		-.0214		-.0092
		(.015)*		(.0082)		(.0119)		(.0135)
740-759 (CONGENTIAL)		.028		.0259		.0140		.0028
		(.034)		(.0174)*		(.0257)		(.0291)
780-796 (ILL-DEFINED)		.0020		-.0082		-.0168		.0135
		(.0183)		(.0095)		(.0139)		(.0159)
800 (ACCIDENTS)		.0081		-.0028		-.0111		-.0277
		(.0148)		(.0077)				
AGE	.0083	.0085	.0017	.0024	.00087	.0024	.0043	.0045
	(.0023)*	(.0023)*	(.0011)*	(.0012)*	(.00172)	(.0017)*	(.0021)*	(.0019)*
CONSTANT	-.194	-.0203	-.024	-.0208	-.0513	-.0536	.029	.029
	(.217)	(.0218)	(.011)*	(.0114)*	(.166)	(.167)	(.020)*	(.019)*
R^2	.025	.083	.038	.062	.018	.072	.033	.202

Notes:
* = t statistic > 1.3.
n.i. = not included because of high collinearity with smoke before 25.

variables PAY and FINSUCCESS on the likelihood of mental illness. We anticipated that individuals who took their first job for financial reasons would have personalities more susceptible to mental illness; in fact, the opposite was the case. For this sample, at least, individuals motivated toward financial success have a significantly lower chance of being diagnosed as psychotic or neurotic. Perhaps PAY and FINSUCCESS are measures of self-confidence as well as motivation.

Our "other variables" included to control for possible omitted effects of X_1 to X_4 on health were generally insignificant. Of some interest to the new theory of family economics is the fact that the effects of mother's education on adult health are apparently fully captured through its effects on childhood health (X_2), educational attainment (X_3), and work attitudes (X_4). When X_2, X_3, and X_4 are included in the health equations, mother's years of schooling is only significant in the psychoses-neuroses equation and almost significant in the arthritis equation—and here the effects are health-reducing! Quantitatively, the impacts are small, however. The presence of both parents and being raised in a large family have no significant effects on adult health, apart from any impact they may have on X_2, X_3, and X_4.[13]

In addition to the significant effects of the behavioral and environmental variables, our results reveal an extremely interesting pattern of disease histories. Rather than simply finding the sick get sicker—in which case all prior illnesses should have a positive or zero effect on the probabilities of recent illnesses—we find that some illnesses are significantly correlated (causation is too strong an inference) with the *absence* of other diseases. For example, diseases of the nervous system (including St. Vitus disease), respiratory system (including childhood asthma), digestive disorders, and infectious diseases are all negatively related to later life chances of arthritis, bronchitis, heart disease, and mental illness. On the other hand, adult heart disease and blood diseases are positively related to each other, as are mental illness, arthritis, and other diseases of the musuclar system. Such disease patterns may be explained by a limited inherited ability to fight the multitude of illnesses. For example, if there are a maximum of n genes to fight n illness types but each individual is endowed with only a small subset m of the full set, then each individual will be potentially immune to only m diseases and susceptible to the remaining n − m. Illnesses will therefore cluster according to genetic patterns. Exposing such adult disease patterns has important implications for medical research and health care policy.[14]

Also of some importance is how little of the variation in the sick/not sick status of the individuals in our sample we are able to "explain" through our long list of X and Y variables and past health. While R^2 as a measure of goodness of fit is bounded between 0 and 1 for continuous dependent variables, this is not the case for 0, 1 dependent variables. Rather the upper bound for R^2 is likely to be substantially less than 1—for example, in the case where the true probability of illness is uniformly distributed across an interval, then the upper bound to R^2 is only .33.[15] While the low R^2 values reported in Table 3 are therefore not unexpected, it does seem reasonable to conclude that for these adult illnesses (except perhaps psychoses/neuroses where the R^2 equals .202) much of the variation in outcomes remains to be explained. Either our analytic framework is inadequate and important determinants of illness are omitted *or* all the potentially important predictors are included but illness outcomes do involve a significant random (or uncontrolled) component. It is important to know which is the case. If potentially important policy variables have been excluded, then knowledge of their marginal effects on adult health is useful. Alternatively, if the incidence of adult illnesses does involve a truly random or, more generally, uncontrollable component, then knowledge of the relative importance of components is necessary for the design of an efficient and fair health insurance policy. If the "controllable" aspects of an illness dominate its incidence, then insurance may involve significant disincentive effects for proper care, yet if the "uncontrollable" aspects dominate, then even sizable compensatory ("pain and suffering") insurance may be socially preferred.[16]

Table 4 focuses attention on the potentially important role of early life environments, X_1 to X_4, on adult health prospects.[17] Based on the results in Table 3 (columns 1, 3, 5, 7) for the "total" effect of X_1 to X_4 on the probability of illness, we have first estimated the probability of illness for an average 40-year-old individual in our sample with a poor childhood health environment. The poor early life health environment is described by the presence of one or more significant childhood illnesses, smoking before 25, 12 years of public school, no preventive health checkups as a child, and the need to take a first job because no other options were available. (Table 4, footnote 1 details the specifications.) The resulting initial probability of illness is recorded in Table 4, column 1. We then

Table 4. Probability of Illness and Early Life Environments

Illness	Initial Probability of Illness[1] (1)	Health (2)	Health, Education (3)	Health, Education, Opportunities (4)
Arthritis	.094	.103	.090	.062
Respiratory (Bronchitis, etc.)	.113	.065	.067	.088
Heart	.081	.015	.019	.045
Psychoses/Neuroses	.078	.053	.027	$0 \simeq$[2]

Notes:
[1] For all illnesses the initial values of the (1, 0) X characteristics include the value 1 for smoke before 25, public school, took first job for reasons of "no option," and lived with both parents. For the childhood disease histories the value 1 was used for croup for arthritis, 1 for hayfever for bronchitis, 1 for asthma and rheumatic fever for heart, and 1 for hayfever for psychoses/neuroses. All other (1, 0) X variables are assigned the value 0. All continuous X variables are given their mean values except schooling, which is set at 12 years; family size, which is set at three brothers and sisters with two older than our "average" sample member; and age, which is set at 40 years.
[2] $0 \simeq$ means estimated probability fell below zero.

examined the change in probabilities resulting from a better early life environment. Removing major childhood illnesses, for example, reduces the probability of adult respiratory, heart, or mental illnesses significantly (column 2).[18] Retaining the good childhood illness record and improving educational opportunities from 12 years to 16 years of schooling reduces illness probabilities further for arthritis and for psychoses/neuroses (column 3). Finally, improving one's reasons for job choice from "no option" to taking the first job because of good pay, the promise of financial advancement and job security, and one likes one's work also adds incrementally to one's adult health prospects for arthritis and mental illness. Interestingly, the chances of respiratory illness and heart disease increase with good job opportunities. Good jobs with the promise of success may help to insure mental health, but only through the simultaneous adoptions of life styles—smoking, drinking, eating behavior induced perhaps by the work environment—which reduce physical well-being. Such results raise the possibility that as a policy matter we may have to choose not a single healthy life style but rather a life style with some preferred mix of health prospects. It is nonetheless encouraging that we have some choice.

IV. CONCLUSION

The United States is at a crossroads in the design of its national health care policy, and the central issue is the appropriate structure for national health insurance. We have learned the lessons of our recent travels well. Significant increases in national insurance coverage promise significant inflation in medical care costs. Yet such inflationary rises in the costs of uncertain illnesses make increased insurance all the more attractive. It is at this juncture that we now stand. Do we pursue increased coverage to reduce the large financial risks of illness, or do we seek to curtail costs with less insurance? Yet in an important sense even this complicated issue of risk coverage vs. cost inflation is too narrowly conceived. For the primary source of insurance-induced inflation is growing medical inputs per case and the widespread adoption of new medical technologies. Surely, this new curative care environment has had some positive impact on the health prospects of patients treated. If so, then the health insurance debate must be decided on three, not two, dimensions: health improvements vs. risk coverage vs. cost inflation.[19] With the inclusion of health outcomes within the health insurance debate it then becomes essential that we understand the causes and consequences of improved population health. It is toward this end that this paper has been written.

The work here is of course preliminary, but nonetheless suggestive. It develops, we feel, the right strategy for including population health changes within health policy debates. First and foremost is the need for empirical work on the behavioral and environmental causes of health. We have provided one example of what we feel is the correct approach, but this analysis must be expanded—

particularly to take advantage of the genetic information available from our twin sample—and replicated with other studies using different populations. Our data are for reasonably healthy, white, male twins and are hardly representative of the population as a whole. Second, an analytic framework is required which can evaluate population health changes within a metric which allows us to trade changes in health against changes in inflation and/or risk coverage. The problems here are even more subtle than those in the much-discussed "value of life" debate. Often significant health changes can only be brought about by altering individual preferences for health-reducing activities such as smoking, drinking, or stressful work. As of now, we know of no compelling basis for considering the welfare implications of such behavioral changes.[20] Yet a normative framework which can successfully deal with this matter is essential. Only with information on the behavioral and environmental causes of illnesses and only through an evaluation mechanism which adequately incorporates health and health behaviors can we begin to design a national health care policy in which health, not just medical expenses, plays the central role.

ACKNOWLEDGMENTS

Inman's initial work began during his visit to Harvard University as a Robert Woods Johnson Foundation Faculty Fellow. We wish to thank Gerard Anderson for his excellent programming assistance with this project and Mark Langer and Alan Feldman for their help in our review of the relevant literature.

The revision of this paper was financed by Grant No. HS 02577 from the National Health Care Management Center, University of Pennsylvania.

An earlier version of our work was presented to the Mathematical Social Science Board Conference on Family Environment and Subsequent Development of the Child, Stanford, California, March 1977. The comments of the participants, particularly Professor Vic Fuchs, were extremely helpful to us in our subsequent revision of this work.

NOTES

1. Fuchs (7) has surveyed recent trends in the U.S. mortality and morbidity rates.
2. For a strong plea that preventive health care policies are needed to improve adult health, see Kristein, Arnold, Wynder (18). The issue is less clear-cut than they suggest, however. See, for example, the discussion of the recent ban on cigarette advertising in Hamilton (11) and the discussion of smoking and illness in Sterling (34) and comments by Weiss (37), Higgins (13), and Bross (4). Two papers by Peltzman are also of interest here. The first critically reviews seat belt and highway safety regulation (27), while the second examines the relative performance of the FDA (26). For an overview of preventive policies, see Fielding (6).
3. Rice estimates the income losses due to death from all causes at about $50 billion for 1963. The most significant losses of course occur for deaths between the ages of 20 and 65. Wages and salaries in that year were $310 billion.
4. See, for example, S. Shavell (32; see the appendix) and P. A. Diamond and J. A. Mirrlees (5).
5. Further, Bartel and Taubman (1, fn. 9) have considered the bias which may result in our analysis from underreporting in later years and consider that it is likely to be small.

6. Other recent work on behavior, environment, and human health which employs microdata bases include Mannheim (21), Grossman (10), and Inman (14). Vic Fuchs has stressed to us that working only with micro samples may be misleading, too, and that work with aggregate data can capture effects of policy not revealed with micro studies. For example, suppose we discover from our micro studies that low-density, mountain living improves health prospects. Do we then recommend that all individuals move to the mountains of Colorado for their health? No, for we will no longer have low-density, mountain living. This is an example of what economists call the "fallacy of composition"—what works for one need not work for many. Macro studies of entire health care systems can reveal these aggregate effects, however.

7. See Grossman (9, 10) and Inman (14) for two alternative economic models of health and health behavior, Pollak (28) for a general survey and bibliography of the economic literature on habits and changing tastes, Berkanovic (3) for a survey of the sociological literature on health behavior (particularly the so-called health belief model), and Solomon and Corbit (33) for a psychological theory of addiction.

8. Notably missing from this list of adult health environments are measures of adult income and the consumption of curative medical services. Our sample lacks data on the utilization of medical services for curative care. While data on adult earnings are available, the recorded earnings information is for years after (1973), not prior to, most of the health histories (1945–1967). Including an income variable would therefore reveal a health-to-income relationship, not the income-to-health connection we seek here. Because measures of adult income and curative medical care will be omitted from our analysis, our estimates of the effects of the included variables may be biased *if* income and curative care consumption are correlated with our included variables *and if* income and curative care have a significant health effect of their own. However, recent research by Newhouse and Friedlander (24), Grossman (10), and Mannheim (21) suggest income and medical care by themselves—holding health habits, job stress, and location fixed—have only negligible effects on the probability of disease contraction. (Note that income and curative care can still ease the burden of illness, once sick, without affecting the chances of becoming ill.) Thus omitting these variables is likely to cause no serious problems for our analysis.

9. Readers familiar with the new "home economics" will recognize this structure. The Y_i (i = 1 . . . 5) equations are demand equations for "market goods" which produce as one of their "commodities" later adult health. See, for example, Becker (2) or more recently Pollak and Wachter (29). Adult health [whether current (H) or lagged (H_{-t})] is defined by the "production technology" of Eq. (6). The recursive specification, made possible by the longitudinal nature of our data, plus our assumption of error independence allows us to identify $\mu(\cdot)$ as a production relationship.

10. A few comments are in order. The NAS-NRC sample lacks data on parents' bequests (X_1) to the members of our sample. However, father's occupation and educational level are reasonable proxy variables to parent wealth, which is the likely causal variable for bequests. See, for example, Ishikawa (15). The variables entitled "Other Variables" are included to hold constant other effects on adult health. There is much hypothesizing, and some evidence, that mother's education and both parents in the home will affect children's health [see, e.g., Grossman (10)], while family size and birth order have been established previously as important determinants of future healthiness [Wray (38)]. The hometown population is included to proxy access to quality medical care during childhood.

11. Formally,

$$\frac{\partial p}{\partial X_i \partial H_{-t}} = \epsilon \beta_i (1 - p) p (1 - 2p)$$

and

$$\frac{\partial p}{\partial Y_i \partial H_{-t}} = \epsilon \delta_i (1 - p) p (1 - 2p).$$

For $p < .5$ (a valid assumption) and $\epsilon > 0$, the comments above follow.

12. See, for example, Goldberger (8, pp. 251–55).
13. But these effects may be quite important. See Inman (14) and Wray (38).
14. The effects of heredity on health have important implications for the statistical estimation of models such as ours, implications we have not yet had time to pursue. A longer version of this paper presented to the Mathematical Social Science Research Conference contained a detailed description of the appropriate statistical methodology, and that paper is available from the authors upon request.
15. See Morrison (23).
16. On these matters, see Shavell (32).
17. We concentrate on the early life variables. While the Y_1 through Y_5 variables do contribute to the explanatory power of the model, they generally do not have quantitatively significant effects on illness probabilities for reasonable changes (say, 1 s.d.) in their variable values.
18. Though one must interpret the coefficients for the effects of child diseases on adult health with caution, there is nothing in this analysis which excludes the possibility that a third uncontrolled variable causes both the childhood and the associated adult illness and that what we are measuring in Tables 3 and 4 is really a spurious, i.e., noncausal, correlation of the two diseases.
19. A recent paper by Harris (12) outlines a methodology for analyzing such a three-way trade-off within the context of health insurance design. The inclusion of population health changes as a relevant dimension of the health insurance debate also forces us now to broaden the scope of that discussion to include health promotion policies generally [see Inman (14)]. For example, preventive care policies may have desirable risk-reducing and cost-controlling properties *because of* their favorable effects on health. The gains from improved health can be substantial (see Introduction) and are certainly on par with costs and net benefits envisioned in the risk vs. cost debates.
20. See the discussion in Pollak (28) and Marschak (22). The theoretical literature on "consistent planning" developed in another context may be a useful starting point; see Marschak (22) for references.

REFERENCES

1. Bartel, A., and P. Taubman (1979) "Health and Labor Market Success: The Role of Various Diseases," *Review of Economics and Statistics*, 61:1–8.
2. Becker, G. (1965) "A Theroy of the Allocation of Time," *Economic Journal* 75:493–517.
3. Berkanovic, E. (1976) "Behavioral Science and Prevention," *Preventive Medicine* 5:92–105.
4. Bross, I. (1976) "Comment on Sterling," *American Journal of Publich Health* 66:161.
5. Diamond, P. A., and J. A. Mirrlees (1978) "A Model of Social Insurance with Variable Retirement," *Journal of Public Economics* 10:295–336.
6. Fielding, J. E. (1978) "Successes of Prevention," *Milbank Memorial Fund Quarterly* 56:274–302.
7. Fuchs, V. (1974) *Who Shall Live*, New York: Basic Books, Inc.
8. Goldberger, A. (1964) *Econometric Theory*, New York: John Wiley and Sons, Inc.
9. Grossman, M. (1972) *The Demand for Health: A Theoretical and Empirical Investigation*, New York: National Bureau of Economic Research.
10. Grossman, M. (1975) "The Correlation Between Health and Schooling," in *Household Production and Consumption*, (ed.), N. Terleckyi, New York: National Bureau of Economic Research.
11. Hamilton, J. (1972) "The Demand for Cigarettes: Advertising, the Health Scare, and the Cigarette Advertising Ban," *Review of Economics and Statistics* 54:401–411.
12. Harris, J. (1979) "The Aggregate Coinsurance Rate and the Supply of Innovations in Hospital Sector," mimeo, M.I.T.
13. Higgins, G. (1976) "Comment on Sterling," *American Journal of Public Health* 66:159–161.
14. Inman, R. (1976) "The Family Provision of Children's Health," in *The Role of Health Insurance in the Health Services Sector*, (ed.), R. Rosett, New York: National Bureau of Economic Research.

15. Ishikawa, T. (1975) "Family Structures and Family Values in the Theory of Income Distribution," *Journal of Political Economy* 83:987–1008.
16. Jablon, S., et al. (1967) "The NAS-NRC Twin Panel: Methods of Construction of the Panel, Zygosity Diagnosis and Proposed Use," *American Journal of Human Genetics* 19:133–161.
17. Kitagawa, E. M. and P. M. Hauser (1968) "Education Differentials in Mortality by Cause of Death, United States, 1960," *Demography* 5:318–353.
18. Kristen, M. M., C. B. Arnold, and E. L. Wynder (1977) "Health Economics and Preventive Care," *Science* 195:457–462.
19. Lave, L. and E. Seskin (1970) "Air Pollution and Human Health," *Science* 169:723–733.
20. Luft, H. (1975) "The Impact of Poor Health on Earnings," *Review of Economics and Statistics* 57:43–57.
21. Mannheim, L. (1974) "Smoking, Education, and Health: Some Preliminary Results on Male Mortality," mimeo.
22. Marschak, T. A. (1978) "On the Study of Taste Change Policies," *American Economic Review* 68:386–391.
23. Morrison, D. F. (1972) "Upper Bounds for Correlations Between Binary Outcomes and Probabilistic Predictions," *Journal of the American Statistical Association* 67:68–70.
24. Newhouse, J., and L. J. Friedlander (1976) "The Relationship Between Medical Resources and Measures of Health: Some Additional Evidence," RAND mimeo.
25. Neyman, J. (1977) "Public Health Hazards from Electricity Producing Plants," *Science* 195:754–758.
26. Peltzman, S. (1973) "An Evaluation of Consumer Protection Legislation: The 1962 Drug Amendments," *Journal of Political Economy* 81:1049–1091.
27. Peltzman, S. (1975) "The Effects of Automobile Safety Regulations," *Journal of Political Economy* 83:677–723.
28. Pollak, R. (1978) "Endogeneous Tastes in Demand and Welfare Analysis," *American Economic Review* 68:374–379.
29. Pollak, R., and M. Wachter (1975) "The Relevance of the Household Production Function and Its Implications for the Allocation of Time," *Journal of Political Economy* 83:255–277.
30. Rabkin, J. G., and E. Struening (1976) "Life Events, Stress, and Illness," *Science* 194:1013–1020.
31. Rice, D. (1966) *Estimating the Cost of Illness,* Public Health Service Publication, No. 947-6, Washington: U.S. Government Printing Office.
32. Shavell, S. (1978) "Theoretical Issues in Medical Malpractice," in *The Economics of Medical Malpractice,* (ed.), S. Rottenberg, Washington: American Enterprise Institute, pp. 35–64.
33. Solomon, R., and J. D. Corbit (1978) "An Opponent-Process Theory of Motivation," abstracted in *American Economic Review* 68:2–24.
34. Sterling, T. (1975) "A Critical Reassessment of the Evidence Bearing on Smoking as the Cause of Lung Cancer," *American Journal of Public Health* 65:939–953.
35. Truett, J., J. Cornfield, and W. Kannel (1967) "A Multivariate Analysis of the Risk of Coronary Heart Disease in Framingham," *Journal of Chronic Diseases* 20:511–524.
36. Vaupel, J. (1976) "Early Death: An American Tragedy," *Law and Contemporary Problems* 40:73–121.
37. Weiss, W. (1975) "Smoking and Cancer: A Rebuttal," *American Journal of Public Health* 65:954–955.
38. Wray, J. D. (1971) "Population Pressures on Families: Family Size and Child Spacing," in National Academy of Science, *Rapid Population Growth,* Baltimore: Johns Hopkins Press.

SUMMARY AND DISCUSSION OF PART II

David S. Salkever

The four papers in this part deal with various aspects of the determinants of health levels at different stages in the life cycle: birth, childhood, adolescence, and adulthood, Proceeding chronologically, let us begin this summary with Lewit's analysis of prenatal care and infant birthweight. Lewit details a theoretical model in which family utility depends upon the number of surviving children and consumption of other goods and services. The number of surviving children is the product of the number of pregnancies and the probability of success of a pregnancy. Each pregnancy involves fixed costs such as loss of maternal time from market work or other pursuits and variable costs such as expenditures on prenatal care. The probability of a successful outcome is positively related to these variable costs. In addition, family resources are required to meet the maintenance costs of surviving children. The family's decision problem is to maximize utility subject to an income constraint by choosing the optimal number of pregnancies and the optimal level of variable costs per pregnancy. Within the context of this model, the demand for prenatal care (represented by the optimal level of variable cost per pregnancy) is a derived demand; prenatal care is not desired for its own sake but rather because it increases the probability of successful outcomes.

Lewit's empirical work is divided into two segments. The first presents empirical estimates of the derived demand for prenatal care. He finds that demand is (1) positively related to maternal and paternal education; (2) lower for nonwhites, Puerto Ricans, and illegitimate births; and (3) nonlinearly related to maternal age. Other variables associated with higher risks of poor outcome (such as previous fetal loss) tend to increase demand. Availability measures are also included for clinic services, obstetricians, and the presence of a Maternal and Infant Care (MIC) project. Only the first of these has a positive effect on demand, while the MIC variable shows significant negative effects.

The second portion of Lewit's empirical work examines the effect of prenatal care on birth weight. He finds that the number of prenatal visits has a positive and significant effect on birth weight. Other significant independent variables include gestational age, parity, prior fetal loss, illegitimacy, and maternal age. Parental education, surprisingly, has no perceptible effect.

The principal policy conclusion emerging from Lewit's results is that measures to increase consumption of prenatal care will yield important health benefits. In fact, the benefits may be even larger than he estimates if prenatal care early in a pregnancy tends to increase gestational age. On the other hand, Lewit's demand function results do not clearly indicate how this increased consumption can be brought about. The lower demand levels for certain population groups (such as nonwhites) might suggest that programs targeted to these groups would be most effective. However, the mixed results for his availability measures (and particularly for the MIC variable) raise doubts about the effectiveness of simply augmenting the supply of providers in the local area. It would be interesting to see if these results also hold for demand functions estimated from data for nonwhites and other "underserved" population groups.

Other avenues of possible future research, as suggested by the author, include the analysis of more recent data and of data from more nationally representative samples to test the generalizability of his findings. He also points out the need to estimate birth weight "production functions" with more detailed input data on medical care, maternal diet, smoking habits, and alcohol consumption. To the extent that prenatal care improves pregnancy outcomes by influencing maternal behavior, this more detailed analysis would help to explain the strong effects of prenatal care reported here.

In their analysis of the growth and health of young children, Chernichovsky and Coate examine three types of empirical relationships: a production function for growth and health in which nutrients appear as inputs, a demand function for nutrients conditional on the child's weight, and reduced-form relationships in which growth, health, and nutrient intake are related to socioeconomic characteristics and health endowment variables (e.g., parental size, infant birth weight, and birth order). The production function results indicate that protein intake is significantly related to physical growth and that parental education and income are negatively related to poor health as measured by the number of colds in a 6-month period. The demand function estimates show that weight, household size, and parental education are all positively related to protein and caloric intake. The reduced form estimates show that birth weight and parental height and weight are significant determinants of growth, as are income and race. Household size and parental schooling have no consistent effects. Reduced-form nutrient demand equations show the opposite pattern: household size and parental education are more significant (and positive), while other variables (except for age and sex of the child) are not important. In the reduced-form health regression, only race is significant.

The authors note that protein and calorie intake levels in their data tend to be well above recommended standards and that no major differences exist in dietary intake by economic class. An interesting question is whether this situation exists because of the effectiveness of diet supplementation programs (food stamps and school lunch subsidies) or would hold even in their absence. Hopefully, future empirical research incorporating such policy variables into the framework developed by Chernichovsky and Coate will address this question.

The approach of separately estimating technical relationships (production functions) and behavioral relationships (derived input demand functions) is also employed in Edwards and Grossman's study of adolescent health and preventive care. Production functions are estimated for physical health and dental health, while demand functions are estimated for dental and physical preventive care (checkups). In the estimated physical health production functions, maternal education tends to have a significant and positive effect on health, while paternal education does not. The authors interpret this as evidence that home health production (by the mother) is an important determinant of health levels. In general, most other variables, including the variable for medical checkups, are not significant in these production function estimates. In the dental health production functions, maternal education is again associated with better health and more significant than paternal education. Moreover, the significantly negative effect on dental health of a mother working full time reinforces the authors' point about home health production. Dental checkups and water supply fluoridation also have significant health-improving effects.

In the demand analyses, as in Lewit's work, the variables of most policy interest are measures of availability—dentists per capita and pediatricians per capita. Both these variables have significantly positive effects on preventive care demand in their respective regressions.

On the whole, the authors present a rather convincing case to show that (1) home health production is important; and (2) preventive care measures are more important for dental health than for physical health. But several limitations of the analysis caution against unqualified acceptance of these conclusions. First, as the authors state, they confine themselves to aspects of physical health (obesity, poor vision, and anemia) which are most amenable to home interventions (such as dietary practices). Other aspects of poor health, such as long-term effects of respiratory diseases, may be more effectively prevented by medical rather than maternal interventions. Second, the measure of medical inputs (a dummy variable for whether or not a physical exam was obtained in the last year) is obviously not a complete representation of medical services obtained by the individual.

One must also treat the results of the demand analysis with care. The data set used by Edwards and Grossman is interesting because of its detailed health information, but it is not ideal for demand studies. Potentially important predictors such as insurance coverage are not contained in the data. (The same comment applies to Lewit's prenatal care demand analysis.)

In contrast to the more structured empirical models in the other papers, Boardman and Inman's study of adult health is frankly exploratory. Empirical relationships are sought between indicators of adult health and nine other groups of variables measuring parental social class, childhood health experiences, and health behavior, education, attitudes toward work and prospects for success upon entering the work force, work success, social support, smoking behavior, drinking behavior, and preventive care behavior. The first four of these categories of variables describe the individual's early life environments while the last five represent adult life environments and behavior.

One of the stronger and more consistent findings of their analysis is that poorer childhood health experiences are associated with poorer adult health. Controlling for these childhood health experiences, they find that higher levels of parental education tend to be associated with poorer adult health. Effects of the individual's own education on his health tend to show that better-educated persons are better able to maintain their own health.

The reader should note that several of these findings are reinforced by evidence from an earlier work of Grossman (1). In that study, overall self-reported health status in high school and years of schooling were both positively and significantly related to adult health. Moreover, these effects were observed even when measures of physical and intellectual ability were entered in the analysis. The precise reason for the link between schooling and health is unclear. Fuchs (2) has suggested it may be due to differences in time preference, and hence in willingness to invest in one's own health, between more and less educated persons. However, Boardman and Inman's schooling effect is measured while controlling for several types of health-affecting behaviors (smoking, drinking, and preventive care use).

One methodological issue which arises in interpreting Boardman and Inman's results for the "adult life environments" variables concerns the direction of causation. Their cross-sectional analysis cannot always distinguish between effects of such variables on health or effects of health on these variables. An example is their result that changes in employment are significant predictors of poor health. Is this because employment changes reflect a lack of job success (e.g., being fired), or because people change jobs to accommodate declines in health status, or because people who change jobs are climbing the ladder to success too rapidly (or too far) for their own health? Clearly, one would need longitudinal data and more detailed information about the reasons for changes in employment to sort out these various explanations.

Taken together, the empirical results of these four studies provide the basis for several general observations about the health production process. First, they demonstrate that health production is indeed an intertemporal investment process in the Fisherian sense of foregoing current consumption to increase future consumption opportunities. As examples of this intertemporal aspect of health production, note that birth weight, which is Lewit's measure of infant health, is one of the most significant determinants of children's growth in the results reported by Chernichovsky and Coate. Similarly, the occurrence of childhood illnesses is significantly associated with illnesses in adult life according to the findings of Boardman and Inman. Edwards and Grossman show a less consistent link between infant and adolescent health measures, but this may be due in part to their particular choice of dependent variables.

Second, these studies demonstrate that health is produced by several different types of inputs besides purchased medical goods and services. We have already noted the "home production" effects reported by Edwards and Grossman. A similar interpretation could apply to Chernichovsky and Coate's estimated effect of parental education on their measure of young children's health (number of colds in the last 6 months). For adults, life style characteristics such as diet, exercise, and use of alcohol and tobacco are important aspects of home health production. Boardman and Inman's results for alcohol and tobacco use confirm this well-known fact. In addition to "home production" effects, the impact on health levels of features of the individual's social and physical environment are also identified in these studies. Examples are the effect of water fluoridation on adolescent dental health reported by Edwards and Grossman and the effects of job-related variables on adult health reported by Boardman and Inman.

Third, while it is currently fashionable for health policy makers to be skeptical about the ability of purchased medical services to affect population health levels, the empirical results presented here warn against sweeping generalizations. As noted above, Lewit reports important benefits of prenatal care, while Edwards and Grossman find that dental checkups contribute significantly to oral health. Thus, it would appear that at least certain types of professional services have an important role to play in the current campaigns for health promotion and disease prevention.

These studies also have much in common from a methodological standpoint. In particular, they point out some of the problems and limitations of the production function concept in health research. Although the concept is in theory purely technical and precise, in practice it is not. Typically, some inputs cannot be measured directly (e.g., quantity and quality of parental time inputs) so that proxies for these inputs (such as parental education or income level) are used instead. Interpretation is further complicated by the fact that even "measurable" inputs can only be crudely depicted. Numbers of physician visits, for instance, can be entered into the analysis, but the quality and content of these visits are more difficult to capture. In the face of these problems, as the authors of these papers recognize, empirical health production function estimates serve as a point of departure. They cannot explain *why* prenatal care increases birth weight or *why* better-educated mothers have healthier adolescents. Instead they can establish relationships which may be validated and more fully understood in subsequent more detailed and clinically oriented research.

Finally, the variety of empirical measures and concepts used in these papers to address the general issues of health production and the determinants of health levels is also of interest. While earlier economic research in this area tended to define health in general terms and to focus on one or two very broad measures of health status, the trend illustrated by these papers is toward analysis of more specific indicators of particular aspects of health. We believe that this trend is beneficial in that it explicitly recognizes the highly multidimensional nature of health and the fact that different components of overall health are not jointly determined by precisely the same set of factors and processes. More importantly, the use of more specific health measures facilitates the linking of economic research with the findings from clinical and epidemiological studies. One can reasonably hope that interdisciplinary communication is thereby fostered and that as a result the quality of economic and noneconomic research and its impact on health levels are enhanced.

REFERENCES

1. Grossman, Michael (1975) "The Correlation between Health and Schooling," in *Household Production and Consumption*, (ed.), N. Terleckyj, New York: National Bureau of Economic Research.
2. Fuchs, Victor (1980) "Time Preferences and Health: An Exploratory Study," Paper presented at the National Bureau of Economic Research Conference "Economic Aspects of Health," Palo Alto, California, July 30–31, 1980.

PART III
HEALTH IN DEVELOPMENT

CORRELATES OF LIFE EXPECTANCY IN LESS DEVELOPED COUNTRIES

Robert N. Grosse and Barbara H. Perry

ABSTRACT

Analyses were performed to investigate several hypotheses concerning the multiple determinants of levels of life expectancy in developing countries in recent decades and some possible explanation for the observed variations in amount of gain in life expectancy from the 1950s to the 1970s. The findings were significant. For level of life expectancy, the results of this present work conform by and large to results of other scholars in this area, although the present work is unique in that only developing countries were included. From the 1960s to the 1970s there was a shift in the relative importance of economic indicators and general social indicators in favor of the social indicators. During the period 1960–65, some 70 percent of the variation in levels of life expectancy was associated with per capita income and literacy rates in a ratio of about 3 to 2 in favor of the economic variable. By 1970–75 the ratio has become 6 to 1 in favor of literacy. In addition, the multivariate model showed that the sanitation variables began to appear as significant correlates of levels of life expectancy in the more recent time period, playing a larger role than level of income per capita. Work pursued as part of a separate but concurrent project explored explicitly this three-way interaction between literacy, life expectancy, and sanitation.

For change in life expectancy from 1950 through 1970, associations were quite different. Per capita income was not associated with the absolute change in life expectancy, and the associations with literacy were much smaller than earlier observed with level of life expectancy at a point in time. In the multivariate model the primary correlates with change were the sanitation variables and health personnel as represented by population per midwife. Tests for such associations with variations in amount of gain in life expectancy have not been found in other literature and comparison with other findings can therefore not be made directly. The present work suggests that lower skill levels of health manpower and activities in sanitation may be the main correlates in a multivariate model of absolute change in life expectancy.

I. INTRODUCTION

Attempts at international comparisons of levels of health and well-being are fraught with all the complexities of trying to understand any elusive phenomenon. A nation's level of health is the result of many interrelated causes. It may have been reached by methods tangible and intangible, direct and indirect, and through multiple levels of social and cultural conditions. Difficulties of definition and interpretation because of the lack of adequate data notwithstanding, in an era of increasing recognition of limited resources worldwide, it becomes ever more important to try to understand the possible reasons why levels of well-being may improve in some countries and remain the same or improve much less quickly in others. That life expectancy at birth has risen about 10 years in the past, 20 in most of the less developed countries, does not mean that simply having time elapse will bring about improvements in the mortality and morbidity in these countries that will bring them all to desired levels. Continuing improvement is likely to be retarded if the marginal productivity of some forces drops because some complementary activities do not move in balance. The parallel rises in economic levels, social well-being, and health service investments that took place in the more developed countries may not repeat themselves in any automatic fashion in the less developed countries.

It appears to be useful to try to identify the forces associated with the courses of mortality experience in less developed countries. In this paper we report on statistical correlations recently performed in an effort to understand what factors seem to determine the levels of life expectancy at birth at particular times and changes in those levels over time. It is recognized that what really is desired is a measure of the health of a population. Unfortunately there is no single measure that will encompass all aspects of health of a population. Life expectancy is chosen as the more positive measure of the mortality experience in a country and relatively more reliable for international comparisons than other measures that might have been used such as age and disease-specific mortality or morbidity. Life expectancy at two points in time was used, 1960–65 and 1970–75, as well as the absolute change from the 1950s to 1970s and the level of infant mortality in 1974.

The following report on the results of the analysis includes first a brief discussion of historical trends and empirical studies, a description of the countries and independent variables used, a section on the indications from zero-order correlations, a section on the multivariate model results, and finally discussion and conclusions.

II. HEALTH LEVELS—HISTORICAL TRENDS AND RELATIONSHIPS

Since the end of World War II, and probably extending as far back as 1930, there has been a substantial decline in mortality in the poorer countries of the world. Table 1 shows the levels and changes in life expectancy at birth from the period 1950–55 to 1970–75 by region of the world. One manifestation of this mortality decline has been labeled the population explosion, in that the large increase in the numbers of people has been associated with drops in deaths, rather than increases in fertility. There has been considerable debate about the causes of the mortality decline; some participants have argued it has been the result of increasing success in the application of communicable disease control technologies through public health activities (9); others argue that it has been the result of general economic

Table 1. Life Expectancy at Birth, 1950–55 to 1970–75

	1950–55	1970–75	Absolute Gain (years)	% Gain
Eastern Africa	34.7	43.8	9.1	26.2
Middle Africa	35.2	41.9	6.7	19.0
Western Africa	32.0	40.9	8.9	27.8
Southern Africa	43.2	50.8	7.6	17.6
Northern Africa	42.0	52.0	10.0	23.8
Eastern South Asia	40.4	50.6	10.2	25.2
Middle South Asia	38.6	48.0	9.4	24.4
Western South Asia	43.9	53.8	9.9	22.6
China	45.0	61.6	16.6	36.9
Other East Asia (exc. Japan)	48.2	61.1	12.9	26.8
Caribbean	52.9	63.1	10.2	19.3
Middle America	49.6	61.5	11.5	23.2
Tropical South America	51.9	60.5	8.6	16.6
Less developed regions	41.6	52.2	10.6	25.5

Source: United Nations, Population Studies No. 60, 1977.

improvement, particularly in the elimination of famines and improved nutrition available to the people of the less developed countries (4).

The relationship of health status to social and economic well-being has long been recognized, but argument has it that the association between economic levels and health status has weakened in recent decades, with the increasing capacity of social or health programs to affect health. George Stolnitz (9, p. 233) in a paper of the 1974 Bucharest World Population Conference argued:

> The evidence appears overwhelming that levels of living and life-style can be greatly offset or even dominated by what might be called programmed disease control, defining the last broadly to include all public health programmes, hospital plus all other medical resources, and sanitary facilities. In areas where mortality is especially high, as in Africa and parts of Asia, public health and sanitation are almost surely the main prime movers needed for achieving rapid initial change. In less developed areas with already reduced, but still high mortality, hospital and other medical facilities have tended to become the strategic main factors for further change once public health and sanitation subsystems have come into being. The same has been true of less developed areas which have begun to penetrate current or recent ranges of mortality in developed areas. In all of these cases, the onset of rapid changes, both in mortality and major causes of death, is found to be closely related with programmes instituted by governments, international agencies and other health related institutions....

In their study of Latin American mortality trends, Arriaga and Davis (1) concluded that after 1930 low- and high-income countries

> have registered a similar and very rapid increase of life expectancy at birth, regardless of the mortality level already reached. The mortality level and its change—principally the latter—can no longer be seen as dependent on economic development. There is no doubt that other factors are at work. Among them, probably the most important is the improvement of public health facilities and medical care.

Samuel Preston (6, p. 81) also argues:

> It seems to have been predominantly broad guaged public health programs of insect control, environmental sanitation, health education, and maternal and child health services that transformed the mortality picture in less developed areas....

This generality of improvement in life expectancy is found to pertain to almost all countries of the developing world since 1950. The argument over whether economic growth or social and health programs have accounted for the health improvements has by no means been conclusively decided in favor of the health programs. There is evidence in the work of Celeste Smucker (8) of the importance of socioeconomic variables over indicators of access to medical care in improvement in mortality levels.

It is puzzling to accept the stated role of health services as the transforming factor in a general trend insofar as there have been substantial differences in the resources devoted to public health programs among developing countries, often with very similar health status results. This alone implies little relationship between results and expenditures. Also there are continual reminders from WHO and developing countries that governmental health programs have been concentrated on a relatively small portion of the population, usually the urban elite.

III. EMPIRICAL STUDIES OF LIFE EXPECTANCY AND SOCIOECONOMIC WELL-BEING

Studies that have been reviewed show mixed results for answers to the question, "What has been most associated with the increases in levels of life expectancy in recent decades and throughout the twentieth century?" Simple correlations performed on data available by WHO show high correlations between such measures as medical density, per capita income, and bed/population ratio with life expectancy at birth for 68 countries (3). While warning about reservations in the quality of the data, the data nevertheless support recognition that the relationships vary greatly in magnitude between richer and poorer countries.

In a most thorough multivariate study by Preston, the change in influence of economic factors was investigated (6). His findings showed that from the period of the 1930s to the 1960s the relationships between income level and life expectancy have taken an upward shift, that there remains an upward slope to the relationship, but that by the 1960s comparable attainment in level of life expectancy cost approximately one-third more than the same level in the 1930s. In terms of multivariate analysis, Preston suggested in his early work that literacy was probably not a major contributor to levels of life expectancy (6, p. 83) but in a separate study (5) finds that indeed literacy does play a major explanatory role. The importance of education levels has also been shown to be the surest correlate of health in a recent study by Fuchs on postindustrial nations (2).

All of these studies, however, used a group of countries as the base which included developed as well as developing countries. Findings show, as referenced above in the WHO report and in the work of Preston, that relationships differ for countries based on their levels of income or other measure of development. This is represented by the often-used logarithmic transformation when using income and comparing levels of life expectancy—Figure 1 from the work of Preston illustrates this.

Other measures of level of health have also been analyzed for their associations with various socioeconomic indicators. In a study of infant and childhood mortality in four developing countries by Sloan (7) female literacy and nutrition were shown to have definite impact. Sanitation variables and variables that characterize the housing stock of the region were found to explain none of the variation in mortality rates.

For the present study, to try to maintain some homogeneity in countries

Figure 1. Relationship between life expectancy at birth (e) and per capita income for nations in the 1900s, 1930s, and 1960s (8).

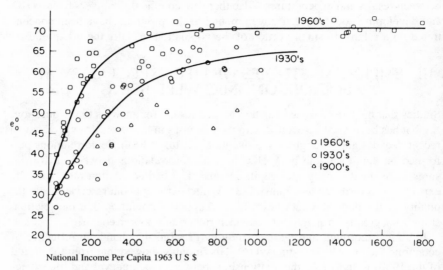

represented, an attempt was made to analyze only developing countries. The United States, Canada, Japan, Australia, New Zealand, and all of Europe were excluded.

IV. DATA ON NINETY DEVELOPING COUNTRIES

A. The Dependent Variables

The data used for the present analysis were assembled from a number of sources.[1] There was at least some information on each of 140 nations classified as developing countries. A distinction was made between those countries above or below population of 1 million with those countries under population 1 million eliminated from the correlation analysis. By including only countries larger than 1 million population it was hoped that the possible overrepresentation of atypical situations would be avoided. Appendix 1 lists the 90 countries included in the current study and their levels on the five proxy measures for level of health. The mean life expectancy for the three periods of time 1950s, 1960s, and 1970s are 42, 47, and 51 years, respectively, with a range in the 1950s of from 30 years in Angola to 66 years in Uruguay. By the 1970s, the range became the low of 36 in Bangladesh to the high of 72 in Puerto Rico. Adequate numbers of observations for infant mortality were available only for 1974, and showed an average value of 117, a range of 45 in Sri Lanka and a reported 17 in Singapore to a high of 200 in Angola.

These five indicators were taken as dependent variables and analyzed both singly and using multivariate techniques to find clues for the major areas of contribution to levels of health in these developing countries.

B. Five Groups of Independent Variables

Over 200 measures were available for use as independent variables. These covered a broad range of activities which can be hypothesized to influence the health of a country, some directly, others indirectly. Since the ultimate aim for this and probably most studies of correlates of levels of health is to assist in eventual policy decisions among alternative and often competing projects, it was felt useful to group the numerous independent variables into familiar groups of broad categories. They were grouped and analyzed as five interacting areas, these being three health input groups—health expenditures, health personnel and facilities, and sanitation activities—and two general social well-being categories—economic level indicators and social and educational status measures. The following is a general definition of the scope of each group:

- Sanitation (power and utilities usage and availability)
- Health expenditures (including various pharmaceutical imports)
- Health personnel and facilities (health manpower, hospitals, health centers)
- Economic indicators (revenues, expenditures, employment, GNP/capita, exports and imports, income distribution)
- Social indicators (includes literacy, nutrition, transportation and communication)

Appendix 2 lists all the independent variables considered, grouped into five broad categories.

V. ZERO-ORDER CORRELATIONS

For many years there has been argument over whether economic growth or social and health programs accounted for the large gains in life expectancy that were observed in developing countries. The most accepted hypothesis—that economic level is losing its historic power to determine levels of mortality—was explored, and analyses were performed to determine the relative contributions of those potentially relevant factors about which we could secure information.

Of initial interest was the relationship between economic level alone with level of life expectancy and with change in life expectancy. Simple zero-order or bivariate correlations demonstrated that in developing countries for the data we used, there was a strong and significant correlation between per capita income and level of life expectancy, but that there was no correlation between absolute *changes* in life expectancy and level of income per capita (see the accompanying figures).

Table 2.

Independent Variable	Dependent Variable	r	r²	Significance Level
Per capita income 1960	Life expectancy at birth 1970–75	.72	.52	$p < .01$
Per capita income 1970	Life expectancy at birth 1970–75	.66	.44	$p < .01$
Per capita income 1960	Life expectancy at birth 1960–65	.75	.56	$p < .01$
Log of per capita income 1960	Life expectancy at birth 1960–65	.79	.63	$p < .01$
Log of per capita income 1970	Life expectancy at birth 1970–75	.77	.59	$p < .01$
Per capita income 1960	Change in life expectancy 1950–55 to 1970–75	−.15	.02	N.S.
Per capita income 1970	Change in life expectancy 1950–55 to 1970–75	.02	.004	N.S.

Some of the variables examined and their correlations are listed in Table 2. The column headed "r" notes the correlation coefficient, and "r²" the proportion of variation explained.

Percentage changes in life expectancy and percentage changes in income were also compared. There was a significant negative correlation between level of income and percent of change in life expectancy but not between percent change in income and percent change in life expectancy.

From these correlations, we could conclude that levels of mortality at any one time are related to average levels of income, but that variations in changes in mortality over time are not.

The logarithms of income per capita were used to express the curvilinear shape of its relationship to life expectancy. That is, as higher incomes and life expectancies are examined, the ratio of income to life expectancy becomes less (a flattening of the curve on arithmetic scale). These findings conform to the suggestions of students of mortality. A further investigation was performed to examine more closely the changes in the influence of income on level of life expectancy.

A comparison of the regression lines between life expectancy in 1960 and in 1970 as compared to income per capita (in 1970 U.S. dollars) is shown in Chart 1. The slopes of the lines are different, with that for 1960 being somewhat steeper. This might indicate that in 1970 increases in life expectancy as a function of economic level required higher economic level per year added than in 1960. In 1960–65, 10 years of additional life expectancy at birth "required" a gain of $300 per capita. For 1970–75, a gain of 10 years "required" a rise of $500 per capita, a 60 percent increase.

While there is a very strong relationship between levels of life expectancy and income per capita, other variables are also associated with life expectancy.

Chart 1. Life expectancy at birth vs. per capita national income (1970 U.S. dollars)

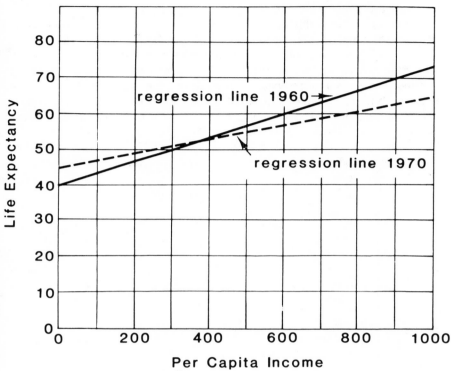

Simple correlations with variables from each of the five groups showed in fact that levels of life expectancy are correlated with a number of factors. Amongst the group of health expenditure variables several indicators were correlated with health measures, but in general not very highly. The total amount of cleansing agent imports displayed a correlation of between .62 and .65 with level of life expectancy in each of the three time periods, but no correlation with the absolute amount of change in life expectancy. The amount of pharmaceutical goods imported other than medicaments was associated with the gain in life expectancy, with $r = .33$. Since these trade data are gathered in different years, all that can really be said is that a positive association is probably affected by activities associated with higher availability of cleaning products.

The association with per capita health expenditures was less strong, with correlation coefficients of between .36 and .41 with level of life expectancy and no significant correlation with amount of change in level of life expectancy over time. Level of total health expenditure also showed positive correlation with level of life expectancy but to a lesser degree ($r = .21$ for life expectancy in the

1950s and .28 for life expectancy in the 1970s). Also, while per capita health expenditure showed no correlation with change in life expectancy, the amount of total health expenditure did, with correlation of r = .31.

With the group of health facilities and health personnel variables, the measures of population per general hospital bed had significant correlations with life expectancy, on the order of .5 to .6. Correlations with population per physician were also significant, ranging from .6 to .7. A somewhat surprising observation was the correlation observed between population per midwife and change in life expectancy from the 1950s to the 1970s. This was −.47 using the 1960 figures and −.48 using the 1970 figures. This was of the same magnitude as correlations between population per physician in 1960 and 1970, −.47 and −.44, respectively.

The utilities and sanitation variables showed very high correlations with levels

Chart 2

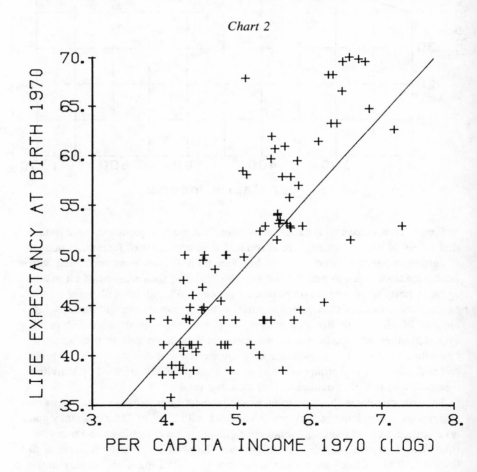

of life expectancy. For example, correlation between level of life expectancy in the 1970s and the percent of urban population with household public water taps was .72. The percent of urban total water coverage, tap or standpipe, in 1975 had a positive correlation with gain in life expectancy for 1950–55 to 1970–75, of r = .31.

Among the social indication variables, as is expected, all are positively correlated with level of life expectancy, but none so highly as the schooling and literacy measures, with correlations as high as .88. Per annum growth rate in food demand showed positive correlation with both levels of life expectancy and absolute change, but not high enough to be more significant than the associations observed with literacy and schooling indicators. A most curious observation for the nutrition components of social well-being was their general pattern of posi-

Chart 3

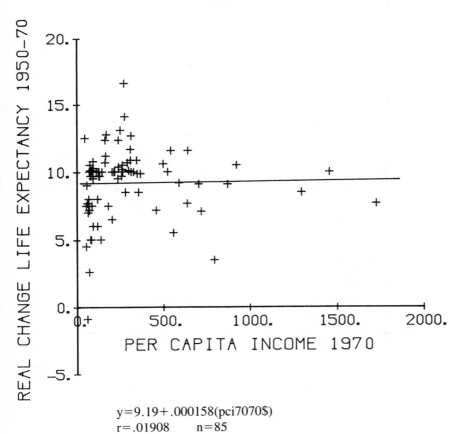

y=9.19+.000158(pci7070$)
r=.01908   n=85

tive correlation with levels of life expectancy and yet negative correlation with changes over time. This seems contrary to what logic would predict.

Plot diagrams for some of the more interesting bivariate regressions are shown in the following charts, namely, the effects of some of the health manpower and social indicators singly on changes and levels of the health measures. To ask questions about the relative magnitude of all these different correlates of life expectancy requires their joint consideration in a multivariate model. The next section addresses this issue.

Chart 4

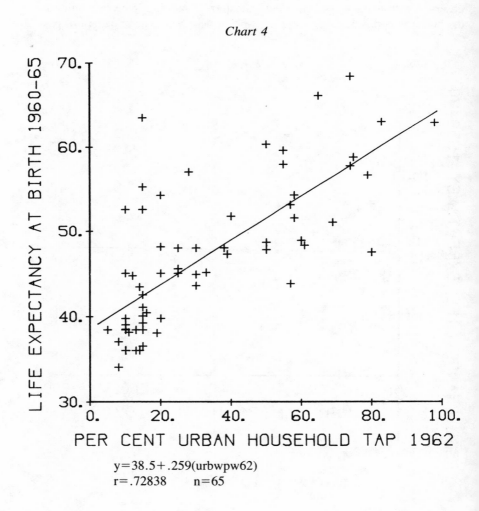

$y = 38.5 + .259(\text{urbwpw62})$
$r = .72838 \quad n = 65$

VI. MULTIPLE REGRESSION ANALYSIS

With some understanding that there are problems with the very measures we wish to associate with one another, and for the time being accepting that the U.N. measures of life expectancy are reasonable estimates, the major measurement errors are in the limited numbers of observations for the same country of the many variables of interest. We have attempted therefore to include the maximum number of countries at each step of the analyses with the intent that the final model will contain observations for as many countries as possible.

Chart 5

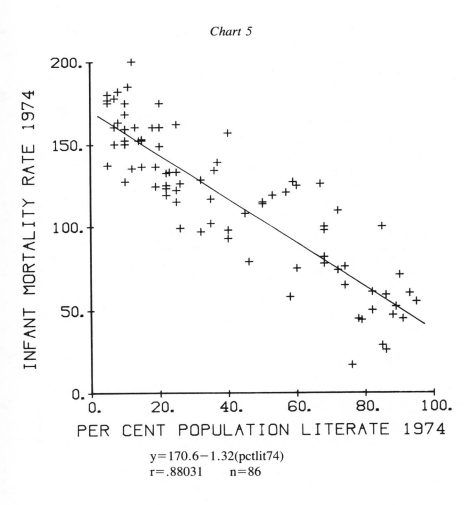

$y = 170.6 - 1.32(\text{pctlit74})$
$r = .88031 \quad n = 86$

The multivariate analysis was done by first selecting those variables which had the highest zero-order correlations with significance levels of $P \leq .05$ when compared with the individual health indicators, and then performing a two-level process of stepwise multiple regression. The most significant correlates within each of the five categories of explanatory variables were selected and the strongest correlates of all those five categories were then combined to show the relative importance of sanitation, health expenditures, health personnel and facilities, economic and social indicators in the presence of the other explanatory variables in terms of their associations with levels of expectancy, with changes in levels of life expectancy over time, and with infant mortality rates.

Chart 6

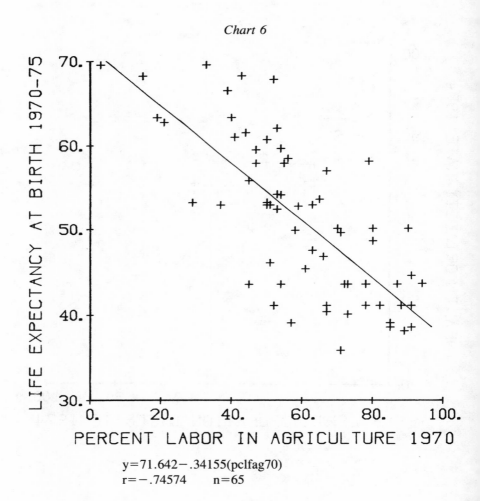

$y = 71.642 - .34155(\text{pclfag70})$
$r = -.74574 \quad n = 65$

Tables 3 through 6 illustrate the process by which representative proxies for each of the five categories of variables were selected. These tables show the results of the stepwise regressions. The starred variables are the resulting selections for the five intragroup analyses. In some cases merely combining these five groups of variables as they were selected provided a very small number of countries for which all of these data were available. Where this was the case, subsets of variables were combined (indicated by □ and #) to provide a larger number of observations for the final regression equations. An explanation of each final result follows.

Chart 7

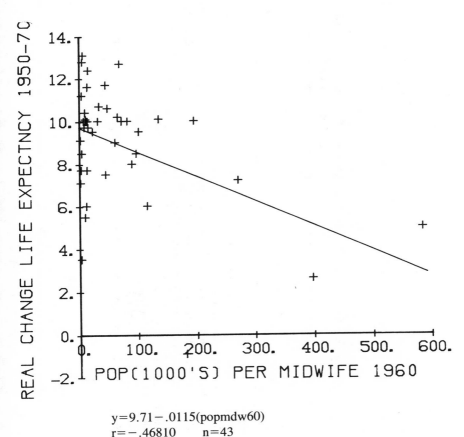

$y = 9.71 - .0115(\text{popmdw}60)$
$r = -.46810 \quad n = 43$

Table 3. Simple and Multiple Correlation Coefficients for Association with Level of Life Expectancy in the 1960–65 Period

Independent Variable	Simple Correlation Coefficient			1st level Multiple Regression (R)	2nd Level Multiple Regression (R)
Utilities (Power and Sanitation):					
5004 Per capita electrical consumption (kwh)	.750	(67)	*		
5006 Percent of urban population with household public water taps	.728	(65)	*	.830	
5011 Percent of urban population with access to standpipe	−.291	(65)		(49)	
5014 Percent change, urban access to standpipes	−.262	(57)	*		
5016 Urban total water coverage (tap or standpipe)	.648	(65)			
Health Expenditures:					
3079 Total expenditure on health	.262	(86)	*	.355	
3081 Per capita health expenditure	.361	(85)		(54)	
8001 S.I.T.C. 541 total imports (Medicinal and pharmaceutical products)	.408	(56)			

Health Personnel and Facilities:
6000 Total number of hospital beds	.286	(58)				
6008 Population per general hospital	−.346	(50)				
7001 Population per doctor	−.551	(64)	*			

Economic Variables:
3006 Per capita national income	.750	(59)	*	.564	(44)	*
3020 Percent of labor force in agriculture	−.789	(58)	*			
3022 Percent of G.D.P. derived from agriculture	−.461	(47)		.793	(35)	*
3045 Total Imports and N.F.S.	.280	(54)				
3049 Merchandise Exports	.307	(54)				
3053 Total exports of goods and N.F.S.	.362	(54)				

Social Indicators:
2006 Primary school enrollment as a percent of children 6–11 years old	.532	(74)	*					
2008 Secondary school enrollment as a percent of children 12–17 years old	.745	(77)	*	.915	(52)	*	.915	(29)
2012 Radios per 1,000 population	.525	(78)						
2014 Automobiles per 1,000 population	.690	(63)	*				.841	(38)
2019 Per capita available kilocalories per day	.564	(76)						
2022 Percent of required calories	.317	(65)						

Table 4. Simple and Multiple Correlation Coefficients for Association with Level of Live Expectancy in the 1970–75 Period

	Independent Variable	Simple Correlation Coefficient			1st Level Multiple Regression (R)		2nd Level Multiple Regression (R)
Utilities (Power and Sanitation):							
5004	Per capita electrical consumption (kwh)	.717	(67)	*			
5007	Percent of urban population with household public water taps	.612	(74)		.855	*	
5008	Percent of urban population with household public water taps	.721	(56)	* #	(38)	□	
5017	Urban total water coverage (tap or standpipe)	.278	(73)				
5033	Percent rural population with some form of excreta	.399	(58)				
5024	Percent of total population with water	.563	(69)				
5027	Percent urban population with in-house public excreta disposal system	.483	(62)				
Health Expenditures:							
3079	Total expenditure on health	.276	(86)	*	.486		
3081	Per capita health expenditure	.362	(85)	#	(51)		
8001	S.I.T.C. 541 total imports (Medicinal and pharmaceutical products)	.407	(56)	* □		*	.970
8009	S.I.T.C. 554 total imports (Soap, cleansing & polishing preparations)	.478	(53)	*			(30)

Health Personnel and Facilities:					
6000 Total number of hospital beds	.275 (58)		□		.948 (47)
6008 Population per general hospital	−.361 (50)		#		.857 (63)
6019 Population per hospital bed	−.369 (85)			.719 (21)	
7004 Population per physician	−.676 (85) * □	#			
7035 Population per midwife	−.281 (51)				
7037 Population per nurse	−.303 (69)				
7040 Population per lab technician, lab assistant	−.266 (63)				
Economic Variables:					
3007 Per capita national income	.661 (85)		□	.746 * (65)	#
3021 Percent of labor force in agriculture	−.746 (65) * □	#			
Social Indicators:					
2004 Percent of population literate	.881 (85) * □	#			
2005 Percent of children 5–19 in school	.772 (85)			.910 * (66)	#
2013 Radios per 1,000 population	.597 (67)				
2021 Per capita available kilocalories per day	.642 (85) *				

Table 5. Simple and Multiple Correlation Coefficients for Association with Absolute Change in Life Expectancy, 1950–55 to 1970–75

Independent Variable	Simple Correlation Coefficient		1st Level Multiple Regression (R)	2nd Level Multiple Regression (R)
Utilities (Power and Sanitation):				
5000 Percent of dwelling units without piped water	.532 (33)	*	.569 (21) ⎫	
5018 Urban total water coverage (tap or standpipe)	.308 (56)	# □	⎬ □	
Health Expenditures:				
3079 Total expenditure on health	.310 (86)	# □	.328 (50) ⎫	
8008 S.I.T.C. 541.9 total imports (Pharmaceutical goods other than medicaments)	.324 (51)	*	⎬	
Health Personnel and Facilities:				
6019 Population per hospital bed	−.243 (85)		⎫	# .527 (26)
7004 Population per physician	−.330 (85)		⎪	
7008 Population per nurse	−.273 (52)		⎬ .681 (25) #	
7010 Population per midwife	−.468 (43)	*	⎪	□ .317 (53)
7035 Population per midwife	−.304 (51)		⎪	
7037 Population per nurse	−.263 (69)		⎭	
Economic Variables:				
3078 Total expenditure on education	.283 (86)	* # □	.283 (86) ⎫	
Social Indicators:				
2002 Percent of population literate	.247 (64)	*	⎪	
2003 Percent of population literate	.220 (82)	# □	⎬ .386 (56)	
2006 Primary school enrollment as a percent of children 6–11 years old	.345 (74)		⎪	
2009 Secondary school enrollment as a percent of children 12–17 years old	.260 (79)		⎭	

Table 6. Simple and Multiple Correlation Coefficients for Association with Level of Infant Mortality, 1974

Independent Variable	Simple Correlation Coefficient			1st Level Multiple Regression (R)	2nd Level Multiple Regression (R)

Utilities (Power and Sanitation):

5005	Per capita electrical consumption (kwh)	−.696	(67)	*		
5007	Percent of urban population with household public water taps	−.585	(74)	□		
5012	Percent urban population with access to standpipe	.337	(70)		.715 (39)	
5016	Urban total water coverage (tap or standpipe)	−.613	(65)			
5017	Urban total water coverage (tap or standpipe)	−.290	(73)			
5024	Percent of total population with water	−.583	(69)			
5027	Percent urban population with in-house public excreta disposal system	−.488	(62)	*		
5033	Percent rural population with some form of excreta	−.312	(58)			

Health Expenditures:

3079	Total expenditure on health	−.242	(86)		.348 (52)	
3081	Per capita health expenditure	−.282	(86)			* .867 (31)
8009	S.I.T.C. 554 total imports (Soap, cleansing and polishing preparations)	−.348	(52)	* □		

Health Personnel and Facilities:

6008	Population per general hospital	.363	(49)			
6019	Population per hospital bed	.290	(86)		.556 (21)	□ .866 (42)
7031	Population per doctor	.351	(75)	*		
7035	Population per midwife	.283	(51)			
7010	Population per midwife	.397	(40)			

(continued)

Table 6. Continued

	Independent Variable	Simple Correlation Coefficient	1st Level Multiple Regression (R)	2nd Level Multiple Regression (R)
Economic Variables:				
3007	Per capita national income	−.543 (82)		
3021	Percent of labor force in agriculture	.664 (65) * ☐		
3023	Percent of G.D.P. derived from agriculture	.461 (65)	.679 (45)	
3043	Merchandise imports	−.292 (58)		
3046	Total imports and N.F.S.	−.430 (71)		
3054	Total imports of goods and N.F.S.	−.358 (71)		
Social Indicators:				
2003	Percent of population literate, 1970	−.793 (81)		
2004	Percent of population literate, 1974	−.880 (86) * ☐		
2005	Percent of children 5–19 in school	−.752 (86) *	.903 (65) * ☐	
2013	Radios per 1,000 population	−.616 (67) *		
2015	Automobiles per 1,000 population	−.561 (67)		
2021	Per capita available kilocalories per day	−.576 (86)		
2025	Percent of required calories	−.613 (86)		

A. Level of Life Expectancy in 1960–65 and in 1970–75

For level of life expectancy in the 1960–65 period, in the 49 countries which had recorded information on the five selected sanitation variables (percent of population with access to electricity plus the four water supply indicators), once the effects of the three starred variables (those representing access to electricity, percent with household taps, and percent increase in urban access to standpipes) were accounted for in the regression model, no additional explanatory power then came from any of the remaining two sanitation indicators: total urban water coverage or urban access to standpipes. This is not to say that the latter two variables are insignificant as far as their impact on life expectancy is concerned. It is rather to say, first, that they most definitely *do* have relationships with life expectancy (this was shown by their significant zero-order correlations), but second, that the direct relationship may well be through other variables with which they are also related.

The combination of electricity, urban household taps, and percentage increase in standpipes yielded a multiple correlation coefficient (r) of .83, and thus in the absence of any of the variables from the other four groups, these were associated with 70% (r^2) of the variation in life expectancy in 1960–65. Because of this, subsequent selection of sanitation proxies came only from this group of three, and in fact a final stepwise regression was performed with only the urban house tap variable consistent with our efforts to recapture as many countries in the final regression as possible. A similar process was performed for each of the other four categories of independent variables. The final regression model selected for level of life expectancy in 1960 is shown below.

Using as the model for determinants of life expectancy in 1960,

$$LE1960 = B_0 + B_1(\text{Sanitation}) + B_2(\text{H.exp.}) + B_3(\text{H.fac.}) + B_4(\text{Eco.}) + B_5(\text{Social})$$

% urban household taps	Total expenditure on health	Population per physician	Per capita income	% Primary school enrollment

the stepwise regression results were

$$LE1960 = 35.36 + (\text{Sanitation}) + (\text{H.exp.}) + (\text{H.fac.}) + .027(\text{Eco.}) + .126(\text{Social}),$$
$\qquad\qquad\quad$ [2.000, 17.7]* $\quad$ n.s. $\qquad\quad$ n.s. $\qquad\quad$ n.s. $\quad$ [.0035, 7.58] [.0248, 5.07]

where $\quad$ n = 38
$\quad R^2$ at first step = .49
$\quad R^2$ at final step = .71 (F = 42.4, p = .001, s.e. = 4.64)
partial r^2 econ-social = .62
partial r^2 social-econ = .42.

As explanations of level of life expectancy in the period 1960–65, the economic (as represented by per capita national income) and social (as indicated by

*[s.e., t]

schooling) indicators predominate as the primary correlates. After these were added in stepwise fashion to the multiple regression model for life expectancy for the 38 countries, no *additional* explanatory force could be gained by the sanitation, health expenditures, and health personnel and facilities variables. That is *not* to say these other variables are not correlated, or indeed important in life expectancy. It means that due to the structure of social and governmental systems, the primary or most significant route to life expectancy improvement may have been first by way of improvements in social and economic patterns. It is no coincidence that with these improvements come changes in sanitation, health expenditures, and availability of health personnel and facilities. The partial correlation coefficients further show that in this period the economic variable was a stronger correlate than the social variable considering each in the presence of the other.

In 1970–75, the impact of economic variables has diminished, and sanitation plays a larger role in the presence of the other variables (Table 2). It is important to recognize that changes in these areas alone, however, may not bring about the desired or expected changes in life expectancy. The model for determinants of life expectancy in the 1970–75 period gave the following result:

$$LE1970 = B_0 + B_1(\text{Sanitation}) + B_2(\text{H.exp.}) + B_3(\text{H.fac.}) + B_4(\text{Eco.}) + B_5(\text{Social})$$

| % urban household taps | Total expenditure on health | Population per physician | % labor force in agriculture | % literate |

$$LE1970 = 44.01 + .103(\text{Sanitation}) - .083(\text{Economic}) + .180(\text{Social}),$$
$$[3.106, 14.2] \quad [.0197, 5.21] \quad [.0336, -2.49] \quad [.0231, 7.81]$$

where $n = 47$
R^2 at first step (social) = .78
R^2 at second step (social, sani) = .88
R^2 at final step (social, sani, econ) = .90 (F = 126.97, p = .000, s.e. = 2.99)
partial r^2 (soc, sani, econ) = .59
partial r^2 (sani, econ, soc) = .39
partial r^2 (econ, sani, soc) = .13

A total of 90 percent of the variation in life expectancy in the period 1970–75 is explained by the three variables literacy, urban water availability, and percent of labor force in agriculture, in that order of importance. After the inclusion of these three indicators, health expenditures and health facilities and personnel variables did not add to the amount of variation explained. Total expenditure on health was used as an independent variable rather than per capita because for the 51 countries with all four items listed under health expenditures in Table 2, total health expenditure was a stronger correlate. Analysis with per capita expenditure did not change the result of the final model. Percent of labor force in agriculture was used as the economic indicator because it was a stronger correlate than per capita income. Had per capita income been used, the contribution of the economic indicator would have been even less.

If we modify the model slightly by substituting electrical consumption for percent of urban population with public water taps as the sanitation and utilities variable, at the cost of using a slightly less powerful sanitation component, we gain 16 additional countries in the sample. The multivariate results then become:

$$LE1970 = 49.83 - .00009(H.fac.) - .106(Eco.) + .215(Social),$$
$$[2.679, 18.60] \quad [.000037, -2.35] \quad [.0332, -3.18] \quad [.0241, 8.94]$$

where
$$n = 63$$
$$R^2 \text{ at first step (social)} = .80$$
$$R^2 \text{ at second step (social + econ)} = .84$$
$$R^2 \text{ at final step (social + econ + H.fac.)} = .86 \ (F = 118.02, p = .000,$$
$$\text{s.e.} = 3.63)$$
$$\text{partial } r^2 \text{ (social-econ, H.fac.)} = .58$$
$$\text{partial } r^2 \text{ (econ-social, H.fac.)} = .15$$
$$\text{partial } r^2 \text{ (H.fac.-social, econ)} = .08$$

Still literacy dominates, with the economic indicator and health personnel variables playing significant, though smaller, roles.

The significant correlation of the sanitation variable with level of life expectancy in 1970–75 prompted a further look at the three-way interrelationship between sanitation, literacy, and level of life expectancy. Figure 2 shows how controlling for literacy can show the effect of higher versus lower sanitation investments with regard to associated levels of health.

B. Absolute Change in Life Expectancy from 1950 to 1970

In a similar manner, health personnel (as represented by population per midwife) and sanitation indicators (as represented by urban water coverage) dominate as explanatory variables for variations in *changes* in life expectancy from 1950 to 1970, though able to explain a much smaller percentage of the variation, 28 and 10 percent, respectively. It is true that most of the variation in *change* in life expectancy remains unexplained by variables included in the multivariate model, but what is also true is that a significant part of the variation is correlated with the sanitation indicator and the lower-level health manpower indicator, population per midwife. Per capita health expenditures were not at all correlated, nor is level of per capita national income, and once the sanitation or health personnel variable is in the model no additional explanatory power comes from the already weak representatives of the other variables.

The first model for change in life expectancy is:

Figure 2. Interaction between literacy, life expectancy, and level of sanitation: high—percent urban total water coverage > 55%; low—percent urban total water coverage ≤ 55%.

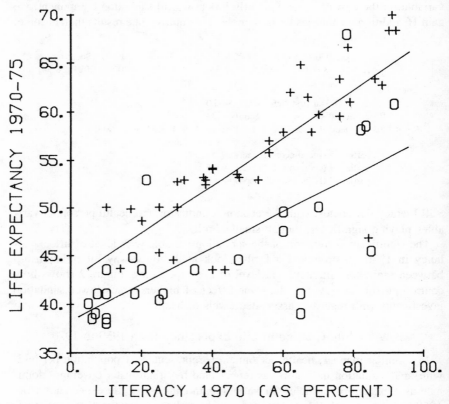

* High: % Urban Total Water Coverage > 55%
   Low: % Urban Total Water Coverage ≤ 55%
   + = countries with high* levels of sanitation
   o = countries with low* levels of sanitation

Source: Hillel Shuval, robert Tilden, and Robert Grosse, "The Health Benefits of Sanitation Investments: A Threshold-Saturation Theory," University of Michigan, School of Public Health, May 1979.

LE1950–1970 = B_0 + B_1(Sanitation) + B_2(H.exp.) + B_3(H.fac.) + B_4(Eco.) + B_5(Social)

| Urban water coverage | Total expenditure on health | Population per midwife | Education expenditures | % literate |

LE1950–1970 = 9.935 − .000014(H.fac.),
[.667, 14.90] [.000004, −3.04]

where n = 26 and R^2 at first step (population per midwife) = .28 (F = 9.24, p = .006, s.e. = 2.97).

Partial correlation of each of the other four were all less than .01 and did not increase the overall amount of variation explained. As was done with level of life expectancy, subsequent analyses were performed to expand the number of countries included as much as possible. By taking the weaker health personnel variable, population per physician, the sample size is increased to 53 and the following results are obtained:

LE1950–1970 = B_0 + B_1(Sanitation) + B_2(H.exp.) + B_3(H.fac.) + B_4(Eco.) + B_5(Social)

| Urban water coverage | Total expenditures on health | Population per physician | Education expenditures | % literate |

LE1950–1970 = 6.205 + .040(Sanitation),
[1.35, 4.59] [.017, 2.39]

where n = 53 and R^2 at first step (% total urban water coverage) = .10 (F = 5.72, p = .021, s.e. = 2.76).

Again, partial correlations of each of the health expenditure, health facilities, economic, and social variables were all under .01 and not significant. It is interesting, however, that the next most significant variable among the remaining four would have been population per midwife.

In conclusion, for the analysis looking at changes in life expectancy over time, economic indicators, health expenditures, and social indicators play less of a role in terms of showing correlation in this multivariate model. The primary indicators seem to be low-level health technology (population per midwife) or levels of sanitation (urban water coverage).

C. Levels of Infant Mortality

In the multivariate analysis of levels of infant mortality the social indicator variables, specifically percent of population literate, dominated as the chief determinant. While no direct causal link can be proven by this high association, it is certainly plausible that the various forms of communication of which literacy levels may be a surrogate are playing a role here.

The regression equation resulting from this stepwise linear regression process is

IMR1974 = B_0 + B_1(Sanitation) + B_2(H.exp.) + B_3(H.fac.) + B_4(Eco.) + B_5(Social)

| | Per capita electrical consumption | Pharmaceutical imports | Population per physician | % labor force in agriculture | % literate |

IMR1974 = 170.24 − 1.355(Social indicator),
[6.61, 25.75]   [.124, −10.96]

where n = 42 and R^2 at first step (% literate) = .75 (F = 120.23, p = .000, s.e. = 21.96).

The other independent variables did not add to the 75 percent variation already explained by literacy level. These results very strongly support the notion that communication and education activities do have an influence on the level of well-being in a region. Though the findings show literacy dominates in the multivariate model, the high associations between IMR and other communication surrogates such as radios per 1000 population also suggest that the communication process broadly has an influence on the health behavior of populations.

Female literacy has been shown in other studies to be highly correlated with childhood mortality, so the present results are consistent with expectations. Some of the causal links hypothesized between literacy and childhood mortality (7) are the following:

1. Increased awareness of personal hygiene;
2. The propensity of the literate to seek medical care from scientific rather than from folk sources;
3. Cognitive development enabling the consequences of particular acts to be anticipated, particularly those associated with child care; and
4. Greater efficiency in consumption, that is, the literate pay lower prices for the goods and services they purchase and thus secure more goods and services from a fixed amount of income.

VII. DISCUSSION AND CONCLUSIONS

The problem of determining what "explains" mortality and changes in mortality is similar in concept to determining the production function of an industrial or agricultural product. That is, many inputs, resources, and enviornmental conditions are necessary. At any given level of output, all the factors of production make contributions. The question of whether to increase the use of some or decrease the use of others is answered by testing the effects on changes in output of small changes in one or another of the inputs. It is essential to keep in mind that the marginal productivity of any input is related closely to the simultaneous utilization of other factors as well. The relative costs of inputs are significant considerations which must be looked at in relation to outputs as well as the physical relationships. Thus there is a "technical production function" and relative prices of inputs to be considered. In our current work we have sought the "social production functions" that relate actions in society to the desired outputs, in this case represented by gains in life expectancy.

The limited explanatory power of income per capita and its weakening strength have been noted by many writers such as Stolnitz, Preston, and Arriaga and Davis. Most of them asserted that the changes in society which were affecting mortality favorably were disease control programs, medical technology, and the like. While our calculations support the idea that income per capita explains only a fraction of life expectancy variations and is lessening in power (in our data from 1960 to 1970; Preston showed this for 1930 to 1960; Arriaga and Davis saw a turning point in Latin America in the 1930s), the health inputs measured—expenditures, personnel, and facilities—had much less statistical significance in explaining variations in levels of life expectancy. It is the set of social factors, among which adult literacy is the strongest, that explains the largest part of life expectancy variations, with sanitation investments also playing a significant role.

APPENDIX I

Ninety Developing Countries: Life Expectancy at Three Points in Time and Infant Mortality Rates in 1974

Country	LIFX5055	LIFX6065	LIFX7075	RCLX5070	IMR74
West Africa					
Benin	31.3	36.0	41.0	9.7	185.0
Ghana	34.0	39.0	43.5	9.5	133.0
Guinea	31.3	36.0	41.0	9.7	175.0
Ivory Coast	33.5	38.4	43.5	10.0	160.0
Liberia	34.5	38.5	43.5	9.0	159.0
Mali	33.5	36.0	38.0	4.5	168.0
Mauritania	33.5	38.4	38.5	5.0	137.0
Niger	33.5	38.4	38.5	5.0	175.0
Nigeria	31.3	36.0	41.0	9.7	162.0
Senegal	33.5	38.0	40.0	6.5	159.0
Sierra Leone	33.5	38.5	43.5	10.0	136.0
Togo	31.3	36.0	41.0	9.7	127.0
Upper Volta	31.0	34.0	38.0	7.0	180.0
East Africa					
Burundi	31.3	36.0	39.0	7.7	150.0

(continued)

Appendix I (Continued)

Country	LIFX5055	LIFX6065	LIFX7075	RCLX5070	IMR74
Ethiopia	31.0	36.5	39.0	8.0	178.0
Kenya	40.0	45.0	50.0	10.0	115.0
Madagascar	33.5	38.4	43.5	10.0	102.0
Malawi	33.5	36.5	41.0	7.5	119.0
Mozambique	33.5	38.5	43.5	10.0	150.0
Rwanda	33.5	38.5	41.0	7.5	133.0
Somalia	33.5	36.8	41.0	7.5	177.0
Rhodesia	41.8	46.8	51.5	9.7	122.0
Uganda	40.0	45.0	50.0	10.0	160.0
Tanzania	34.2	39.2	44.5	10.3	160.0
Zambia	36.0	41.0	44.5	8.5	157.0
North Africa					
Algeria	43.1	48.1	53.2	10.1	126.0
Egypt	42.4	47.4	52.4	10.0	98.0
Libya	42.9	47.9	52.9	10.0	125.0
Morocco	42.9	47.9	52.9	10.0	149.0
Sudan	38.6	43.6	48.6	10.0	136.0
Tunisia	43.6	48.6	54.1	10.5	128.0
Central Africa					
Angola	30.0	34.0	38.5	8.5	200.0
Central African Empire	33.0	37.0	41.0	8.0	163.0
Chad	31.3	36.0	38.5	7.2	160.0
Congo	33.5	38.4	43.5	10.0	175.0
Cameroon	33.5	38.4	41.0	7.5	135.0
Zaire	38.5	40.0	43.5	5.0	160.0
South Africa					
Lesotho	35.9	40.9	46.0	10.1	114.0
South Africa	44.4	48.6	51.5	7.1	117.0
Tropical America					
Bolivia	40.8	43.8	46.8	6.0	108.0
Brazil	54.2	57.9	61.4	7.2	82.0
Colombia	50.2	56.6	60.9	10.7	76.0
Equador	47.2	54.2	59.6	12.4	78.0
Paraguay	51.5	57.0	61.9	10.4	65.0
Peru	44.8	51.0	55.7	10.9	110.0
Venezuela	54.2	60.2	64.7	10.5	50.0
Middle America					
Costa Rica	58.2	62.8	68.2	10.0	52.0
El Salvador	43.7	51.5	57.8	14.1	58.0

Appendix I (Continued)

Country	LIFX5055	LIFX6065	LIFX7075	RCLX5070	IMR74
Guatemala	41.2	47.2	52.9	11.7	79.0
Honduras	36.9	45.1	53.5	16.6	115.0
Mexico	51.6	59.5	63.2	11.6	61.0
Nicaragua	43.0	47.9	52.9	9.9	121.0
Panama	58.8	62.9	66.5	7.7	44.0
Temperate America					
Argentina	62.7	66.0	68.2	5.5	60.0
Chile	54.1	57.7	62.6	8.5	71.0
Uruguay	66.3	68.3	69.8	3.5	45.0
Caribbean					
Cuba	58.8	65.1	69.8	11.0	29.0
Dominican Republic	45.1	53.0	57.8	12.7	98.0
Haiti	40.5	45.5	50.0	9.5	150.0
Jamaica	57.9	65.8	69.5	11.6	26.0
Puerto Rico	64.4	69.5	72.1	7.7	
East Asia					
People's Republic China	45.0	55.5	61.6	16.6	55.0
Hong Kong	60.9	66.1	70.0	9.1	
Korea (DR)	47.5	55.2	60.6	13.1	100.0
Korea (R)	47.5	55.2	60.6	13.1	47.0
Middle Asia					
Afghanistan	30.2	35.3	40.3	10.1	182.0
Bangladesh	36.7	40.8	35.8	−.9	132.0
Bhutan	31.1	38.1	43.6	12.5	
India	38.7	44.7	49.5	10.8	134.0
Iran					139.0
Nepal	33.1	38.1	43.6	10.5	152.0
Pakistan	39.1	44.9	49.8	10.7	124.0
Sri Lanka	56.6	63.5	67.8	11.2	45.0
South East Asia					
Burma	40.0	45.0	50.0	10.0	126.0
Kampuchea	39.4	43.4	45.4	6.0	127.0
Indonesia	37.5	42.5	47.5	10.0	125.0
Laos	37.8	40.4	40.4	2.6	123.0
Malaysia	48.5	54.2	59.4	10.9	75.0
Philippines	46.0	52.5	58.4	12.4	74.0
Singapore	60.4	65.8	69.5	9.1	17.0

(*continued*)

Appendix I (Continued)

Country	LIFX5055	LIFX6065	LIFX7075	RCLX5070	IMR74
Viet Nam	37.8	41.7	44.6	6.8	100.0
Thailand	45.2	52.5	58.0	12.8	
West Asia					
Iraq	42.7	47.7	52.7	10.0	99.0
Jordan	43.2	48.2	53.2	10.0	97.0
Lebanon	54.0	58.7	63.2	9.2	59.0
Saudi-Arabia	34.7	39.7	45.3	10.6	153.0
Syria	43.8	48.8	54.0	10.2	93.0
Turkey	47.0	51.7	56.9	9.9	119.0
Yemen (R)	34.7	39.7	44.8	10.1	152.0
Yemen (PR)	34.7	39.7	44.8	10.1	152.0

APPENDIX II

Five Groups of Independent Variables

I. HEALTH EXPENDITURES

HLTH 65	14	Health expenditure*	1965
HLTH 70	77	Health expenditure*	1970
HLTH 73	54	Health expenditure*	1973
EXHLM$US	103	Total expenditure on health*	1974
XPCHLT74	102	Per capita health expenditure*	1974
SITC541	72	S.I.T.C. 541 total imports med and pharm products	
SITC5411	35	S.I.T.C. 541.1 total imports vitamins	
SITC5413	39	S.I.T.C. 541.3 total imports antibiotics	
SITC5414	27	S.I.T.C. 541.4 total imports vegetable alkaloids	
SITC5415	27	S.I.T.C. 541.5 total imports hormones (insulin, etc.)	
SITC5416	50	S.I.T.C. 541.6 total imports antisera, vaccines	
SITC5417	68	S.I.T.C. 541.7 total imports medicaments	
SITC5419	62	S.I.T.C. 541.9 total imports pharmaceutical goods (other than medicaments)	
SITC554	69	S.I.T.C. 554 total imports soap, cleansing and polishing preparations	
SITC5541	51	S.I.T.C. 5541 total imports bar soaps and soap products	
SITC5542	56	S.I.T.C. 5542 total imports cleansing agents	

II. HEALTH FACILITIES AND PERSONNEL

TOTBDS60	79	Total number of hospital beds	1960
R HOSP70	36	Number of rural hospitals	1970
POPRH70	36	Population per rural hospital	1970
H CNTR60	23	Number of health centers	1960
POPHCT60	23	Population per health center	1960
H CNTR70	33	Number of health centers	1970

Appendix II—(Continued)

POPHCT70	33	Population per health center	1970
LABTEC A	105	Number of auxiliary health personnel and other	
TRDMWDRA	24	Number of traditional midwives, doctors, etc	
G HOSP60	69	Number of general hospitals	1960
POPGHP60	69	Population per general hospital	1960
G HOSP70	55	Number of general hospitals	1970
POPGHP70	55	Population per general hospital	1970
RHBDS70	33	Number of rural hospital beds	1970
POPRHB70	33	Population per rural hospital bed	1970
HCBDS70	23	Number of health center beds	1970
POPHCB70	23	Population per health center bed	1970
GH13DS60	59	Number of general hospital beds	1960
POPGHB60	58	Population per general hospital bed	1960
GHBDS70	54	Number of general hospital beds	1970
POPGHB70	54	Population per general hospital bed	1970
POPBED74	102	Population per hospital bed	1974
DOCTOR60	93	Number of doctors	1960
POPDR60	92	Population per doctor	1960
DOCTOR70	55	Number of doctors	1970
POPDR70	55	Population per doctor	1970
POPPHYS	102	Population per physician	1974
NURSE 60	77	Number of nurses	1960
POPNRS60	77	Population per nurse	1960
NURSE 70	52	Number of nurses	1970
POPNRS70	52	Population per nurse	1970
MIDWIF60	62	Number of midwives	1960
POPMDW60	61	Population per midwife	1960
MIDWIF70	44	Number of midwives	1970
POPMDW70	44	Population per midwife	1970
PHARM 60	61	Number of pharmacists	1960
POPHN60	60	Population per pharmacist	1960
PHARM 70	50	Number of pharmacists	1970
POPPHM70	50	Population per pharmacist	1970

III. SANITATION ACTIVITIES AND UTILITIES

PCWOPC60	35	Percent of dwelling units without piped water	1960
PCWOPW70	30	Percent of dwelling units without piped water	1970
PCWELEC60	31	Percent of dwellings with electricity	1960
PCWELEC70	27	Percent of dwellings with electricity	1970
ELEC C60	79	Per capita electrical consumption (kwh)	1960
ELEC C70	81	Per capita electrical consumption (kwh)	1970
UWB WH62	66	Percent of urban population with household public water taps	1962
UBW WH70	88	Percent of urban population with household public water taps	1970
UBW WH75	65	Percent of urban population with household public water taps	1975
CUWPW627	61	Percent change, urban population with household water taps	1962–70

(*continued*)

Appendix II—(Continued)

CUWPW705	63	Percent change, urban population with household water taps	1970–75
UBW PS62	66	Percent urban population with access to standpipe	1962
UBW SP70	85	Percent urban population with access to standpipe	1970
UBW SP75	64	Percent urban population with access to standpipe	1975
CUWSP627	58	Percent change, urban access to standpipes	1962–70
CUWSP705	61	Percent change, urban access to standpipes	1970–75
UBTOTW62	66	Urban total water coverage (tap or standpipe)	1962
UBTOTW70	87	Urban total water coverage (tap or standpipe)	1970
URTOTW75	67	Urban total water coverage (tap or standpipe)	1975
CUTWW627	60	Percent change, urban total water coverage	1962–70
CUTWW705	63	Percent change, urban total water coverage	1970–75
RURACW70	82	Percent rural population with access to water	1970
RURACW75	65	Percent rural population with access to water	1975
CRWAW705	55	Percent change, rural with access to water	1970–75
UANDRW70	82	Percent of total population with water	1970
UANDRW75	63	Percent of total population with water	1975
CTOTW705	58	Percent change, total population with water	1970–75
UBWSWG70	70	Percent urban population with in-house public excreta disposal system	1970
UBWSWG75	62	Percent urban population with in-house public excreta disposal system	1975
CUWSW705	49	Percent change, urban population with in-house excreta disposal systems	1970–75
UBTOTS70	65	Percent urban population with some form of excreta disposal system (in-house, privies, pots, etc)	1970
UBTOTS75	53	Percent urban population with some form of excreta disposal system (in-house, privies, pots, etc)	1975
CUTSG705	47	Percent change, urban population with some form of excreta disposal	1970–75
RURWSG70	65	Percent rural population with some form of excreta disposal	1970
RURWSG75	56	Percent rural population with some form of excreta disposal	1975
CRWSG705	38	Percent change, rural with some form of excreta disposal	1970–75
URTOTS70	61	Total population (urban and rural) with some form of sewage system	1970
URTOTS75	51	Total population (urban and rural) with some form of sewage system	1975
CURSG705	44	Percent change, total population with sewage systems	1970–75

IV. SOCIAL INDICATORS

PCTLIT40	17	Percent of population literate	1940
PCTLIT60	62	Percent of population literate	1960
PCTLIT70	85	Percent of population literate	1970
PCTLIT7A	91	Percent of population literate	1970
PCTLIT74	102	Percent of population literate	1974
PCINSCHL	102	Percent of children 5–19 in school	1974

Appendix II—(Continued)

PRSCHL60	87	Primary school enrollment as a percent of children 6–11 years old	1960
PRSCHL70	106	Primary school enrollment as a percent of children 6–11 years old	1970
SDSCHL60	91	Secondary school enrollment as a percent of children 12–17 years old	1960
SDSCHL70	112	Secondary school enrollment as a percent of children 12–17 years old	1970
NPERRM60	45	Average number of persons per room	1960
NPERRM70	48	Average number of persons per room	1970
RADIO 60	125	Radios per 1,000 population	1960
RADIO 70	79	Radios per 1,000 population	1970
CARS 60	75	Automobiles per 1,000 population	1960
CARS 70	79	Automobiles per 1,000 population	1970
QOFLI76	100	Quality of life index (0–100)	1976
POLFRI76	100	Political freedom index (0–100)	1976
CAL 40	13	Per capita available kilocalories per day	1940
CAL 65	86	Per capita available kilocalories per day	1965
CAL 70	91	Per capita available kilocalories per day	1970
CAL 74	100	Per capita available kilocalories per day	1974
PCTCAL60	73	Percent of required calories	1960
PCTCAL65	86	Percent of required calories	1965
PCTCAL70	76	Percent of required calories	1970
PCTCAL74	100	Percent of required calories	1974
PROT 60	74	Daily protein consumption in grams	1960
PROT 65	86	Daily protein consumption in grams	1965
PROT 70	76	Daily protein consumption in grams	1970
PCPTA 60	41	Percent of protein from animal and pulses	1960
PCPTA 70	76	Percent of protein from animal and pulses	1970
FPPTGR65	86	Food production percent growth rate	1965
FDPAGR65	79	Food demand per annum growth rate	1965
GINI60	27	Gini coefficient, early 1960s (1960–63)	
PCI 2060	27	Percent of income held by bottom 20 percent	
PCI 9060	27	Percent of income held by top 10 percent	
GINI65	18	Gini coefficient, mid-late 1960s (1966–69)	
PCI 2065	18	Percent of income held by bottom 20 percent	
PCI 9065	18	Percent of income held by top 10 percent	
GINI70	23	Gini coefficient, early 1970's (1970–73)	
PCI 2070	23	Percent of income held by bottom 20 percent	
PCI 9070	23	Percent of income held by top 10 percent	
PCTI2060	33	Percent of income held by bottom 20 percent	1960
PCTI2070	32	Percent of income held by bottom 20 percent	1970

V. ECONOMIC INDICATORS

GDP 50$	18	Gross domestic product at factor cost*	1950
GDP 55$	27	Gross domestic product at factor cost*	1955
GDP 60$	81	Gross domestic product at factor cost*	1960
GDP 65$	88	Gross domestic product at factor cost*	1965
GDP 70$	88	Gross domestic product at factor cost*	1970

(*continued*)

Appendix II—(Continued)

GDP 73$	83	Gross domestic product at factor cost*	1973
PCI6070$	68	Per capita national income (in 1970$)	1960
PCI7070$	120	Per capita national income (in 1970$)	1970
PTI6070$	68	Percent change in per capita income	1960–70
RCI6070$	68	Real change in per capita income (in 1970$)	1960–70
RGNPI74	70	Real gross national product index (U.S. = 100)	1974
PCTUNE60	31	Percent of labor force unemployed	1960
PCTUNE70	45	Percent of labor force unemployed	1970
PCLFAG60	69	Percent of labor force in agriculture	1960
PCLFAG70	77	Percent of labor force in agriculture	1970
PCGDPA60	63	Percent of G.D.P. derived from agriculture	1960
PCGDPA70	80	Percent of G.D.P. derived from agriculture	1970
EXEDM$US	103	Total expenditure on education (1)	1974
XPCED74	102	Per capita education expenditure (1)	1974
PCGDPA75	57	Percent of G.D.P. derived from agriculture	1975
ECONTYPE	108	Type of economic system	1976
ACGDP607	32	Annual growth rate G.D.P.	1960–70
PCCGDP67	32	Per capita annual growth rate G.D.P.	1960–70
GOVCON67	30	Annual growth rate of government final consumption	1960–70
PRICON67	30	Annual growth rate of private final consumption	1960–70
GFCAPF67	32	Annual growth rate of fixed capital formation	1960–70
XPORT607	26	Annual growth rate of exports of goods and services	1960–70
IPORT607	26	Annual growth rate of imports of goods and services	1960–70
ACGDP05	36	Annual growth rate G.D.P.	1970–75
PCCGDP05	36	Per capita annual growth rate G.D.P.	1970–75
GOVCON05	33	Annual growth rate of government final consumption	1970–75
PRICON05	32	Annual growth rate of private final consumption	1970–75
GFCAPF05	34	Annual growth rate of exports capital formation	1970–75
XPORT705	30	Annual growth rate of exports of goods and services	1970–75
IPORT705	30	Annual growth rate of imports of goods and services	1970–75

Note
*Million U.S.$

ACKNOWLEDGMENT

The authors wish to thank Mr. Robert D. Eckstein and Mr. Robert L. Tilden, both researchers at the University of Michigan School of Public Health, for their assistance in the original data collection and in the production of the graphics work.

NOTES

1. Variables were taken from data tapes of the World Bank, most of which have been published in *World Tables 1976*, published for the World Bank by Johns Hopkins University Press, 1976; *United Nations Statistical Yearbook*, 1977, U.N.: *World Population Prospects as Assessed in 1973*, 1977, U.N.; *Population and Vital Statistics Report*, Series A, Vol. XXIX, No. 4, 1977, U.N.; *World Military and Social Expenditures*, 1977, Ruth Leger Sivard, Rockefeller Foundation, 1977; *Size Distribution of Income, A Compilation of Data*, Shail Jain, World Bank, 1975; *Real G.N.P. Per*

Capita for More than 100 Countries, Irving R. Kravis, Alan W. Heston, Robert Summers, *The Economic Journal,* Vol. 88, Cambridge University Press, June 1978; United Nations Demographic Yearbook, 1977, U.N.: *Community Water Supply and Wastewater Disposal,* Mid-Decade Progress Report, Director General, World Health Organization, May 1976.

REFERENCES

1. Arriaga, Eduardo E., and Kingsley Davis (1969) "The Patterns of Mortality Change in Latin America," *Demography* 6(3):223–242.
2. Fuchs, Victor R. (1979) "Economics, Health and Post-Industrial Society," *Milbank Memorial Fund Quarterly* 7(2):153–182.
3. Gilliand, Pierre, and Rene Gallard (1977) "Outline on International Comparison of Public Health, Based on Data Collected by the World Health Organization," *World Health Statistics Reports* 30(3):227–238.
4. Krishnan, P. (1975) "Mortality Decline in India, 1951–1961: Development versus Public Health Program Hypotheses," *Social Science and Medicine* 9:475–479.
5. Preston, Samuel H. (1974) "Causes and Consequences of Mortality Declines in Less Developed Countries during the Twentieth Century," Conference on Population and Economic Change in Less Developed Countries, September 30-October 2, 1976.
6. Preston, Samuel H. (1976) *Mortality Patterns in National Populations,* Academic Press.
7. Sloan, Frank (1971) *Survival of Progeny in Developing Countries: An Analysis of Evidence from Costa Rica, Mexico, East Pakistan, and Puerto Rica,* Santa Monica, California: Rand Corporation.
8. Smucker, Celeste M. (1975) *Socio-Economic and Demographic Correlates of Infant and Child Mortality in India,* Doctoral Dissertation, The University of Michigan.
9. Stolnitz, George J. (1975) "International Mortality Trends: Some Main Facts and Implications," United Nations, Department of Economic and Social Affairs, *The Population Debate: Dimensions and Perspectives, Papers of the World Population Conference,* Bucharest, 1974, Volume 1, *Population Studies,* no. 57.

THE POWER OF HEALTH

Wilfred Malenbaum

Gradually, over recent years, there has emerged a new acknowledgement of the power of health policy, program, and action to serve national economic and societal goals. Today policies for socioeconomic development in the Third World anticipate interventions on the part of the world's health service establishment. Within the span of a decade or more or less, the composite effects from health sector inputs have been dramatically reassessed. Where originally they were held to provide overall *impediments* to societal progress in poor lands, they are increasingly judged *important and even essential* for progress in these countries.

So striking and rapid a metamorphosis in health policy, program, and action must reflect developments in the health sciences and the social sciences. Have new doctrine and new evidence come to the fore? In fact, of course, pertinent theory and factual appraisal are still suggestive, not yet definitive. If the trend toward health interventions for social change is indeed notable, the explanation must lie not only with what social and health scientists have discovered over the recent past but with the desperate need for world policymakers to uncover new tools to serve economic and social progress in poor lands especially. This juxtaposition of need and scientific discovery requires that a new priority be given to discovering critical interrelations of the health and the social and economic spheres, and particularly the circumstances in which these relationships may become functionally operable over specific time intervals.

The present paper will (1) illustrate that health actions do appear to have a new role in a world concerned with the further progress of man and his society; (2) delineate how underlying doctrine provides health actions with the power for this role transformation; and finally (3) indicate some priority directions of further study and research to improve our assessment of the power of health in social and economic development.

I. NEW TASK FOR HEALTH INTERVENTIONS

Basic doctrine on economic development is in the process of dramatic change. However impressive the actual Third World development experience since midcentury, there is widespread recognition that neither past results nor future prospects meet the world need. The "development decades" have seen advances in total gross national product (GNP) and in average product per person (GNP/population) at rates never previously experienced by the world as a whole. The same is true for the rich lands separately and for the poor lands separately (even with Third World population growth at record levels). Indeed economic gains in poor lands have been accompanied by expanded education systems, higher literacy, improved nutrition and health conditions, increased technological sophistication, industrialization, and greater urbanization. National "numbers" for all these measures reflect unparalleled rates of economic and social advancement over some 25 years throughout the world.

Of course, the less developed lands of 1950 remain poor, both by absolute and relative measures. In the early 1970s average per capita income in the poor nations was but 10 percent of the per capita level in the rich nations. Over the rapid-growth years from 1950 to 1975, this disparity between rich and poor lands seems actually to have widened. In addition *within* many of the world's nations the spread between average per capita income of the top 20 percent families and of the bottom 20 percent has apparently also grown larger. For the countries of the Third World, with more than 70 percent of all the world's people, the absolute level of living of some 20–30 percent (perhaps ¾ billion) of these people may have improved little if at all over the decades from 1950. Notwithstanding the record of rapid economic and social advancement on an overall national basis, poverty, malnutrition, and disease continue to prevail among large populations throughout the Third World.[1]

The world's development experience has thus not yet served to narrow the relative output per person levels between rich and poor lands. Nor have the development decades assured gains in economic well-being to large groups of poor nation populations. Furthermore there is today little evidence to support the view that these growing disparities (even where accompanied by rapid expansion in selective extractive or industrial sectors) are preambles to a long-period development that will progressively benefit an expanding proportion of the total population of the world's nations.

Given this outlook and its differential promise for the rich in contrast to the poor, world policy has in the 1970s turned determinedly to new types of economic development efforts. Guided by the poor nations, world organizations exert persistent and mass influence for a New International Economic Order (NIEO) that seeks a rapid increase in the share of world output generated in the Third World nations. This share was some 16 percent in recent years; NIEO hopes it will reach 25 percent by the year 2000. Per capita income in the poor lands is to increase from 10 percent currently to some 20 percent of rich nation levels in 2000. Redistribution through NIEO will be accomplished primarily through international action to establish new pricing, trading, and financing arrangements for output of goods and services in poor and rich nations.

With respect to income distribution within poor lands and even with respect to growth of their total domestic product, it seems increasingly clear that NIEO-type actions must be supplemented by an additional development focus. A second arm of the revised development strategy calls for direct action to meet "basic needs" of the bottom-income 20–30 percent of the populations, particularly in nations of the Third World. For the most part these poverty groups are in rural areas, primarily engaged in various agricultural and handicraft activities. "Basic needs" encompass food, housing, and education as well as health care.

"Basic needs" programs envision gains through transfers from richer parts of the nation and world—perhaps eased by NIEO benefits. But mainly and fundamentally the gains for the low-income categories are to arise from programs that spurt a differential productivity gain for these workers. In the short run, the achievement goal is an expanded output in agriculture and consumer goods and services, in increments to be produced and in considerable measure consumed by the low-income population in rural areas. In the longer run these gains are expected to spread expansionary influences through other parts of the nation, urban as well as rural. World experience over the past decade especially, and hence international policy, are giving more emphasis to this basic needs focus (and less to any short-period restructuring of world income flows through NIEO) as the fundamental policy tool in achieving the development aspirations of the poor lands. "The shift of emphasis in development towards the concept of basic needs . . . is now accepted in principle by the international community and by the Development Assistance Committee (DAC) of the OECD [Organization for Economic Cooperation and Development]" (5, p. 5).

Finally, it is in this basic needs context that the role of health actions impinges upon the critical development goals of world society today. The professional literature and the political discussion among nations now identify health interventions as a primary component in any basic needs achievement. "Health is coming to be recognized as a fundamental goal of development in its own right, and also as a means of increasing productivity and of economic development.... [Health] affects productivity, mental potential of children, infant and general mortality, and the allocation of resources within a family, community and nation" (5, p. 5). Health policy, program, and action have thus assumed a central role in achieving economic development, a goal that remains a major concern of world operational and policy entities, like nations and world organizations. In but a few years the World Bank, the Development Assistance Committee of the OECD, and several ad hoc meetings of the world's foreign aid donors all seem to have absorbed the theme: health actions constitute an essential force in social and economic development.[2]

Perhaps even before the OECD-DAC emphasis noted above, the World Health Organization has itself played a fundamental role in identifying and promoting this new emphasis. Thus WHO and UNDP (United Nations Development Program) jointly launched in 1976 a Special Program for Research and Training in Tropical Diseases (TDR). These diseases affect "several hundred million people" who are thus "cut off from the main stream of social and economic progress.... [T]hey are not free to choose and plan a better future" (25). A long and effective attack on these diseases, and on malaria, schistosomiasis, and filariasis especially, had not been able to maintain progress. A broader spectrum of scientific involvement was deemed necessary through TDR. An independent group of distinguished scientists, assembled in September 1976 by WHO to exercise their expert interdisciplinary judgment on the plans and programs of TDR, agreed that the Special Program had to "develop a rigorous model of the relationships between health interventions and economic and social progress.... In the light of present knowledge in the social science disciplines ... this would require research on the role of human health in the process of socioeconomic development.... The effectiveness of the Special Program should ultimately be judged in social and economic terms..." (26, p. 12). Soon thereafter the World Bank joined UNDP and WHO as a sponsor of TDR with particular concern about health-economic interdependences. Under this composite sponsorship scientific groups have been assembled since 1976 to help crystallize appropriate research efforts in behavioral sciences and health sciences on both theoretical and applied levels. Formulation of such interdisciplinary research tasks retains high priority in the TDR program today.

Another health theme has also emerged on the current world development scene: the WHO goal of "health for all by the year 2000" (24, p. 8). From its very establishment WHO has accepted "health as a universal right without political, geographic, economic and social discrimination."[3] But reality in WHO and in the world as a whole has for the most part paid lip service to this objective. The new goal now reflects (and strengthens) today's new recognition of the power of health. While the specifics of the program still remain in discussion stage between WHO and Third World countries, there can be little question about program support for extensive increases in "primary health care" centers, a health service organization emphasis that features importantly in current plans for the health role in socioeconomic progress. Again most of the action will need to take hold in rural areas of the Third World nations.

In sum, the persistence of poor health and nutrition, of high infant mortality, and low life expectancy at birth in a large share of the Third World population accompanied the dynamic growth of economic output per person in these countries over the 1950–75 period. It became clear that neither economic betterment nor better health for these hundreds of millions of low-income people could be expected from past types of development action. The record of past development decades invited new policies for the economic advance the world seeks. It also opened new horizons for the role of health in this goal. The health services of the world did increase their earnings by larger percentages than are recorded by the rapid growth of real-world GNP for the years 1950–75. By and large three-fourths or more of all health services in poor lands was provided to the top 20–25 percent of the income groups in those countries. Indeed, at least for the years to 1970, important institutional spokesmen for the medical profession did not support shifts of health services to the poor lands and to the poorer parts of their peoples. Health was considered a handmaiden to *rapid* population growth; low-per-capita-income nations need *lower* rates of population increase. "At least until recently . . . health appears to have been regarded as a low priority by many agencies—only some 3–5% of bilateral aid went to health in 1975—a factor which is to a large extent reflected in attitudes and orientation of personnel in both bilateral and some of the multilateral Agencies'' (5, p. 13). It is obvious that the new role for health interventions will require important changes in world health service delivery. Such changes also imply an altered appreciation of the power of health in world progress even on the part of the health establishment itself.

II. WHAT CAN HEALTH INTERVENTIONS CONTRIBUTE?

One would expect answers, or directive argument, about health and development relationships from our impressive body of economic theory—classical, Marxist, neoclassical, Keynesian, neo-Keynesian, even radical. Or, and in addition, relevant empirical relationships should be found both in the rich economic history now available for past centuries of socioeconomic growth and in the very large number of case studies of the new national growth achievement in this century's developing lands. Actually neither the analytics of theory nor the facts of history (yet) offer the straightforward guides to existing health-economic interdependences that one would presume from the new policy directions of Section I.

Nonetheless, invaluable lessons on this relationship do emerge from a growing body of empirical observation and analysis of micro-type case material, essentially from today's developing nations. Increasingly social scientists are discovering the potential of health actions in poor regions and also the conditions that release the power of health. The health field does appear to offer new policies and programs that deal both with the overall economic growth goals of poor nations and with the income distribution patterns these nations seem to seek. Moreover, these output and distribution targets could well be consistent and indeed mutually reinforcing.

The present state of knowledge calls for further study of micro-type case materials as matters of high priority. The gap between these new lessons on health-economic relationships and the limited lessons provided by theory and history can be expected to narrow. Already, social scientists accept the need to adapt these disciplines better to cope with the fundamental role of the human factor in socioeconomic change. Behavioral attributes of man are but partially reflected in the economic man of traditional theory.

The familiar doctrines of economic theory, and neoclassical economics in particular, have been basic to the growth models used in the development programs in Third World countries. In neoclassical doctrine, changes in national product arise from changes in the two primary factors of production, labor and capital (including land). These inputs govern levels of output with any given state of the art of production. Changes in outlays for health services—or changes somehow achieved in the state of health—do not feature explicitly as inputs in any of these familiar theories of economic growth. Insofar as the theory does consider the health sector, the emphasis tends to be "consumer good," not "producer good." By and large man's health is an appreciated end product of the economy: health and health services fit reasonably well into the body of theoretical doctrine dealing with consumer choice and his satisfaction-generating expenditures determined by the style of his total income. Presumably health outlays, like other expenditures in such important consumption categories as food and clothing and like other expenditures in investment categories, do bear upon a nation's subsequent labor and capital supply. Thus increases in health outlays might reduce savings and thus serve to reduce new capital formation. Increased health outlays could, on the other hand, also serve to expand the man-hours of work by the labor force or permit more output per man-hour. Here is some producer-good, investment-in-human-capital emphasis. On the whole, traditional theory can be held to argue that increased health outlays serve to shift a nation's relative factor endowments toward labor. Since Third World countries are generally held to be amply endowed with labor, this interpretation is consistent with the antipathy often held—perhaps by the medical establishment in particular—against relative improvement in health programs in Third World lands, and especially among the poorer sectors of their populations.

Empirical analysis of actual development experience over a century and more has of course already made the argument that economic growth theory must go beyond such usual and familiar models and their preoccupation with the absolute and the relative quantity of capital and labor inputs. For economic development, theory must provide for independent inputs directed to improvements in the *quality* of the conventional inputs of labor and capital. The history of the development of the rich world cannot be interpreted without explicit regard to inputs of technology, education, motivation, entrepreneurship—and health. The actual development experience in Third World countries over the past decades—noted similarity in Section I—has suggested similar conclusions: there is need for growth models with explicit provision for changes in the quality of conventional factor inputs. The recorded history of economic growth attributes major responsibility for expansion to productivity increases per unit of capital and labor, the traditional factors of production. This intensive route to growth has tended to become progressively more important relative to the extensive route, (i.e., more units of capital and labor each taken to be of constant quality). In any event, this shift of emphasis in accounting for world growth experience pertains for all long-period growth records in today's world. Without exception, productivity gains on a national base are attributable to quality improvements in factors. Whether the improvement is associated with labor or capital, it originates from the skills, motivations, and aspirations of man and of the institutions he creates or maintains.[4]

Since productivity advances have been of such decisive importance in accounting for the economic achievement of the rich lands, considerable attention has been given to study of the forces responsible for the productivity gains. Education and health interventions have been used in such empirical studies, among other factors. Men "are the active agents who accumulate capital, exploit natural resources, build social, economic, and political organizations, and carry forward national development. Clearly, a country which is unable to develop the skills and knowledge of its people and to utilize them effectively in the national economy will be unable to develop..." (8). There is support, in argument and in measurement, for associating interventions in the health area with establishing the forces for attitudinal and motivational shifts in a region seeking economic development.[5]

Before proceeding to the contribution of case study materials to the role of health, note should be taken, however summarily, of a further dimension in the relevant historical and theoretical literature. Any account of economic progress must handle population growth as well as national output growth. Continuously, from the very creation of formal economic doctrine, distinguished social scientists have been immersed in the interplay of these two growth rates and of the factors that influence them. In some degree this question is subsumed in the economic development discussion of the preceding pages. Certainly population growth has intimate ties with labor supply growth. Yet a specifically demographic emphasis broadens the analysis of the more comprehensive subject of economic growth. Moreover, it provides a direct link into the health area. For birth and death rates, morbidity measures, and life expectancy are core variables in population growth; changes in health thus become an integral part of a changing population-product relationship. Over time Malthusian doctrine made room for the more flexible "demographic transition." This last has been questioned persistently as regards its relevance to Third World nations over recent decades: declining mortality rates over a generation and more were not followed by meaningful reduction in birth rates. Today the relevance of the broad transition doctrine to the changing rates of population growth in the rich world over past centuries is also being debated anew. The main reasons are recent research results with bodies of local community data over long periods, from the 1700s and in Western Europe particularly (11). In another vein, the role of progress in health sciences as a basic determinant in morbidity decline and in expanded life expectancy has been confronted by the hard reality of the actual progress achieved through expanding health knowledge. Greater longevity was the result in some small degree of advances on the medical frontier. Mostly it was achieved by the more rapid growth of economic product than of population, even with high and indeed record levels of population growth (13).

In sum, the extensive study of population-output relations with particular regard to the strength and influence of the direct components of population growth (births and deaths) has yet to yield accepted doctrine on the role of health in socioeconomic progress. Important new contributions began appearing in 1976; they offer a broader framework for analyzing population and economic change [(7), (12, ch 5–10)]. The two studies cited emphasize the role of shifts in attitudes and aspirations of component groups in a society in process of change. Concept and operation are multidisciplinary. The specifics of population and output change depend on the behavioral characteristics of pertinent elements of an evolving society. Output growth and population growth involve changes in the health area. Bases for development action rest on knowledge that must come from study of changes in group behavior.

The elusiveness of both theory and history on the nature of the health-output relation notwithstanding, many specific attempts have of course been made to identify and quantify the role of health actions or of changes in health status on economic product. These efforts pursue what seem to be straightforward and logical propositions. With regard to the world's poor nations and particularly to their rural sectors, there "are four principal ways by which health programs can affect the pace of economic development . . . (1) increasing the number of man-hours of work available; (2) increasing the quality and productivity of the existing workforce; (3) making feasible the development of previously unsettled regions; and (4) changing the attitudes of persons toward innovation and entrepreneurship (growth-creating activities)."[6]

On the first two of these categories documentation available for nations, provinces, and even small regions and communities is equivocal with respect to the sequence of causation, the sign of the relationship, the relevant time interval, and more. Perhaps no set of the empirical results offers firm coefficients for use in planning even for the specific region of study to say naught of other regions. Basically such results reflect the problems discussed earlier. The processes of production, especially the labor-intensive, family-dominated activities of the rural areas, embody institutional relationships not well reflected in the economic man optimizing syndromes of our theory. Nor can the relationships be adequately measured with numbers of workers and/or worker vigor as major contributing factors in output. This conclusion about the limited value of items 1 and 2 in the categorization seems now to have been widely accepted. Indeed, important new efforts to probe the role of health actions tend to be very explicit in setting aside the simplistic nature of research in which disease ratios, health-associated work absence rates, and the like are used to explain changes in (discounted) economic product flows.[7]

An expanded "pace of economic development" attendant upon health actions is often documented for case 3. Disease-infested areas in various parts of the world, and in tropical areas particularly, become economically more viable with introduction of disease eradication efforts. Malaria control programs in particular are credited with major economic breakthroughs in Nepal, Thailand, Sri Lanka; limited control of river blindness and of schistosomiasis has discouraged economic expansion in parts of Africa. Where output expanded, the programs encompassed much more than health actions. Transportation, land reform and land purchase programs, and power development, among other efforts, had to accompany disease control [(20, pp. 47–9), (14, pp. 165–73)]. Where output was declining as disease rates increased, the missing inputs were more than the degree of health impairment. Primary among inputs are attitudinal and motivational factors of production. These are always essential supplements to other factor inputs for rural development programs in Third World countries; with high aspirations for progress this factor could itself serve in mobilizing other inputs. In all cases where the disease problem looms large in economic change—whether in the past developments in Sri Lanka and Nepal or in the current problems in Ghana and Upper Volta—major decisions in many areas other than disease control play key roles; action is needed with aggressive implementation. Without such motivational drive, the apparently technical input/output sequence disintegrates. Recent analysis of a colonization zone in a malaria-infected region in eastern Paraguay provides unusual and specific documentation. "Elimination of the disease will not spark any significant cumulative process unless there are accompanying economic opportunities...." "Land acquisition, a network of roads,... were part of the same effort that included malaria control...." "Self-help... was conspicuously present,..." but only after opportunities had been made available over a period of years.[8]

Only in the past decade or so have social scientists considered the role of health interventions upon patterns of individual behavior. Pioneer work had already established important ties between protein intake in the unborn fetus (and indeed in young children below the age of 3) and brain development. Given the growing appreciation of the quality-of-man requirements for economic growth, here was physiological evidence of the importance of nutrition (and health) for behavior consistent with socioeconomic progress.[9] In addition, there now began to appear arguments and some empirical support for a psychological tie between health sector programs and individual self-esteem. Awareness of actions a man could take for his family health led to new awareness that *he* could influence his family's well-being and life prospects [(20, p. 50), (14)]. While systematic quantitative investigations of these ideas were not undertaken, development experience, notably in rural activities in poor lands, did yield cumulating support for the general idea. It is perhaps this route that led to the important application of health-economic relationships reflected in Section I of this paper.

A good illustration is available in an International Labor Organization working report already noted (21, pp. 11–14):

> "The *a priori* considerations adduced foregoing have, in the author's experience, been corroborated by the natural history accounts of field workers—particularly those of agricultural consultants in Pakistan. . . . [There] appears to be a nearly universal lament of these consultants—namely, lack of success in achieving the adoption and propagation of improved, productivity-increasing agriculture technologies of a seemingly simple nature, e.g., water-management techniques, cropping patterns and the like. . . . [These] field workers usually will characterize the apparent health status of the village farmers . . . as abysmally bad . . . and so chronic that the affected populations may fail to recognize them as health problems.[10]" . . .
> "Indeed, in the view of some such observers it is precisely the deteriorated health status of the farmers which constitutes a major bar to the adoption and propagation of readily available improvements in technology which could pay big dividends in increased output. [While] this is partly a matter of physical vigor, . . . it is partly a matter of subjective states—e.g., ability to marshall the attention necessary for forward planning and willingness to assume the risks which are entailed by departures from accepted modes.[11]"

Economic progress in the world's poor lands remains a priority need; the limited achievement of the past three decades and the uncertain future outlook for hundreds of millions of people constitute a challenge to world and national leadership as well as to the scientific community everywhere. A comprehensive review of analytical and empirical research, product of some 10 years of economic and social field investigations in nations of the Third World, was recently published in the American Economic Association's *Journal of Economic Literature*. It concludes: "Fortunately there is now a considerable measure of agreement concerning the fundamental importance of policies and programs that accelerate the growth of agricultural production and which foster broad participation in that growth and more rapid expansion of employment in both agriculture and non-agriculture" (10, p. 900). The review assessed several major alternative approaches in the literature before reaching this conclusion. It also appraised alternative proposals for achieving the output and participation objectives. The position adopted rested on recognition of the primary need: actions that offered some hope for generating new aspirations and motivations among the populous poor in the Third World lands. A comprehensive family health program is recommended with very "broad coverage so that the poorest segments of the population are reached by development activities" (10, p. 899). An integrated local approach can best serve the essential development task. In particular, the health program has a promotion and education focus, rather than basic concern with curative care; it demands community-based health workers as key components of local efforts that integrate many disciplines in the achievement of long-term economic progress.

However competently these conclusions and these implementing guides are derived from the available body of theory and experience, the relevant sources are limited in quantity and in depth of interdisciplinary analysis. "Given the enormous importance of the objectives . . . it is appalling that there has been so little research at assessing the feasibility and effectiveness of alternative strategies for attaining those objectives" (10, p. 901). Multidisciplinary research efforts are required. At the minimum these must augment the available documentation of the record with respect to the socioeconomic effects of interventions in the health area. Within the context of policy and administrative action, more study is needed of possibilities of collaborative policy actions spurred by economists, demographers, health and nutrition specialists, *and* political scientists. Finally, the central position of the health function in these new programs poses new questions of the capacity of the appropriate health establishments to organize and direct basic primary health facilities in local and particularly very poor regions. A few observations on these further needs for basic knowledge are offered in the concluding section.

III. NEW RESEARCH DIRECTIONS

If world development policy is in fact giving new priority to factor inputs from the health field, there must also be a new priority for the discovery of additional knowledge about relevant input-output relationships. This type of statement is always made when responsible policy officials and competent social and medical scientists are moved to recommend health actions for socioeconomic gains in the economies of Third World nations. Thus the special program for research and training in tropical diseases, TDR, is now striving valiantly to establish research programs with a detailed enough basis in health sciences and a comprehensive enough basis in the behavioral sciences to yield realistic parameters to guide new program and policy.[12]

The goal of new research seems to be a better "technological" dimension in the health and economic development sequence. It is hard to object to such further research efforts—more intensive in each discipline, more comprehensive and integrated among disciplines. Nonetheless, the discussion of Section II emphasized that responsible policy/scientific support for an expanded input role from the health fields rested heavily on what might be called "organizational" dimensions of the health and economic development sequence. The critical problem was discovering a structure for delivery which had a chance of reaching the sectors of the economy (mostly the rural poor) where progress was a prerequisite for the current problem of economic development in the world's poor lands. Indeed, according to our Section II, this structure demanded not only primary health personnel but a primary health care program in a rural organizational framework geared to comprehensive development achievement on the part of the rural poor. The essential need is for organizational devices that have a chance to capture the psychological-motivational links that have in recent years begun to be associated with health interventions in very poor areas. The structure of the program would be geared to developing new aspirations on the part of the rural poor and new identification of their own role in personal, family, and community advancement. New technological knowledge—perhaps along lines now being formulated by TDR—may be valuable only in the context of a new organizational structure for the health and development sequence.

Fortunately, significant beginnings toward more knowledge of this organizational dimension are already in hand. It is clear that if health actions are to involve the populous poor particularly, and rapidly, new knowledge is needed about existing health-financing patterns and about alternatives to them. WHO has over some years sought new information both on resources flows into health services production/distribution and on alternative approaches to meeting basic health needs in developing countries (6, 24). Both these WHO research interests are of course geared to the goal of broadening health services availability and to the WHO effort to pioneer ways to deliver primary health care by means which reinforce effective routes to the overall socioeconomic progress of poor lands. The new "basic needs" emphasis in economic development is also reflected in the international financing area where the Development Assistance Committee (DAC) of the OECD has long played a central analytical and operational role. Seemingly, therefore, the many important new research needs associated with a new health area emphasis in economic growth are being pursued by institutions which are aware of the need for using such new information. On the whole this augurs well for future additions to new knowledge.

Yet the research emphasis in such official institutions tends toward national, macro coverage. The nature of the actual organizational (and associated technological) dimension tends rather to be local or regional and micro. Social scientists concerned with the quest for new knowledge about health and socioeconomic development must take the lead in contributing new results for the case study type of source material. The basis for these would be in connection with relatively new specific health or development projects,[13] or from reexamination of data from old projects including those of a continuing nature.[14] Only a persistent flow of such case analyses will permit the eventual presentation of a national program for health inputs that can best serve a nation's development goals.

The preceding observations can be made more concrete by reference to studies associated with financing alternatives in the health area under WHO auspices.[15] On a national basis and in a national accounts conceptual framework the value of final health sector services can be itemized by nature of goods and services (medical supplies, professional services, hospital services, public health services, research, training) and by the final purchasing unit (household, various types of public and private institutions). Similarly, the same total expenditures can be traced by sources of finance as these become available to the final purchaser. This basic framework permits health service accounting through all ministries (i.e., health and others—e.g., military, housing, etc.), and for a wide range of financing flows from personal outlays, insurance arrangements, public funds, charities, etc. It permits inclusion of services provided by traditional healers as well as modern practitioners, and the inclusion of expenditures for drinking water, sanitation, private transport to clinics, and similar outlays relevant to distribution of health services.

From the national level data, further breakdowns can be attempted so that household income tax burden and health service purchases can be seen by population deciles, separately for urban and rural population as well as for the nation. Further, health services purchases by income groups can be identified by the purchaser (households direct, various hospital-type institutions); these can be separated between curative and preventive services and each separately for rural and urban regions. Were all these data available, there would be a clear picture of who gets what services and through what means.

Unevenness in distribution of services and payments would emerge clearly. The tables would reveal the imbalance between urban and rural services per person, the preponderance of curative over preventive services, and the higher burden of costs per low-income person for comparable services. They would suggest straightforward alternative arrangements for making services more readily accessible to the critical groups of rural poor. In case study material, associated economic output data, as well as the particulars of any economic development program, would provide a relevant basis for technological dimensions of the health development sequence. This organizational framework permits posing specific alternatives in resources flows, health services flows, output flows. Such research offers a relevant input-output argument; it is an integral part of an organizational structure for rural development.

In sum, the current state of the art in the development process in poor lands directs interventions toward affecting the attitudes of the populous poor in rural areas. There is a broadening expectation that health actions will thus serve the goals of economic advancement. The additional knowledge required consists fundamentally of the facts about current use of the broad range of health services and facilities and about the specific channels by which these uses are accomplished. The power of health in new policy can only be realized thorough actions which alter these existing use patterns and open the possiblity that the vast body of people in the poor nations can assume greater control of their own economic and social prospects.

NOTES

1. The broad characterization of development achievement is presented in some detail in *The World Bank* (23), especially Chapter 2.

2. Of the many possible citations, the following are most relevant: Development Assistance Committee (4), [the document in reference (5) was a Secretariat submission to this meeting]; International Conference on Primary Health Care (9); and World Bank (22). Notable in the professional literature is the American Economic Association's "Food, Health, and Population in Development," a survey of development literature since 1970 (10).

3. This is the characterization made by Dr. Halfdan T. Mahler, as the new Director-General of WHO, in an address to the WHO Regional Commission for the Americas, Washington, D.C., September 30, 1974.

4. The argument of this paragraph is presented more fully in my "Health and Economic Expansion in Poor Lands" (14, pp. 162–166).

5. Thus, see my 1977 consultancy report "Health and Economics" (13) and C. M. Stevens (21).

6. Discussion and illustrations here draw upon Alan L. Sorkin's *Health Economics in Developing Countries* (20).

7. Thus, see the *Report of the Initial Planning Consultation (IPC) on Socio-Economic Research in the Special Tropical Disease Program* (16). "The Consultation was unanimous in its judgment that the classical benefit-cost calculations attempted so far have been incomplete and highly misleading.... [They] cannot yet produce valid answers to the question of what is the effect of health status on the magnitude and growth of economic product..." (p. 25). But the need for improved understanding of the health-output relationship is great. Considerable field research is needed on community beliefs and associated behavior, on cultural as well as socioeconomic practices that bear upon acceptance and effective use of health interventions. Such new research efforts are being incorporated under the developing Tropical Disease Program.

8. See Gladyn N. Conly (3), pp. 1–5 and 27–82, and especially 83–99. Quotations are from this last section (pp. 96–7) and summarize the detailed survey results. Conly's impressive study is germane also to Sorkin's case (4) in the text: health interventions are related to attitude changes.

9. For a convenient account see Alan Berg, Nevin S. Scrimshaw, David L. Call, Editors (2): articles by J. M. Bengoa, J. Cravioto, Wilfred Malenbaum. In his recent Ph.D. dissertation thesis at the University of Pennsylvania, C. Ross Anthony derived empirical support for this physiological route on the basis of original data assembled for a health study in a development center in Nepal (1).

10. Precisely this last point is documented in detail in the Ross Anthony study (1).

11. These themes are developed throughout the ILO working paper (21) which also emphasizes (as in the present treatment) the interdependence of health inputs with other factor inputs and the importance of a local core for development programs. "There is increasing recognition of the crucial role played by "community participation" in mounting effective health [and other development] programs" (p. 15).

12. See, for example, *Scientific Working Group on Epidemiology, Report on First Session* (19); *Report on the Meeting of the Initial Planning Consultation on Socio-Economic Research* (17); and *Report of the Preparatory Scientific Working Group on Social and Economic Research* (18).

13. E.g., as in the malaria control project in Bendel State, Nigeria, or the schistosomiasis control

project at Lake Volta in Ghana, among others involved in the developing research plans cited by TDR (documents cited in note 12 above).

14. E.g., as in a list of South Asian research studies in the health field; see Malenbaum (14, pp. 174–175).

15. The present discussion follows "Methodology for Exploratory Surveys on Financing Health Services in Developing Countries," a working paper which I prepared in April 1977 for the Sandoz Institute for Health and Socio-Economic Studies (15). This private organization is closely associated with WHO studies on health services financing.

REFERENCES

1. Anthony, C. R. (1979) *Health, Population and Income: A Theoretical and Empirical Investigation Using Survey Data from Rural Nepal,* University of Pennsylvania, Ph.D. Dissertation.
2. Berg, A., N. S. Scrimshaw, and D. L. Call, (eds.), (1971) *Nutrition, National Development and Planning,* Cambridge: M.I.T. Press.
3. Conly, G. N. (1975) *The Impact of Malaria in Economic Development: A Case Study,* Washington, D.C.: American Region Office of WHO.
4. Development Assistance Committee (1978) *Summary Report of the Informal Meeting of Experts on Aid for Rural Development in a Basic Needs Perspective,* Mimeo, DD-578, Revised, Paris, 8 August 1978.
5. Development Assistance Committee (1978) *The Role of Foreign Aid in Meeting Basic Health Needs for Rural Development,* (78) Paris, April 11, 1978.
6. Djukanovic, V. and E. P. Mach (1975) *Alternative Approaches to Meeting Basic Health Needs in Developing Countries,* Joint UNICEF/WHO Study, Geneva.
7. Easterlin, R. A. (1976) "The Conflict between Aspirations and Resources," *Population and Development Review* 2(3): 417–425.
8. Harbison, F. H. (1971) "Human Resources as the Wealth of Nations," *American Philosophical Society* 115(6): 426.
9. International Conference on Primary Health Care. (1978) *Declaration of Alma Ata,* Information Bulletin PHC/14, Alma Ata, USSR, 12 September 1978.
10. Johnston, B. F. (1977) "Food, Health and Population in Development," *Journal of Economic Literature,* pp. 870–907.
11. Knodel, J. and E. van de Walle (1979) "Lessons from the Past: Policy Implications of Historical Fertility Studies," *Population and Development Review* 5(2): 217–245.
12. Leibenstein, H. (1976) *Beyond Economic Man,* Harvard.
13. Malenbaum, W. (1977) "Health and Economics," Geneva, MPD/RCT-TDR.
14. Malenbaum, W. (1973) "Health and Economic Expansion in Poor Lands," *International Journal of Health Services* 3(2): 161–176.
15. Malenbaum, W. (1977) "Methodology for Exploratory Surveys on Financing Health Services in Developing Countries," Sandoz Institute for Health and Socio-Economic Studies, Working Paper, Geneva.
16. *Report of the Initial Planning Consultation on Socio-Economic Research in the Special Tropical Disease Program,* Geneva, 1–7 December 1977, TDR/SER/IPC/77.3.
17. *Report on the Meeting of the Initial Planning Consultation on Socio-Economic Research,* Geneva, 1–7 December, TDR/SER/IPC/77.3.
18. *Report of the Preparatory Scientific Working Group on Social and Economic Research,* Geneva, 1979.
19. *Scientific Working Group on Epidemiology, Report on First Session,* Geneva, 6–10 June 1977, TDR/MPN/SWG-EPIA (1) 77.3.
20. Sorkin, A. L. (1977) *Health Economics in Developing Countries,* Lexington.

21. Stevens, C. M. (1975) "Health, Employment and Income Distribution," World Employment Research Program, Working Paper, Geneva: ILO.
22. World Bank. (1975) Health Sector Policy Paper.
23. World Bank. (1978) *World Development Report, 1978,* Baltimore: Johns Hopkins University Press.
24. World Health Organization. (1978) *Financing Health Services,* Report of a WHO Study Group, Geneva.
25. World Health Organization. (1976) *Tropical Diseases,* Geneva.
26. World Health Organization/United Nations Development Programme. (1976) *Report of the First Technical Review Group,* Geneva, TDR/TRG/WP/76.3, unpublished document.

HEALTH PLANNING IN THE SUDAN

Ronald J. Vogel and Nancy T. Greenspan

I. INTRODUCTION

The World Health Organization has defined health as being "... a state of complete physical and mental and social well-being and not merely the absence of disease or infirmity."[1] This definition is not as rigorous and precise as are many of those in economics, yet it does reflect the ambiguity of the state "healthy," or its antithesis "unhealthy." A more economically operational concept of health might place it on a continuum between a pure investment good and a pure consumption good. In a pure investment model, the economist would view health as one of many investments which the individual could make in himself in order to increase his stock of human capital; likewise, in a pure consumption model, health would be one of the many commodities which the consumer might choose, given the relative prices of the commodities and given his budget constraint.

The first section of this chapter examines the theoretical underpinnings of the concept of health capital and the demand for health. In that section, we will also formulate the rationale for health planning, particularly in the context of developing nations. Having established a theoretical basis for health planning, and having offered some suggestive empirical evidence on the effects of good health on economic growth, we will then present Sudan in a comparative setting within the context of the World Health Organization's definition of health and health care. The third section of the chapter will commence with a brief description of the Sudanese economy so that the reader will have an economic perspective on the resource constraints involved in that part of the world. Then planning and budget trends will be discussed and major health problems and possible solutions to them will be presented. Finally, the concluding section will seek to draw some health and planning lessons, not only for Sudan but also for those developing nations which find themselves in an economic situation similar to that of Sudan.

II. THE ECONOMIC THEORY OF HEALTH INVESTMENT AND PLANNING, AND SOME EVIDENCE

Although there have been a number of interesting studies exploring health differentials among and within countries, none of the results are useful in establishing the basic patterns of and relationships between a population's health status and its environment.[2] That is, "these studies have not," as Grossman (1972) points out, "developed behavioral models that can predict the effects that are in fact observed."[3] However, a model constructed by Grossman, (1970, 1972(a), 1972(b)) which implicitly refers to investment in health within the context of a developed country, is sufficiently rich to be generally applicable to any stage of economic development.

The central proposition of Grossman's model is that "health can be viewed as a durable capital stock that produces an output of healthy time,"[4] Following Becker (1965), Lancaster (1966), Muth (1966), Michael (1969), Becker and Michael (1970) and Ghez (1970), he views healthy time as a "commodity" whose production function has as its arguments goods and services. The demand for goods and services is, therefore a derived demand. Medical care would be one of those services which enters into the production of "healthy time" and, as a consequence, the demand for medical care is a derived demand. Likewise, housing, diet and environmental variables enter into the production function for the output of healthy time. Grossman writes the gross investment production function for health in period i as:

$$I_i = M_i g(t_i; E_i) \qquad (1)$$

where I is gross investment, M is medical care, t is healthy time divided by medical care, and E is the stock of human capital.

The marginal monetary rate of return on an investment in health is given by:

$$\gamma_i = (W_i G_i)/\pi_{i-1} \qquad (2)$$

where γ is the monetary rate of return on an investment in health, or, the marginal efficiency of health capital (MEC), W is the wage rate, G is the marginal product of health capital and π is the marginal cost of gross investment in health.

Setting the marginal utility of healthy days or the marginal disutility of sick days equal to zero, we can write the optimal amount of health capital in period i as:

$$\frac{W_i G_i}{\pi_{i-1}} = \gamma_i = r - \pi_{i-1} + \delta_i \qquad (3)$$

where r is the rate of interest and δ is the rate of depreciation of the stock of health capital. The relationship in equation (3) is illustrated in Figure 1, where H is the stock of health in period i.

MEC is the demand curve which shows the relationship between the stock of health and the rate of return on an investment in health. The supply curve shows the relationship between the stock of health and the cost of health capital. The supply curve is posited as being infinitely elastic, because the cost of capital is independent of the stock.

This model contains a number of important properties. It begins by assuming that individuals inherit an initial stock of health capital that is subject to depreciation over the life-cycle and that can be increased by investment. The MEC curve slopes downward because of the diminishing marginal productivity of health capital. Most importantly, the model explains "variations in both health and medical care among persons in terms of variations in supply and demand curves for health capital"[5]

Within its context, increases in the rate of depreciation on the stock of health with age cause upward shifts in the supply schedule, while increases in the wage rate and in education cause outward shifts in the MEC schedule. The model yields three important predictions:

1. The quantity of health capital demanded declines over the life cycle as the rate of depreciation increases with age.
2. Demand for health and medical care are positively correlated with the wage rate.
3. The more educated would demand a larger optimal stock of health, if education increases the efficiency with which gross investments in health are produced.

While this model offers a particularly satisfying explanation of the demand for good health, or "healthy time", the set of phenomena which Grossman sets out

Figure 1. The Stock of Health Capital*

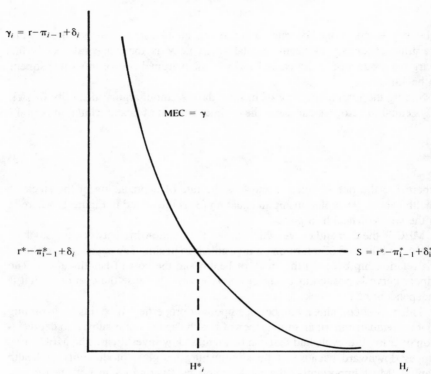

*From: Michael Grossman, "On the Concept of Health, Capital and the Demand for Health," *Journal of Political Economy*, March/April 1972, vol. 80, no. 2 p. 232. Copyright © 1972 by the University of Chicago.

to explore with it would seem to have more relevance for a developed economy. Let us now explore its applicability to a set of phenomena more appropriate to a developing economy.

We begin by examining the determinants of shifts in the MEC schedule. Grossman[6] includes, as direct inputs into the household production function for health, the own time of the consumer and market goods such as medical care, diet, exercise, recreation and housing. He also allows the production function to depend on certain environmental variables, "that influence the efficiency of the production process," the most important of which "is the level of education of the producer."[7] From the perspective of a developing nation, Grossman's list of "environmental variables" could be expanded to include the purity of the supply of drinking water, bathing water, and sanitation facilities in general, the existence of endemic and infectious diseases and all of those non-market health and other health related goods which fall under the broad heading of "externalities."

Figure 2. The Stock of Health Capital*

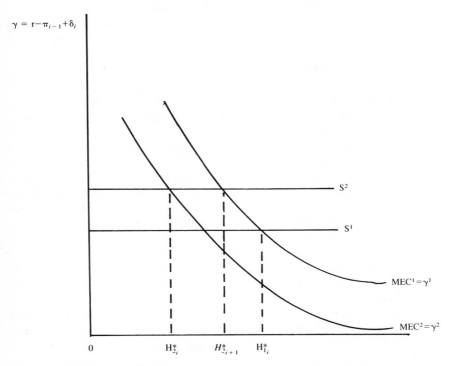

*Adapted from: Michael Grossman, "On the Concepts of Health Capital and the Demand for Health", *Journal of Political Economy*, March/April 1972, vol 80, No. 2, page 242. Copyright © 1972 by the University of Chicago.

To the extent that negative externalities exist and enter negatively into the production function for good health and "healthy time" we would expect them to shift the MEC function inward, even though rising levels of education in the developing country might tend to have the opposite effect. On the supply side, any phenomenon which would cause r or δ to rise, such as a shortage of capital in general or a small initial endowment of health stock due to a mother's poor prenatal care, would cause the supply schedule for the health stock to rise. Given the severe negative effects of many of the "environmental variables," in a developing economy, we would expect the stock of health which an individual holds, in any given period, to be lower in a developing economy than in a developed economy, all other things being equal. Figure 2 depicts such a case.

In Figure 2, we initially posit an individual living in a developing economy in Africa, who has limited or no access to clean drinking water, who lives in a village where both human and animal fecal matter do not have adequate and sanitary disposal, where malaria and schistosomiasis are endemic, where housing consists of mud huts, and where the level of his diet is small in quantity and low in nutritional value. Because of these environmental variables, his MEC function is MEC_2, and his health stock in period i is H_{2i}^*. If he could change the above environmental variables his MEC function in period $i + 1$ would move outward to MEC_1 and his health stock would be H_{2i+1}^*, because his efficiency in producing "healthy time" would have been increased. The problem is that these environmental variables are public goods (or bads) in the classical sense and the individual does not, therefore have sufficient incentive to change them himself. If he were in the mainstream of a developed economy where such conditions no longer exist, and where r and δ might be lower, his health stock would be H_{1i}^*.

A horizontal aggregation of the MEC functions for all of the inhabitants in a developing country might appear as MEC_2 in Figure 2. Because of the public goods nature of the environmental variables, in this instance, there *would be* an aggregate incentive to remove or ameliorate the environmental variables and move to MEC_1, making the achievement of health stock H_{2i+1}^* possible.[8] The only constraint on this prescription is whether manipulation of some other policy variable, such as increases in the level of education or growth in GNP, might achieve an aggregated level of the health stock of H_{2i+1}^* at less cost. This is an empirical issue, the evidence for which is scanty, but partially addressed by the work of Malenbaum (1970), which focuses on the relationship between health and growth in GNP.

Malenbaum tried to find "support for the hypothesis that programs for health improvement are important components of a program for the expansion of output per man in poor lands."[9] Table 1 summarizes his econometric efforts to determine the direction of the causality between health and output. In three of his five

Table 1. Determinants of Output in Agriculture

Case	Output	R^2	Percentage of Covariation Associated with Inputs		
			Economic	Health	Other
Developing nations, 22 countries	Changes in agricultural output	.62	20	79	1
Mexico	Output per worker in agriculture				
1940		.66	1	28	71
1960		.63	20	40	40
Thailand, 50 provinces	Output per worker in agriculture	.62	85	5	10
India, 20 blocks	Agricultural output	.73	85	14	1

Source: Wilfred Malenbaum, "Health and Productivity in Poor Areas," in Herbert E. Klarman ed. *Empirical Studies in Health Economics,* (Baltimore: Johns Hopkins University Press, 1970), p. 49. Copyright © 1970 by the Johns Hopkins Press. Reprinted with permission.

equations, health related variables, such as vaccination programs, cleaner drinking water, and improvement in malarian death rates explained a relatively large percentage of changes in output. Malenbaum provides three reasons why better health might increase output: (1) an increase in the energy potential of labor; (2) making additional land accessible for use, by eliminating external polluting elements (such as the tsetse fly), which has an indirect and longer run influence upon human health; and (3) an improvement in the motivation and/or attitudes of labor, with Malenbaum emphasizing the last factor as potentially the most important.

Support for the latter emphasis can be elicited from past production function studies. As Malenbaum points out, most production function studies have used some variant of the Cobb-Douglas production function:

$$O_t = A_t L_t^b K_t^{(1-b)} \qquad (4)$$

where O_t, L_t and K_t are, respectively, total output, labor and capital inputs in year t, where A_t is an index of total factor productivity in that year, and where b and $(1-b)$ are elasticities of output with respect to capital and labor. These studies, whether focused on Europe, on the United States or on the developing nations, leave a large residual variance in the explanatory power of the equations, leading one to suspect that the traditional inputs into the production function fail to capture important quality dimensions which may be interactive upon themselves. Although we know that adjustments for capital depreciation have been made in some studies, and that, from Grossman's work, the notion of a health stock and the variables which influence it in the production of "healthy time" certainly imply a quality dimension, many quality aspects of the labor component of production functions for aggregate output have not yet been isolated and measured empirically to complete satisfaction within the economics profession.[10] This lack of specificity leaves a great vacuum in our knowledge about the degree of importance of human capital in labor's contribution to the growth process. However, as Malenbaum asserts, it is not unreasonable to speculate that the qualitative dimension is important, given the size of the residuals in most studies. Moreover, the interactions among the quality of the labor force, the quality of capital and economic growth and progress, have their own effects upon the motivation and attitudes of labor, thus producing further progress. Although the stock of health capital is but one measure of the quality of the labor force, Malenbaum's empirical results give some degree of credence to the fact that improved health has a positive impact upon economic performance and growth rather than vice versa.[11] In addition, examination of the degree of intercorrelation between his equations' independent variables, health inputs and the country's literacy rate showed that there was little causality between the two. If these assertions concerning the relationship between health and policy variables such as economic growth and levels of literacy are in fact true, and if the health stock in developing economies is lower than in the developed economies primarily because of Grossman's "environmental variables," these "environmental variables" must be modified or changed. Because the environmental variables are largely public goods (or bads), careful identification of the most perniciously important variables is first necessary, and then careful planning and investment by government must follow.

III. COMPARATIVE DATA ON HEALTH AND WELL-BEING

In Table 2, the industrial countries [12] are compared with Africa[13] and with the Sudan alone, using some of the indices that could reasonably pertain to the World Health Organization's definition of health. Most of the indicies are available for the years 1960 and 1970. The data for Sudan are broadly similar to the data for the rest of Africa, although life expectancy at birth in Sudan is a little more than seven years greater than for all of Africa. Other areas of relatively small difference are population density per square kilometer of land and of agricultural area, the percentage of the labor force in agriculture, the percent of women in the labor force, school enrollment ratios, radio receivers,[14] and electric power consumption. When one compares the industrial countries with either Africa or Sudan, he finds wide discrepancies in all of the indices. Because the crude birth rate is so much higher than the crude death rate, the population growth rate is much higher in Sudan than in the industrial countries. Infant mortality is 140 per 1,000 live births, whereas it is only 17.1 in the industrial countries. Given the age structure of the population, the dependency ratio in Sudan is almost twice as high as that of the industrial countries which have 63.3 percent of their population in the working ages of 15–64 years. Thus, there are relatively more non-working mouths to feed in Sudan than in the industrial countries. If output per worker in agriculture is low and consequently output per capita is not higher than in the industrial countries, relatively less food per capita will be available. The comparative data on per capita calorie supply and protein supply suggest that this is the case, even though 80% of Sudan's labor force is employed in the agricultural sector. Schooling is much less advanced in Sudan than in the industrial countries at both the primary and secondary levels, and, as a consequence, the literacy rate is 19% in the former and 98.7% in the latter. Finally, the last four indices in Table 2 give some comparative data on the general availability of information and transportation, both of which can be used for preventive and curative health purposes.

The general impression conveyed by comparisons of the data in Table 2 is that the health and well-being of the average African and Sudanese citizen is well below that of the average citizen in an industrialized country. In part, this is a normative judgement because it is impossible to compare the benefits of a longer lifespan to a shorter lifespan of (say) viewing the sun rise each morning over the River Nile or over Mount Kilimanjaro. Nevertheless, from an economic standpoint, and all other things being equal, a person with a longer life has the

Table 2. Africa and Industrial Countries, Social and Health Indicators, 1960 and 1970

Indices of Well-Being	Industrial Countries		Africa		Sudan	
	1960	1970	1960	1970	1960	1970
Crude birth rate (per thousand)	18.3	16.7	47.2	44.2	52.0	49.0
Crude death rate (per thousand)	9.6	9.6	21.9	19.9	19.0	18.0
Infant mortality rate (/1,000 live births)	27.6	17.1	134.9	N.A.	N.A.	140.0
Life expectancy at birth (years)	69.4	71.7	N.A.	40.6	N.A.	48.0
Gross reproduction rate (%)	1.1	1.2	N.A.	3.1	N.A.	3.4
Age structure (% of population)						
0–14 years	26.0	24.5	42.8	44.5	47.0	45.0
15–64	64.4	63.3	53.5	51.8	51.0	52.0
65 years and older	10.3	11.8	3.5	3.6	2.0	3.0
Population growth rate (%)						
Total	0.9	0.9	2.3	2.5	2.9	2.9
Urban	1.1	2.1	N.A.	5.1	4.0	5.0
Population density per sq. km. of total land	126.9	136.4	11.7	14.8	5.0	6.0
Population density per sq. km. of agr. area	216.7	239.5	28.9	34.8	32.0	47.0
Urban population (% of total)	66.1	72.6	14.0	18.2	8.0	12.0
Percent of labor force in agriculture	20.3	12.9	73.9	68.0	86.0	80.0
Percent of labor force in industry	39.3	39.0	6.7	N.A.	6.0	N.A.
Percent of women in total labor force	32.6	34.8	24.0	24.8	10.0	10.0
Dependency ratio	0.9	0.9	1.3	1.4	1.6	1.7
Per capita calorie supply (% of requirements)	118.7	118.7	88.1	92.7	81.0	91.0
Per capita protein supply						
Total (grams per day)	90.4	94.4	57.7	60.5	55.0	63.0
From animals (grams per day)	49.8	55.0	N.A.	19.0	N.A.	24.0
Adjusted school enrollment ratio						
Primary	112.0	107.0	32.5	48.8	17.0	25.0
Secondary	63.2	73.6	3.2	6.5	5.0	9.0
Adult literacy rate (%)	N.A.	98.7	N.A.	16.3	N.A.	19.0
Radio receivers per 1,000 population	257.7	388.4	9.7	38.1	1.0	80.0
Passenger cars per 1,000 population	87.0	255.6	2.8	4.0	1.0	2.0
Electric power consumption (KWH per capita)	1,888.9	3,644.7	30.0	66.2	8.0	25.0
Newsprint consumption (per capita Kg/year)	13.5	20.4	0.2	0.2	0.1	0.2

Source: World Bank, World Tables, 1976, (Washington and Baltimore: The World Bank and The Johns Hopkins University Press, 1976), pp. 506–527. N.A. = Not Available.

Table 3. Percentage Distribution of Deaths by Cause in Selected Model Populations

	Model Developing Country	Model Developed Country
All Causes	100.0	100.0
Infectious, parasitic and respiratory diseases	43.7	10.8
Cancer	3.7	15.2
Diseases of the circulatory system	14.8	32.2
Traumatic injury	3.5	6.8
All other causes	34.3	35.0

Source: World Bank, Health Sector Policy Paper, March 1975, p. 10, as adopted from Population Bulletin of the United Nations, pp. 111–112, particularly Table V.33.

opportunity to produce a greater quantity of goods for consumption, which is, after all, the end of all economic activity.

Data for the purpose of comparing specific diseases by country are difficult to obtain, especially for developing nations, but the data in Table 3 show a marked contrast in disease patterns, at least as evidenced by causes of death in model developing and developed countries. The predominant killers in developing nations are the infectious, parasitic and respiratory diseases, while cancer and disease of the circulatory system are the most lethal diseases in the developed nations. Combining the information in Tables 2 and 3, we might derive a simple health scenario for developing nations which goes as follows: birth rates in developing nations are high, but, because of the high incidence of infectious, parasitic, and respiratory diseases, many new-born infants and children do not survive to become adults. Malnutrition compounds the problems because it makes those at risk to these diseases more susceptible to contracting them, and, once contracted, makes them less able to survive the pernicious debilitative

Table 4. Health Status and GNP Groups

	Countries Grouped by 1970 GNP Per Capita	Crude Birth Rate (per thousand)	Crude Death Rate (per thousand)	Infant Mortality Rate (per thousand live births)	Life Expectancy at Birth (years)
Group 1	$100 and Below	44.9	23.2	159.5	43.7
Group 2	$101 to $200	48.2	18.5	113.1	46.5
Group 3	$201 to $375	37.6	12.6	100.2	55.9
Group 4	$376 to $1000	32.9	9.8	55.6	61.4
Group 5	$1001 and Above	17.1	9.3	19.1	71.6

Source: World Bank, World Tables, 1976, (Washington and Baltimore): The World Bank and The Johns Hopkins University Press, 1976) pp. 506–508.

Health Planning in the Sudan

affects of the infection or parasite. It would appear that in a developing nation, a smaller percentage of those ever born survive to suffer the primarily degenerative diseases of older age which occur in a developed nation.

Table 4 reveals the relationship, which Table 1 implies, between levels of per capita Gross National Product and health status. On a macro level, high levels of life expectancy at birth and low crude death rates, low crude birth rates, and low infant mortality are associated with high levels of GNP per capita. These relationships are to be expected in light of Malenbaum's findings. A debilitated, high-dependency ratio economy would have difficulty in pursuing high levels of GNP per capita.

IV. COMPARATIVE DATA ON HEALTH INPUTS

Levels of health care are usually measured in terms of input variables, such as physician-population ratios; this practice does not make much economic sense and can only be defended on the grounds that health outcomes are difficult to measure precisely and that most of the available data are in input form. For comparative purposes, however, input data can give some information about the relative intensity of the health effort. Whether that effort is pointed in the right direction is an issue which we will explore later.

Table 5 contains data on three measures of health expenditures for Sudan and for a few of its neighbors, plus data for the United Kingdom and for the United

Table 5. Health Expenditures in Selected Countries 1971

	Health Budget as a Percent of the National Budget	Health Budget as a Percent of Gross National Product	Government Expenditures Per Capita ($)	GNP Per Capita (1973) ($)
Central African Empire	8.4[a]	1.9[a]	2.81[a]	160[e]
Chad	NA	NA	NA	80
Zaire	6.1	1.8	4.82	140
Uganda	9.6	1.7	2.24	150
Kenya	6.4	1.7	0.14	170
Ethiopia	6.9	0.8	0.67	90
Libya	5.8	2.4	35.00	3530
Egypt	8.4	1.6	3.91	250
Sudan	4.9[b]	1.6[b]	1.49[b]	130
United Kingdom	9.5	4.3	105.16	3060
United States	20.9[c]	7.6[d]	368.25[d]	6200

Source: [a] World Bank, *Health Sector Policy Paper*, March 1975, pp. 74–75.
    [b] World Bank, *World Tables 1976*, pp. 218 & 373.
    [c] *Statistical Abstract of the United States*, 1974, p. 242.
    [d] Robert M. Gibson & Charles R. Fisher, "National Health Expenditures, Fiscal Year 1977," *Social Security Bulletin*, July 1978, p. 5.
    [e] World Bank, *World Tables 1976*, pp. 70, 72, 246, 236, 138, 94, 152, 90, 219, 280, 282.

States. On the first two measures, Sudan ranks at or near the bottom, although, because of its smaller population, it spends more than Ethiopia and Kenya, on a per capita basis. All of these countries, with the exception of oil-rich Libya, simply do not have the financial resources to devote as much money to health care as do the United Kingdom and the United States, whose GNP per capita is much higher.

This lack of resources is reflected in the data in Table 6 on health personnel and hospitals. Anyone familiar with such data for the United States will immediately recognize how much lower all of the measures are for these selected neighboring African countries. On many of these conventional physical input measures, Sudan again ranks near the middle or the bottom, although it consistently does better than its poorer neighbors, Chad, Tanzania, and Zaire. The few pieces of data available in the last two rows of Tables 6 indicate that the distribution of physicians is even more skewed than in the United States and in other developed countries. Again, the relative poverty of these countries explains why the measures are all, with the possible exceptions of Egypt and Libya, lower than in the developed countries.

The World Bank (1975) has stated that the developing nations do not spend optimally on health care. In a 1975 report, *National Health Programme*, the Sudanese Ministry of Health discussed its country's health status and health policy:

> Communicable endemo-epidemic diseases are highly prevalent in the Sudan and affect particularly the child population. In general, diseases resulting from poor environmental sanitation constitutes the bulk of the health problems which the Sudanese Government has to deal with. Unfortunately, it would seem that until now the real expansion of health facilities which has taken place since the last eight to ten years was not directly and specifically related to the solution of these problems.
>
> Furthermore, some of the economic programs which have been accomplished during the last decade by the Sudanese Government in the field of agricultural development have aggravated certain disease problems, such as malaria and schistosomiasis, which would result in decreasing the economic output expected of the projects. More generally, any disease should not be looked at just as a medical but as a social and economic problem...
>
> It is to be noted that until very recently the development of curative medical services has received priority attention with the result that a large proportion of the budget of the Ministry of Health is presently committed to the maintenance and operation of facilities and services which are not necessarily the best and most effective way of controlling endemo-epidemic diseases largely associated with poor environmental conditions.[15]

This is not simply a rhetorical statement, but is contained in a document which is the result of a careful study undertaken jointly by the Sudanese Ministry of Health and the World Health Organization between October 25, 1974 and April 24, 1975. The document itself, as its name implies, is an analysis of health problems in the Sudan, but, equally importantly, it is a plan to deal with those problems. Before discussing the essence of *National Health Programme*, its

Table 6. Health Personnel and Hospitals North East and East Africa, Varying Years

Category	Chad	Egypt	Ethiopia	Kenya	Libya	Sudan	Uganda	Tanzania	Zaire
Health Personnel									
Year	1972	1971	1972	1972	1972	1972	1971	1972	1972
Physicians per 10,000 population	0.2	5.5	0.1	1.7	7.9	0.8	1.2	0.5	0.4
Population per physician (000)	63.2	1.8	69.3	5.8	1.3	13.1	8.7	21.5	28.0
Dentists per 10,000 popultion	0.0	0.7	0.0	N.A.	0.6	0.0	0.1	0.1	0.0
Population per dentist (000)	1,895.5	13.6	1,080.5	N.A.	16.0	203.6	198.6	177.0	714.4
Nurses per 10,000 population	0.9	6.6	N.A.	7.7	21.9	5.4	1.9	2.4	0.9
Population per nurse (000)	11.7	1.5	N.A.	1.3	0.5	1.9	5.2	4.2	11.7
Nurses & Midwives per 10,000 population	1.2	11.3	0.4	11.1	24.2	7.7	3.1	N.A.	1.4
Population per nurse & midwife (000)	8.3	0.9	22.3	0.9	0.4	1.3	3.2	N.A.	7.2
Pharmacists per 10,000 population	0.0	2.0	0.0	0.1	1.5	0.2	0.1	N.A.	0.1
Population per pharmacist (000)	631.8	5.1	231.5	84.4	6.7	57.1	168.8	N.A.	174.5
Hospitals									
Population per bed	1,070	460	3,080	760	230	1,040	640	700	320
Beds per 10,000 population	9.4	21.7	3.2	13.2	43.6	9.6	15.6	14.3	31.4
Admissions per 10,000 population	N.A.	N.A.	N.A.	N.A.	N.A.	N.A.	N.A.	N.A.	N.A.
Bed occupancy rate (%)	N.A.	N.A.	N.A.	N.A.	N.A.	N.A.	N.A.	N.A.	N.A.
Distribution									
Physicians per 10,000—Capitol	N.A.	N.A.	1.7	7.5	6.2	N.A.	3.9	N.A.	N.A.
Physicians per 10,000—Rural areas	N.A.	N.A.	0.1	0.2	4.1	N.A.	0.1	N.A.	N.A.

Source: World Health Organization, *World Health Statistics Annual, 1972*, Vol. 3, *Health Personnel and Hospital Establishments*, (Geneva: World Health Organization, 1972), pp. 47, 51, 55, 60, 64, and 201.

N.A. = Not Available.

import and its prospects, we must view it within the twin contexts of Sudanese planning and the Sudanese budgetary process.

V. PLANNING AND BUDGETING IN THE SUDAN

An understanding of the planning process in general and of health planning in the Sudan would be incomplete if divorced from the context of Sudan's modern economic history since independence from Great Britain. The first part of this section describes the recent development of the economy and the growth of the various economic sectors.

Sudan's independence was declared in January 1956. Before that date, Sudan had been an Anglo-Egyptian domain since 1899, with Britain the senior partner throughout the period. During the early part of the "condominium" (1899–1956), most of the developmental emphasis was placed upon the construction of railways, harbors, and river steamer services. However, the greatest change in the economy was caused by the introduction of modern irrigation schemes, as witnessed by the Gezira scheme, still the largest farm in the world. The introduction of pump schemes along the Nile and its tributaries was another major development in laying the basis for Sudan's present economy.

The parliamentary system, which had been adopted after independence, lasted until a military coup in November 1958. Civilian rule and the parliamentary system were again restored in October 1964. The years 1956–1969 can be characterized as years of much political instability, of intermittent civil war in the South and of a basically poor, free enterprise, laissez-faire economy. However, in 1960, an Economic Planning Secretariat was formed in the Ministry of Finance and Economics,[16] and its task was to prepare a comprehensive plan of economic development. The initial period of the Plan was seven years which was later extended to ten years, but political instability forestalled the full implementation of the Plan beyond 1964. The relative political stability that prevailed and continues to prevail since the 1969 May Revolution has set the stage for a more effective utilization of the country's resources and a faster rate of economic and social development. The May Revolution of 1969 also enunciated a socialist political and economic philosophy. Consistent with that philosophy, the entire economy was nationalized, and in order to ensure the management of the socialist economy, a team of Russian experts was brought in to begin work on a five year plan. For the first time in the history of the country, a fully-fledged Ministry of Planning was created and in 1970, the 1970/71–1974/75 Five Year Plan was officially begun.

As was shown in Table 2, Sudan's social and health indicators are broadly comparable to those of the rest of Africa and very dissimilar to those of the industrial countries. Sudan's economy is based upon agriculture, with approximately 80 percent of the population engaged in agricultural or livestock activities; it has been classified by the United Nations as one of the twenty five

Table 7. Gross Domestic Product Per Capita at 1970 Prices
(1961 and 1971 in U.S. Dollars)

Country	1961	1971	Percent Change (1961–1971)
Nigeria	112	141	26%
Egypt	171	220	29%
Morocco	182	232	27%
Ghana	283	286	1%
Ethiopia	59	69	17%
Kenya	88	149	69%
Uganda	79	121	53%
Tanzania	61	98	61%
Gambia	96	131	36%
Sudan	120	122	2%

Source: United Nations, Economic Commission for Africa, *African Economic Indicators*, (Addis Ababa: United Nations, 1973), p. 6.

poorest countries in the world. In 1973, its per capita income was measured by the World Bank to be $130.[17]

Table 7 gives relative rates of growth of Gross Domestic Product per capita in the ten years of the First Plan. Sudan's rate of growth in GDP per capita was the second lowest of the group. This is one indication of the success of the first Ten Year Plan (1960–1970). The results of the first Five Year Plan (1971–1975), which was extended two more years until 1977, are not yet available. Finally, the government has recently begun the Six Year Plan of Economic and Social Development (1977–1983). Coincident with the beginning of the Six Year Plan, the government's philosophy has shifted from socialism to capitalism, nationalized industries are being returned to their former owners, and now there are more Americans than Russians in Khartoum, the capital.

Table 8 is interesting because it gives a basis for comparing what was actually planned during the Ten Year Plan and the Five Year Plan and what actually happened. During the Ten Year Plan, actual investment in agriculture and irrigation exceeded planned investment, 39% vs 32% of the budget. The opposite was true for social services and public administration, 23 percent vs 32 percent of the budget.[18] The actual investment in the other two categories, industry and mining, and transport and communication both exceeded planned investment by one percentage point. Therefore, the sole net loser was social services and public administration. During the succeeding Five Year Plan, actual investment in agriculture and irrigation was 13 percentage points less than planned investment as a percent of the total budget. Investment in transport and communications was almost double the planned investment and actual allocations for the social services and public administration, from a planned 22 percent to 20 percent of the total budget. Columns three and six give aggregate figures for the fifteen year

Table 8. Sudan Public Sector Allocations 1961–70 and 1970–75 (Millions of Sudanese Pounds)[a]

	Planned Investment						Actual Investment (Current Prices)					
	Ten Year Plan (1960–1970)		Five Year Plan (Original) (1970–1975)		Two Plans Total (1961–1975)		Ten Year Plan (1961–1970)		Five Year Plan (Original) (1970–1975)		Two Plans Total (1961–1975)	
	Amount	Percent	Amount	Percent	Amount	Percent	Amount	Percent	Amount	Percent	Amount	Percent
Agriculture & Irrigation	90	32	81	39	171	35	113	39	66	26	179	33
Industry, Mining & Power	42	15	49	24	91	19	46	16	73	29	119	22
Transport & Communications	63	21	30	14	93	19	65	22	62	25	127	24
Social Services & Public Administration	90	32	45	22	135	27	65	23	49	20	114	21
Total[b]	285	100	205	100	490	100	289	100	250	100	539	100

Notes:
[a] 1 Sudanese Pound, £S 1 = $2.50 at the time of the Plan.
[b] Excluding foreign technical assistance.
Source: Ministry of National Planning, The Democratic Republic of Sudan, *The Six Year Plan of Economic and Social Development*, (Khartoum: Ministry of National Planning, 1977) p. 40.

period 1961–1975. The largest gap between planned and actual allocations was in the social services and public administration sector, with £S 135 million planned and £S 114 million actually allocated. Analysis of the trends shown here reveals that relatively greater emphasis was placed upon agriculture and irrigation during the Ten Year Plan (1961–1970); during the Five Year Plan (1970–1975), relatively greater emphasis was placed on industry, mining and power, and especially upon transport and communications. The public sector shares that are envisaged for the Six Year Plan (1977–1982) are given in Table 9. For purposes of this analysis, columns (2), (3), and (4) are the most revealing. Under proposed investment in new projects, the social services would garner 31 percent, the largest share of the budget. Because resources expected to be available for public sector investment were limited to £S 1570 million and because on-going projects were expected to consume £S 370 million, only £S 1,200 million were available for investment in new projects. Of this, the social services will receive £S 235 million, or 20 percent of the total new investment budget for the six years. The 77 percent reduction in social services between proposed and allowed investment in new projects is the largest reduction for any sector. However, there is a reserve of £S 225 in column (3), and taking account of "past experience, existing commitments, future investment proposals, resource constraints, and the overall development strategy," the reserve will probably be distributed so that final shares appear as follows:[19]

Sectors	Percent Share	Change From Column (4) Table 9 (percentage points)
Agriculture & Irrigation	32	+5
Industry, Mining, Tourism and Power	25	+4
Transport and Communications	24	+4
Social Services and Public Administration	19	+2

Again, the social services suffer, relative to the other sectors, as the difference column shows. A comparison of Tables 8 and 9 shows that the share of the social services sector in the total budget is lower in the *proposed* Six Year Plan than it has been in either the planned investment or in the actual investment of the two previous Plans (1961–1975), albeit that the absolute amount of funds to be spent in that sector has climbed dramatically. Nevertheless, comparison of the relative shares in the three Plans gives some indication of the priorities of the Ministry of Planning, which does have the final word in budgetary matters.

As regards investments in health *per se*, very little is said in the 174 page planning document for the Six Year Plan. Mention is made that between 1977 and 1994, "major expansion and improvement will take place in all the basic services and particularly in the fields of education and health . . .",[20] and that

Table 9. Sudan, Public Sector Allocations, On-Going Projects, Proposed Investments, and Allowed Investments, Six Year Plan (1977–1982) (Millions of Sudanese Pounds)[a]

Sector	(1) On-Going Projects		(2) Proposed Investment in New Projects		(3) Allowed Investment in New Projects		(4) Total Allocations (1) + (3)	
	Amount	Percent of Total	Amount	Percent of Total	Amount	Percent of Total	Amount	Percent of Total
Agriculture & Irrigation	90	24	900	27	335	28	425	27
Industry, Mining, Tourism and Power	160	43	600	18	175	15	335	21
Transport & Communications	90	24	730	24	230	19	320	20
Social Services, Housing & Public Administration	30	8	1030	31	235	20	265	17
Reserve	—		—		225	19	225	14
Total	370	100[b]	3310	100[b]	1200	100[b]	1570	100[b]

Notes:
[a] Sudanese Pound £S 1 = $2.50 at the time of the Plan.
[b] Totals may not add to 100% because of rounding.
Source: Ministry of National Planning, The Democratic Republic of Sudan, *The Six Year Plan of Economic and Social Development* (Khartoum: Ministry of National Planning, 1977), pp. 41–42.

one of the Plan objectives is: "providing more social services and upgrading their standards, particularly in the fields of health and education."[21] However, in the summary statement, "General Strategy" at the end of Chapter Three, "Salient Features of the Six Year Plan," health *per se* is not even mentioned, only the ideas of "ensuring adequate social services" and "promotion of equity and the widest distribution."[22]

Chapter VII, "Human Resources and the Development of the Labour Force" is the largest chapter in the plan document. Yet, health is not mentioned once. The strategy for the development of human resources is divided into five headings: (1) demographic aspects, where hygiene is recognized, (2) employment and the labour force, (3) productivity and technological methods, (4) training, and (5) the labour market and the administration of labour matters.[23] The essence of the chapter is concerned with education and labour market disequilibrium. Pages 142–145 give some consideration to proper nutrition, in a chapter entitled, "Consumption". Finally, p. 168 mentions "the distribution of general services of education, health and social welfare etc., in a proper manner, and in the way which would eradicate social and economic differences in the various parts of the country."[24] Close study of the total document makes it clear that the development of "human resources" is almost synonymous with "education and training", from a labour market entry perspective.

Table 10 shows the yearly trend in expenditures for Sudan's central government for 1970–1976. The top half of the Table gives the sums in current Sudanese Pounds, and the bottom half expresses them as percentages of the total. We have broken down the social services sector into its components so that the reader might have a better feel for the relative weights of each of the social services over time.

Before analyzing the social services themselves, it might be well to discuss another aspect of relative shares within the budget. Under "Other Expenses," the share of local government increased from 14 percent in 1969/70 to 21.2 percent in 1975/76, with the sharpest jump occurring in 1973/74. During the same period, defense expenditures declined from 25.8 percent of current expenditures to 16.5 percent. This relatively large shift in budgetary policy was the result of the 1971/72 cessation of the sixteen year civil war between North and South Sudan. The predominantly black-animist southern third of Sudan which contains a fourth of the population had tried to secede from the Arab-Moslem North. The dispute was settled in the Addis Ababa Accord of 1972 with the proviso that the local southern government would be semi-autonomous and that the central government in Khartoum would channel a relatively greater share of resources to its poorer subunit in the South. Hence, the shift in budget expenditures in 1973/74, once the administrative and political details of the Accord had been resolved. Also, the "settlement of loans" sector has almost doubled as a percent of expenditures during the same time period. As has been true for most developing nations, the rapid increases in the price of oil and petroleum prod-

Table 10. Current Expenditures of the Central Government Sudan, 1970–76 (Millions of Sudanese Pounds)[a]

Sector	1969/70	1970/71	1971/72	1972/73	1973/74	1974/75	1975/76 July/Feb.
ECONOMIC SERVICES	18.3	18.5	21.1	23.6	23.4	28.4	20.5
SOCIAL SERVICES	19.8	20.1	21.7	23.4	16.8	18.7	13.0
EDUCATION	9.3	9.9	10.0	10.9	8.3	8.6	5.4
HEALTH*	7.9	7.8	8.4	8.5	4.6	5.2	4.3
INFORMATION	1.6	1.3	1.4	1.8	2.4	2.8	1.6
YOUTH	0.3	0.5	1.1	1.1	0.7	1.3	1.0
COOPERATION	0.2	0.2	0.2	0.6	0.2	0.1	0.1
LABOUR	0.4	0.3	0.3	0.3	0.4	0.4	0.4
HOUSING	0.1	0.1	0.2	0.2	0.2	0.3	0.2
SETTLEMENT OF LOANS	9.2	11.8	14.7	16.1	19.0	27.6	22.6
DEFENSE AND SECURITY**	37.1	46.2	47.6	45.7	41.2	41.9	30.5
OTHER EXPENSES	59.5	49.8	48.2	63.1	89.2	147.4	98.2
LOCAL GOVT.	20.1	17.7	18.6	15.6	37.9	48.1	39.1
TOTAL	143.9	146.4	153.3	171.9	189.6	264.0	184.8

* Including Social Welfare Department
** Including Civil Aviation

Percent of Total

ECONOMIC SERVICES	12.7%	12.6%	13.8%	13.7%	12.3%	10.8%	11.1%
SOCIAL SERVICES	13.8	13.7	14.2	13.6	8.9	7.1	7.0
EDUCATION	6.5	6.8	6.5	6.3	4.4	3.3	2.9
HEALTH	5.5	5.3	5.5	4.9	2.4	2.0	2.3
INFORMATION	1.1	0.9	0.9	1.1	1.3	1.1	0.9
YOUTH	0.2	0.3	0.7	0.6	0.4	0.5	0.5
COOPERATION	0.1	0.1	0.1	0.4	0.1	.04	0.1
LABOUR	0.3	0.2	0.2	0.2	0.2	0.2	0.2
HOUSING	0.1	0.1	0.1	0.1	0.1	0.1	0.1
SETTLEMENT OF LOANS	6.4	8.1	9.6	9.4	10.0	10.5	12.2
DEFENSE AND SECURITY	25.8	31.6	31.1	26.6	21.7	15.9	16.5
OTHER EXPENSES	41.4	34.0	31.4	36.7	47.1	55.8	53.1
LOCAL GOVT.	14.0	12.1	12.1	9.1	20.0	18.2	21.2
TOTAL	100.0	100.0	100.0	100.0	100.0	100.0	100.0

Note:
[a] Sudanese Pound £S 1 = $2.50 at the time of the Plan.
Source: Ministry of Finance, Planning & National Economy, Sudan, *Economic Survey 1975–76*, (Khartoum: Ministry of Finance, Planning and National Economy, 1977) Appendix No. 5.

ucts, including fertilizers, has put greater pressure on an already precarious balance-of-payments position in Sudan. Mounting pressures have necessitated a greater reliance on loans, many of them being short-term.

Given the shifts that have occurred in the last three of the five expenditure sectors because of these factors, it is not surprising that economic services and social services have suffered a relative decline. The greatest *relative* and *absolute* decline has been in the social services, from 13.8 percent of current expenditures in 1969/70 to 7.1 percent in 1974/75 and from £S 19.8 million to £S 18.7. The data on the components of the social services sector show that education and health have borne the brunt of the consequences of shifts in external factors, such as the increasing price of oil, and of the internal political factors, with the end of the civil war. Between 1969/70 and 1974/75 current expenditures on education and health declined 49 percent and 64 percent respectively as a percent of total current expenditures. There is no complete statistical record of how the newly increased resources are being used in the South of Sudan. Given the physical ravages caused by the civil war, it might be safe to assume that very little emphasis is being placed upon the social services and that the "current expenditures" sent to the South become "capital expenditures" once they arrive.[25]

The Sudanese government's practice is to classify expenditures as "development" or "recurrent." Table 11 contains the breakdown for the health budget for the years 1971–1975. In comparing Table 10 and Table 11, Column (3), which are from different sources, it can be seen that the data on recurrent (or current) expenditures for health are somewhat inconsistent, especially for 1973/74. We can only guess at the reason for the large difference for that year, but again it probably results from the confusions involved in the transfer of semi-autonomy to the Southern region after the civil war. Although the total amount allocated to health from all sources (Table 11, Column 7) increased from £S 15.4 million to £S 22.3 million, most of this increase was absorbed by wage and salary inflation.[26]

In order to assess the allocation of resources within Sudan, Table 12 presents the recurrent and development expenditures for all services in the Southern region for the only year available, FY 1973. In the recurrent budget, almost 100 percent of the funds came from the central government, while 94 percent of the funds in the development budget came from the central government.[27] Even though 18% of current expenditures are devoted to health, 70 percent of that money was spent on wages and salaries. For fiscal 1973, the recurrent government health budget was £S 15.6 million in Table 11; the expenditure of £S 1.3 million on health in the South in that year represents 8.3 percent of current expenditure. The South has 25 percent of the population of Sudan. Health development expenditure in the South amounted to 16 percent of total health development expenditure.

Let us summarize the contents of this section. The main current planning document, the *Six Year Plan of Economic and Social Development,* barely men-

Table 11. Government Health Budget 1970/71 to 1974/75 (Millions of Sudanese Pounds)[a]

Year	Development Expenditure Ministry of Health		Recurrent Expenditure				Total	
			Ministry of Health		Ministry of Local Government			
	(1) Amount	(2) Percent	(3) Amount	(4) Percent	(5) Amount	(6) Percent	(7) Amount	(8) Percent
1970/71	1,470	9.6	7,911	51.4	6,000	39.0	15,381	100
1971/72	1,650	9.9	9,030	54.1	6,000	36.0	16,680	100
1972/73	2,213	12.4	9,631	54.0	6,000	33.6	17,844	100
1973/74	2,418	12.4	11,023	56.7	6,000	30.9	19,440	100
1974/75	3,053	13.7	4,771	21.4	14,522	65.0	22,345	100

Note:

[a] Sudanese Pound £S 1 = $2.50 at the time of the Plan.

Source: Ministry of Health, Democratic Republic of Sudan, *National Health Programme, 1977/78–1983/84*. (Khartoum: Khartoum University Press, 1975) Annex I, p. 16, Table 2.14 (a).

Table 12. Southern Region, Expenditures by Type of Service, FY 1973 (Millions of Sudanese Pounds)[a]

Recurrent Expenditures		
	Amount	Percent
Administration	2,721.7	39.0
Economic Services	1,093.4	15.7
Social Services	3,157.6	45.3
Health	1,254.9	18.0
Education	660.7	9.5
Information	536.6	7.7
Housing & Public Utilities	528.2	7.6
Labor	177.2	2.5
Total	6,972.7	100.0
Development Expenditures		
Agriculture	1,026.0	20.5
Industry	63.0	1.3
Transport & Communications	1,719.2	34.3
Education	204.0	4.1
Health	351.7	7.0
Housing	525.0	10.5
Information	460.7	9.2
Public Utilities	157.0	3.1
Administration & Other	505.6	10.0
Total	5,012.2	100.0

Note:
[a] Sudanese Pound £S 1 = $2.50 at the time of the Plan.

Source: Ministry of Finance, Planning, and National Economy, Southern Region, The Democratic Republic of the Sudan, *Southern Region Budget* (Juba: Ministry of Finance, Planning, and National Economy, Southern Region, 1973).

tions health or its importance for economic development. Expenditure priorities, as revealed in the budgets spanning the period 1960–1983, give the social services a declining share. Although the data in Table 10 indicate that current health expenditures at the central government level have fallen as a percent of the central budget, it is difficult to see what has happened at the local level because the data are not sufficiently refined. Table 12 indicates a maldistribution of health spending because the South contains 25 percent of the population of Sudan but does not have as large a share of money for expenditures on health. Finally, because the data are in current Pounds, they mask the effect of inflation, which may have reduced the real amount of funds spent on health, much of which goes for wages and salaries. Of course, these figures present only a partial picture because, under the circumstances, money spent on agriculture may increase nutritional levels, and money spent on transport may make it easier to build better housing, both of which enter into the production function of the commodity, health.

VI. THE NATIONAL HEALTH PROGRAMME

The document *National Health Programme* (NHP) is a plan, but not one in an economic sense. Since it was done separately at the Ministry of Health and not integrated with the more general and main plan, the *Six Year Plan of Economic and Social Development* drawn up at the Ministry of Planning, the NHP does not really have a budget constraint that would naturally flow from the main planning document itself. Lacking a budget constraint, the NHP could be viewed as a shopping list of what the most perceptive physicians in the Ministry of Health and at the World Health Organization would like to have. On the more positive side, the NHP is a valuable piece of work because it represents a detailed compilation and ranking of Sudan's health problems and does present estimates of what it might cost to ameliorate or eradicate them. As far as the authors are aware, most countries of the developing world have not done this in as great detail.

In the terms of reference given to the study group which compiled the NHP, the Minister of Health laid down the following tasks:

1. Study and analyze the documentation and data related to health in the fields of demography, economy, health and overall socio-economic policy of the government.
2. Identify the health problems of the country and relate them to the social, cultural, political, economic and demographic components.
3. Propose feasible programmes and projects to deal with the identified problems.
4. Draw alternative strategies for choice by decision makers.
5. Define methods of implementation of projects in relation to material and human resources available.
6. Identify programme areas for external assistance.[28]

And, the following priorities were stated:

1. Preventive and Social Medicine are considered as the top priority especially in the fields of control or eradication of endemic and epidemic diseases and improvement of environmental health conditions. In this respect, special attention is to be given to Maternal and Child Health and School Health Services.

2. Strengthening of rural health care facilities to ensure complete health coverage and fair distribution to the entire population with basic health care.
3. Provide training facilities for all levels of professional, technical, and auxiliary health manpower.
4. Consolidate existing curative health care facilities to provide better services for the population and allow for some expansion in curative health care facilities in the less developed areas.
5. Direct medical research towards health problems according to their priorities.[29]

Table 13 sets forth a concise summary of the essence of almost one hundred pages of the NHP. At the moment, malaria and bilharzia are the most difficult problems. Malaria is an anomalous problem, for, as the reader can see under heading (2), in Table 13, Sudan's extensive irrigated agricultural schemes are creating much of the problem. One of the problems faced by countries in the tropics is that there is no yearly cycle of change in weather conditions. That is, there is no winter to freeze and kill off the various forms of insect life, which, if left unattended, proliferate in geometric proportions throughout the year.[30] If a tropical country exacerbates the climatic difficulty by providing man-made pools of water, then natural breeding areas are enlarged and the difficulty becomes compounded. The anomaly of the situation in Sudan is that it has vast natural resources in the River Nile and in the rich agricultural land surrounding the river, particularly in the triangular areas, south of Khartoum, and delineated by the Blue and White Niles (the Gezira). Sudan has done much, and has ambitious plans for the future, to siphon off the waters of the Nile, and using gravity irrigation, to feed these potentially rich soil lands for the production of much-needed export crops such as cotton, and, for the production of domestically-consumed wheat, groundnuts (peanuts) rice and sugar. As the "Objectives" column of Table 13 suggests, there is a tradeoff to be made. Man-made malaria can only be completely eradicated by eliminating the irrigated agricultural projects, projects which are responsible for 40 percent of Gross Domestic Product and 90% of export earnings and generate over half of total employment. These earnings, in turn, produce a surplus which is plowed back into further economic growth for investments in physical capital and in human capital, such as investments in education and in health care. However, from the viewpoint of productivity, malaria takes its toll. During the prime agricultural periods of planting and harvesting, thirty-three working days are lost per worker, which in 1977 caused a labor shortage resulting in difficulties in getting the cotton crop picked on time. The group which produced the NHP concluded that a workable solution lay in controlling man-made malaria in the areas where it exists, at technically and economically feasible levels.

The other type of malaria listed under heading (1) can be mainly eliminated as

Table 13. Health Program Measures & Objectives, Sudan, 1975

Program	Present Level of Problems	Objective for 1984
(1) Malaria, Nation-Wide	Estimated as affecting 20% of population.	Reduce morbidity from 20% to 5%.
(2) Malaria, Man-Made	An average of 33 working days lost/worker/year in irrigated agricultural areas, during planting and harvesting seasons.	Man-made malaria technically difficult if not impossible to eliminate, so, control malaria at technically & economically feasible levels.
(3) Primary Health Care		
(a) Primary health care services	Three well-to-do provinces have reasonably good care; others do not.	The building of 1,247 additional primary health care service units to provide 100% coverage of rural population.
(b) Public lack of health information and lack of hygienic habits	90% of the population is illiterate and 60% of the illiterate population has not acquired hygienic habits.	Reduce the percentage of people lacking health information from 60% to 40% and instill hygienic habits in population through better information.
(c) Communicable diseases preventable by immunization	25% of eligible children in Khartoum province are immunized by DPT and polio vaccines. Other parts of Sudan, insignificant immunization.	80% annual coverage of eligible infants in each accessible primary health care population.
(d) Protein calorie malnutrition	One of the major health related problems in the Sudan; 50% of children, age 0–4 years, suffers from some degree of malnutrition.	50% reduction in malnutrition.
(e) Gastro-enteritis dysentery	8,255/100,000 population are inflicted.	Reduce the incidence of gastroenteritis by 70% of level expected in 1984.
(f) Tuberculosis	Incidence of 0.12%/year (120/100,000 population).	Increase the cure rate from 20% to 50%. Decrease incidence in 15–29 year age group by 40% and decrease incidence in 30+ age group by 10%.
(g) Sleeping Sickness (Southern Region)	Extent not completely known but may be of epidemic proportions in some areas of South.	Re-establish control of sleeping sickness in the South by 1980.

(continued)

Table 13—(Continued)

Program	Present Level of Problems	Objective for 1984
(h) Kala Azar (Visceral Leishmaniasis)	Present and expected levels not known. Large population remain to be surveyed in affected areas.	Control of the disease through diagnosis and treatment. Assessment of the magnitude of the problem.
(4) Bilharzia, Man-Made (Schistosomiasis)	At 3 years of age, 11% prevalence, at 5 years of age, 35% prevalence, at 7 years of age, 60% prevalence in the Gezira and Managil areas.	Reduce the prevalences in the following groups: 3 years: from 11% to 0% 5 years: from 35% to 0% 7 years: from 60% to 10%.
(5) Safe Water Supply	900 shallow wells belonging to the government & 30,000 privately owned shallow wells & water holes need facilities for preventing human and/or animal contamination.	To provide facilities to prevent contamination of all 1,900 government wells expected by 1984.
(6) Environmental health	Need for a detailed intersectoral programme formulation.	Will be done before the beginning of the next socio-economic development plan period.
(7) Food (Maize) in certain regions of Sudan	60% of the population concerned (1,785,000) are affected by the lack of maize.	100% of the population concerned will be provided with this basic commodity.
(8) Onchocerciasis (Blindness and debility)	40% of males in a random sample of school children (5–14 years) in a specific geographic area skin or nodule positive.	To reduce the percentage of skin and/or nodule positives of the age group (5–14) males by 25%.

Source: Ministry of Health, *National Health Programme, 1977/78–1983/84.* (Khartoum: Khartoum University Press, 1975), pp. 49–130.

was done in Louisiana, Florida and Bermuda: with use of pesticides, such at DDT, and the use of well-known antilarval measures. The constraints for this kind of program are threefold: (1) lack of hard currency for the purchase of the chemicals, (2) lack of transportation for the chemicals because of difficulties in the road system, and, (3) in the Western Provinces, an inadequate supply of water for spraying purposes, which is also linked to the transportation problem.[31]

A related health problem can be found under heading number (4) in Table 13. Bilharzia or schistosomiasis is a disease caused by a microscopic parasite which lives in the water and enters the skin and, subsequently, the blood stream. It kills later in life via irreversible tissue damage. Like malaria, its main economic consequences are its debilitating effects on workers. The disease is spread through the larvae of the parasite which are fostered and born in a type of snail (*Schistosomae Mansoni* or *Schistosomae Haematobium*), living in slowly moving water, such as that in irrigation ditches and in water which does not change its level. Human defecation feeds the snail. Unlike malaria, bilharzia could be better controlled by spraying with molluscides, by inducing humans to discontinue defecating in the irrigation ditches, and by providing alternate facilities for such human needs. Thus, the solution to the bilharzia problem is basically one of adequate information and education, rather than medical. For those who are already ill with the disease, treatment with the drug Hycanthone is sometimes successful if the course of the disease has not become too prolonged.

The second gravest set of problems listed in Table 13 pertains to the distribution of health manpower in the Sudan as delineated in Table 14. By western medical standards, the number of physicians per 10,000 inhabitants is low in the Sudan, but the physician-population ratio in Khartoum relative to elsewhere throughout Sudan is high.[32] Likewise, there is a much greater concentration of the other categories of health professionals in Khartoum and in the provincial capitals than in the rural agricultural area where the majority of the population of Sudan live. The economic and sociological incentives for these health professionals to live in Khartoum and in the provincial capitals are so strong, that it is difficult to imagine their doing otherwise in the absence of compulsion or strong economic counterincentives. Most professional people want to work in areas where they can earn the highest income, where they can associate with and learn from fellow professionals, and where the best and latest capital equipment is available, where they might associate socially with other educated and urbane professionals, and where there is a reasonable probability that their own children will receive a decent education to enable their children to have the eventual possibility of achieving a similar professional status. In most countries, whether developed or in the process of developing, rural areas simply do not have the attractions that most professional people voluntarily seek. The problem is exacerbated in developing nations because their rural areas relative to their urban areas have even less to offer to their professionals' economic, cultural and social

Table 14. Specialty Distribution of Health Manpower, Sudan, January 1975

Classification	Number	Classification	Number
1. Medical Doctors		3. Medical Assistants	
(a) Headquarters	18	(a) General	1,049
(b) Public Health	36	(b) Lab Assistants	212
(c) Research	14	(c) Theatre Assistants	153
	68	(d) Dental Assistants	66
(d) Specialists		(e) Eye Assistants	75
Gynaecological	39	(f) Anaesthesia Assistants	67
Pediatric	16	(g) Psychiatric Assistants	9
Physician	29	(h) Pharmacist Assistants	14
Surgery	48	Total	1,645
Phys. Medicine	1	4. Technicians	
Eye	23	(a) Radiographers	107
Chest	16	(b) Lab Technicians	88
Psychiatry	11	(c) Refractionists	84
Skin	7	(d) Sisters	195
Anaesthesia	10	(e) E.C.G. Technicians	7
X-Ray	14	(f) Physiotherapists	3
	214	(g) Other Technicians	30
Fellowship (residents)	113	Total	514
Total	395	5. Pharmacists	42
(e) Registrars	76	6. Health Visitors	307
(f) General Doctors	506	7. Health Statisticians	23
(g) Housemen	161	8. Dental Technicians	50
(h) University	102	9. Social Workers	27
(i) Armed Forces	47	10. Dental Mechanics	4
	892	11. Nutritionists	3
Grand Total	1,287	12. S.P.H.D.	19
2. Dentists		13. P.H.D.	62
(a) Specialists	4	14. Public Health Officers	151
(b) Dentists	50	15. Sanitary Overseers	271
Total	54	16. Nurses	11,120
		17. Village Midwives	4,438

Source: Ministry of Health, *National Health Programme, 1977/78–1983/84*, Khartoum: Khartoum University Press, 1975), pp. 30–31, Annex I.

aspirations than in the developed countries. But, this is only part of the explanation for difficulties in providing adequate primary health care.

The reader will note from Table 14 that the 214 specialist and the 113 fellowship (residents in a specialty) medical doctors in Sudan comprise almost a third of all physicians working in the Sudan. Although there is nothing wrong with specialization *per se,* policymakers in most countries have now begun to realize that funds spent on primary health care seem to be more effective in realizing and maintaining good health status than money spent on medical specialization. However, there remain powerful economic and professional incentives for the individual to specialize: (1) higher incomes than general practitioners, (2) government subsidies to pursue specialization and (3) the enhanced professional esteem often associated with specialization.

The availability of enough primary care physicians, the introduction of information, and the eradication of unhygienic habits would solve many of the health problems enumerated under heading (3) in Table 13. However, given the patterns of distribution of the physicians, it is probably futile to rely on medical doctors to provide primary health care in all but the urban areas of a country. What is needed is someone with less training than the expensive-to-produce physician. For example, the Chinese have successfully employed the concept of the "bare-footed" physician,[33] and the Russians, the "felcher". These "barefooted physicians" and "felchers" represent inexpensively trained peasants who, for many reasons, will never leave or aspire to leave their own geographical areas where they were born. They cannot perform complicated procedures but are sufficiently trained to provide a host of services basic to the maintenance of health, such as immunizing and providing health information. For maximum effect, these substitute providers of primary health care would have to be bolstered by imaginative programs for disseminating information, because persuading people to act on information is often more difficult than providing the information. In order to facilitate the use of the information, it may be necessary to demonstrate to rural people the effectiveness of certain preventive measures upon their health, even though the perception of the more subtle health changes, such as a five percent increase in one's ability to work, may be difficult to inculcate. Teaching people to boil their water is a simple example. If most people in a village are constantly plagued by diarrhea or dysentry, and many children die from dehydration caused by it, then an effective demonstration of the results of boiling water may produce far-reaching effects for the health of the whole village. Such results could be demonstrated easily and cheaply through the use of an audio-visual presentation which is geared to an illiterate and unsophisticated audience, or, through establishing a "control group" within a village who boils its water and serves as an example for the village. Similarly, hygienic habits can also be taught. The problem is choosing the right medium for the transmission of this information.[34]

Heading number (5) in Table 13 is also of crucial importance and is related to all of the problem areas contained under heading (3). Sudan will have to devote large sums of money to drilling new wells and insuring the security of both old wells and newly-dug wells from the seepage of human and animal fecal material. Again, the informational aspect of this problem is important. People must be made aware of and convinced of the health consequences of defecation by them and their animals in inappropriate places.

Finally, headings (6), (7), and (8) in Table 13 relate to somewhat different problems, but again, they are of a public health nature: (1) lack of food, although arising from difficulties with the transportation system, and basically a question of income redistribution for certain geographical areas, results in less resistance to disease and is, consequently, a health problem; and (2) little is presently known about the extent of environmental health problems and the prevalence of onchoceriasis, which is a major problem in some parts of west central Africa.

Table 15 presents a summary of the estimated costs of intervention in the eight program areas in 1975 Sudanese Pounds.[35] Average annual development expenditures over the seven years of the Ministry of Health Programme will be £S

Table 15. Summary Table of Costs of Programmes
(Millions of Sudanese Pounds)[a]

	Programme			
No.	Title	Development Cost in £S.	Training Cost in £S.	Recurrent Avge. Cost in £S.
1.	*Malaria Nation Wide*	320,000	—	1,305,000
2.	*Malaria Man Made*	—	—	2,529,000
3.	*Primary Health Care*			
	3.1 Primary Health Care Services	16,786,644	207,300	6,758,760
	3.2 Public Lack of Health Information and Lack of Hygiene Habits	243,650	123,750	47,600
	3.3 Communicable Diseases Preventable by Immunization	Unit Cost: 20 Piasters Per Immunized Child		
	3.4 Protein Calorie Malnutrition	—	—	1,219,000*
	3.5 Gastro-Enteritis	2,532,000	—	227,000 [b]
	3.6 Tuberculosis	—	—	184,630
	3.7 Sleeping Sickness	55,000	—	45,000
	3.8 Kala Azar (Phase II)	31,100	—	18,765
4.	*Bilharzia Man Made*	—	—	874,200
5.	*Safe Water Supply*	6,750	—	7,740 [c]
6.	*Environmental Health*	Subject to Programme Formulation		
7.	*Lack of Food (Dura) in Certain Regions of the Sudan*	31,750,000	—	—
8.	*Onchocerciasis*	236,000	—	115,000
	Totals	51,961,144	331,050	13,381,695

Notes:
*Optional Component.
[a] Sudanese Pound £S 1 = $2.50 at the time of the Plan.
[b] Cost of Plan for production of parental rehydration fluids and recurrent cost for the same Plan per year. The actual cost for rehydration fluids by this programme if imported by the Medical Supplies Department will amount to £S 3,359,300 per year.
[c] Pilot Project for research only.
Annual Recurrent cost of Hospitals £S 10 275 993
NB. All these costs are at 1975 rates. An annual inflation rate of 6.5% is officially expected for the plan period.
Source: Ministry of Health, The Democratic Republic of the Sudan, *National Health Programme 1977–78/1983–84*, (Khartoum: Khartoum University Press, 1975), p. 58.

7,423,021. Annual training costs will be £S 331,050 and average annual recurrent costs will be £S13,381,695. Further, annual costs for the hospital system are £S 10,275,993, with a plan to update three provincial hospitals and twenty three rural hospitals at an average annual development cost of £S 287,382 and recurrent costs of £S 1,724,292. In addition, we estimate that it costs about £S 700,000 annually to conduct the Khartoum University Medical School,[36] although one might argue that this item should be placed in the education budget. The sum of these figures, gives a planned total average annual outlay of £S 34,123,433. Referring to Table 9, we see that the Ministry of Planning has allocated £S 265 million for the sector, "Social Services, Housing and Public Administration" for the entirety of the Six Year Plan of Economic and Social Development, or an average annual expenditure of £S 44 million. Therefore, if the Ministry of Health could succeed in spending the £S 34 million per annum, only £S 10 million would remain each year for all of the other social services, housing, and public administration, unless the Ministry could siphon funds from the contingency reserve shown in Table 9, or rely on large infusions of foreign aid. Given the historical trends that we have identified in the previous Sudanese plans and budgets, it is not likely that more money will be allocated to the social services sector and to health; rather it is more likely that less will be allocated. Although the NHP (p. 129) lists six of the eight program areas as "areas appropriate for external assistance," it gives no hint of either how much or from where it expects to receive external assistance. Furthermore, if inflation in the Sudan continues at its present rate, internal resources, such as those which hospital salaries and wages buy, will become more expensive throughout government and there may have to be a reshuffling of government budget priorities, unless the government continues to print more money.

It is legitimate to question whether the study group for the *National Health Programme* did meet its terms of reference. A careful study of the entire document leaves the reader impressed with the amount and quality of work done: the emphasis of the NHP is placed directly on those problems, whose neglect is so often bemoaned by scholars on development and by scholars on the economics of health. The NHP is, thus, a good example of where the emphasis in health planning should be laid. On the other hand, the obvious lack of coordination with the Ministry of Planning and the consequent apparent financial discrepancies between the budget of the Ministry of Health and that of the Ministry of Planning leave us somewhat dubious about the ultimate outcome of the *National Health Programme*.

VII. CONCLUDING SECTION

The World Bank (World Bank, 1975) has been highly critical of health policies in developing nations. The introduction to its health sector policy paper had this to say:

> In many developing countries, health policies are inefficient and inequitable. Too large a proportion of public expenditures on health are allocated to impressive, but expensive modern hospital facilities and sophisticated medical manpower. The allocations have resulted in a bias toward inpatient, curative care—a practice which barely touches health problems in areas beset with serious environmental hazards to health. Furthermore, the resources are typically concentrated on the needs of urban areas...
>
> ... services would focus primarily on improving environmental and public health, personal health practices and nutrition. Reforms in the service offered to the poorer people should concentrate on improving health at the community level. The objectives should include changes in the living habits and attitudes, as well as household and community activities to improve water supply and sanitation. While the demand for curative care would not be denied, a more economical balance would be struck between measures to treat disease and measures to control its incidence.[37]

Moreover, it is the perception at the World Bank that, despite officially declared policy in planning documents, most governments do not have the political will to provide effective health care for the bulk of the population.[38] This view has been produced, no doubt, by bitter experience in countries where the World Bank has done considerable work. However, as the International Labour Organization points out, "... Sudan is very distinct from other developing countries and in some respect presents opportunities for development more favorable than those available elsewhere."[39]

Sudan's greatest asset is that it has tremendous agricultural potential. Presently, only 15 million feddans of a possible 200 million feddans[40] of arable land, with access to irrigation, are under cultivation, land which is also in the sections of the country where there is an extremely favorable climate for plant growth.[41]

Also, Sudan does not have a population problem, so that, in contrast to its neighbor, Egypt, Sudan's land can accomodate the rising population growth exhibited in Table 2. Indeed, Sudan may even be underpopulated with its present population of approximately 17 million. Another positive aspect of Sudan's present situation is its relative political stability. Although it is hazardous to make a prognosis in the area of African or Middle-East politics, (witness the now defunct East African Community or recent events in Iran) the Sudanese Socialist Union, the overwhelmingly majority party, has matured considerably and now accommodates a wide range of views. The President, Gaaffar El-Numeiry, who is in his second six year term, enjoys widespread popularity, support and respect. Even though the country has now returned to a Capitalist economy, the Socialist ethic concerning the need for equity remains strongly on the official agenda. On the other hand, Sudan has all of the other textbook development problems associated with underdevelopment, plus the fact that, being the ninth largest country in the world, with one million square miles of land, it has a more aggravated transportation problem than most developing nations. For instance, Port Sudan, its artery to the outside world, is six hundred miles from Khartoum, the capital, and the whole country has only a little more than 150 miles of paved roads.

However, given Sudan's favorable agricultural potential, it does not seem to be overly simplistic to say that massive infusions of capital, both human and otherwise, could go a long way in solving problems that might not be solvable in some other developing countries.

With the help of the WHO, the Ministry of Health in Sudan has begun to identify the mistakes of the past in the health care sector, and to specify future projects more consistent with today's ideas on investment in health and health care in developing countries. One would hope that, in the future, the Ministry and WHO would use even more sophisticated methods for planning such as the linear programming method developed by Feldstein.[42] However, until the Ministry of Health coordinates its plan and budget constraint with those of the Ministry of Planning, it is difficult to see how health planning in Sudan can achieve its desired effects for environmental improvement and primary health care without an expansion of the health budget. Much of the present curative type medical structure, which consumes fairly large financial resources, would have to be abandoned, if environmental health and primary health care are to be pursued with more vigor. However, the existence of the medical lobby and pressure from the upper classes who are the primary beneficiaries of the present system make retrenchment from existing health care facilities unlikely. This circumstance means that the health budget must be expanded, either through a shift in budget priorities, which appears difficult, given all of Sudan's other pressing investment needs in agriculture, transportation, and education, or through reliance on foreign aid. Given Sudan's strong commitment to equity, one would hope that these foreign funds would be spent on the programmes identified and enumerated in the *National Health Programme*. Whether the foreign funds will come is difficult to predict. Sudan's present balance of payments position is precarious, and those foreign aid donors who do contribute seem to prefer investments in the traditional investment sectors such as agriculture and transportation. However, if the stock of health capital is as important as the stock of educational or physical capital in the process of economic growth, relative neglect of important areas in the health sector by the Sudanese government and by foreign donors of aid could constitute a serious mistake in the long run.

ACKNOWLEDGMENT

Portions of this paper were written while Vogel was on a two-year loan to the International Monetary Fund, as Training Advisor in the Ministry of Finance and National Economy, Democratic Republic of the Sudan. Greenspan's work with the World Health Organization in Geneva significantly aided her in the co-production of this paper. Any views expressed herein represent the opinions of the authors and should not be interpreted as official views of the International Monetary Fund, World Health Organization, Health Care Financing Administration, or U.S. Department of Health and Human Services.

NOTES

1. World Health Organization (1946), p. 1.
2. See Adelman (1963); Fuchs (1965); Larmore (1967); Newhouse (1968); Auster, Leveson, and Sarachek (1969); Sorkin (1975); and World Bank (1975).
3. Grossman (1972), (a) p. 247
4. Grossman (1972), (a) p. 223.
5. Grossman (1972), (a) p. 247.
6. Grossman (1972), (a) p. 225.
7. Grossman (1972), (a) p. 225.
8. Given the interaction between the environmental variables and the service "medical care," the MEC function would probably tend to become more elastic once the environmental variables were ameliorated, i.e. the treatment of protein deficiency is probably more effective on a patient who is not weakened simultaneously with malaria than on one who is.
9. Malenbaum (1970), p. 31.
10. The educational component of this quality of dimension has been explored to a greater extent than the health and other human capital components. See: Blaug (1967); Johnson and Stafford (1973); Ribich and Murphy (1975); Dougherty and Psacharopoulos (1977), and the accompanying references contained in these four papers. The Dougherty and Psacharopoulos paper is the most interesting and complementary paper to this Chapter, given the perspective of this Chapter and given that they point out the human capital returns to one quality dimension (education), but also show the allocative inefficiency and possible distributional inequity of present educational expenditures in most developing nations. (See Sections VI and VII of this Chapter for the complementarity.)
11. In his comment upon Malenbaum's paper, Richard Goode believes that the unimpressive results of the health variables for Thailand and India may be primarily due to the faulty specification of the variables arising from data limitations.
12. The World Bank defines "Industrial Countries" as: Australia, Austria, Belgium, Canada, Denmark, Finland, France, West Germany, Iceland, Ireland, Italy, Japan, Luxembourg, Netherlands, New Zealand, Norway, South Africa, Swedan, Switzerland, United Kingdom, United States. World Bank (1976) p. 549.
13. The World Bank defines "Africa" as: Algeria, Angola, Benin, Botswana, Burundi, Cameroon, Central African Empire, Chad, (People's Republic of the) Congo, (Arab Republic of) Egypt, Equatorial Guinea, Ethiopia, Gabon, Gambia, Ghana, Guinea, Ivory Coast, Kenya, Lesotho, Liberia, Libyan Arab Republic, Madagascar, Malawi, Mali, Mauritania, Mauritius, Morocco, Mozambique, Niger, Nigeria, Rhodesia, Rwanda, Senegal, Sierra Leone, Somalia, Sudan, Swaziland, Tanzania, Togo, Tunisia, Uganda, Upper Volta, Zaire, Zambia. World Bank (1976) p. 549.
14. Some readers may well wonder what connection "radio receivers" have with health. Later in this chapter, we will discuss the importance of verbal and audiovisual information for a largely illiterate rural population.
15. Ministry of Health, The Democratic Republic of the Sudan (1975), p. 25.
16. In the following pages, it is hoped that the reader does not become confused with the various appellations for the Ministries. As governments and philosophies change in this part of the world, so do the names of Ministries. The Ministry of Health has always had that name. However, the finance and planning functions in the Sudan have undergone bureaucratic and organizational permutations. Thus, since independence, Sudan has had, in order,: (1) a Ministry of Finance and Economics, (2) a Ministry of Finance and Economics *and* a Ministry of Planning, (3) a Ministry of Finance Planning, and National Economy, and (4) at this writing, a Ministry of Finance and National Economy *and* a Ministry of National Planning. Further footnotes, sources for the Tables and the References will reflect these changes in names, depending upon the time period analyzed.
17. World Bank (1976), p. 218.
18. "Social Services and Public Administration" is the most refined budgetary classification available for these comparisons on human capital and, more specifically, health expenditures.

19. Ministry of National Planning, The Democratic Republic of the Sudan (1977), p. 42.
20. Ministry of National Planning, The Democratic Republic of the Sudan (1977), p. 24.
21. Ministry of National Planning, The Democratic Republic of the Sudan (1977), p. 32.
22. Ministry of National Planning, The Democratic Republic of the Sudan (1977), p. 45.
23. Ministry of National Planning, The Democratic Republic of the Sudan (1977), pp. 81–124.
24. Ministry of National Planning, The Democratic Republic of the Sudan (1977), p. 168.
25. In June of 1977, Vogel spent six days in the Southern capitol of Juba and its environs on a familiarization mission. Visual and verbal evidence indicated that large sums were being spent on the construction of fine, large, local government office buildings and on vehicles for local government officials. Interviews with local Ministry officials indicated a perception of the need to alleviate local health problems, but little activity was observed on that front.
26. Ministry of Health, The Democratic Republic of the Sudan (1975), p. 9.
27. Ministry of Finance, Planning, and National Economy, Southern Region, The Democratic Republic of the Sudan (1973), p. 15.
28. Ministry of Health, The Democratic Republic of the Sudan (1975), p. 1.
29. Ministry of Health, The Democratic Republic of the Sudan (1975), pp. 1–2.
30. Thus, "... the speed with which the fly multiplies depends on temperature. At 16°C it takes forty-four days for the fly to develop from egg to adult. The time decreases to sixteen days at 25°C and ten days at 30°C." Kamarck (1976), p. 81.
31. Another constraint is that extensive use of antimalarials in some areas has brought about new strains of highly resistant forms of malaria, now more difficult to treat and control.
32. Fifty percent of all generalists and specialists are concentrated in the two provinces of Khartoum and Gezira. Ministry of Health, The Democratic Republic of the Sudan (1975), p. 33.
33. Wen and Hays (1975).
34. The most effective way to disseminate such information may be through religious leaders and religious teachers. In general, these leaders and teachers enjoy high prestige and credibility, while government functionaries are usually distrusted for a variety of reasons. In a Moslem country, such as Sudan, where women have little contact with religious leaders, cheaply trained midwives could provide similar information.
35. For the Plan period, an annual inflation rate of 6.5% is officially anticipated. See: Ministry of Health, The Democratic Republic of Sudan (1975), p. 58. However, lately, the rate of inflation in the Sudan has been averaging around 20% per annum. See: Bank of Sudan (1977), p. 16.
36. Ministry of Health, The Democratic Republic of the Sudan (1975), p. 128 states that the cost of graduating a medical student is £S 4,200. Each year recently, the medical school has taken in 180 students and graduated 160.
37. World Bank (1975), p. 4.
38. World bank (1975), p. 60.
39. International Labour Organization (1976), p. XVI.
40. 1 feddan = 1.038 acres.
41. On the importance of this last factor, see Kamarck (1976), Chapter 2 and 3.
42. Feldstein (1970), pp. 139–163.

REFERENCES

Adelman, Irma, "An Econometric Analysis of Population Growth," *American Economic Review*, Vol. 53, June, 1963 pp. 314–349.

Ali, Abdel Gadir Ali, "A Note on the Brain Drain in the Sudan," *Bulletin No. 49*, Economic and Social Research Council of Sudan, National Council for Research, 1976.

Auster, Richard D., Irving Leveson and Deborah Sarachek, "The Production of Health: An Exploratory Study," *Journal of Human Resources*, Vol. 4, Fall 1969, pp. 411–436.

Bank of Sudan, The Democratic Republic of the Sudan, *Seventeenth Annual Report* (Khartoum: Bank of Sudan, 1977).

Becker, Gary S., "A Theory of the Allocation of Time," *Economic Journal*, Vol. 75, September, 1965, pp. 493–517.

Becker, Gary S., and Robert T. Michael, "On the Theory of Consumer Demand," unpublished paper, 1970

Blaug, Mark, "The Private and Social Returns on Investment in Education: Some Results for Great Britain," *Journal of Human Resources*, Vol 2, Summer 1967, pp. 330–346.

Dougherty, Christopher and George Psacharopoulos, "Measuring the Cost of Misallocation of Investment in Education," *Journal of Human Resources*, Vol XII, Fall 1977, pp. 446–459.

El-Hassan, Ali Mohamed, (ed.), *An Introduction to the Sudan Economy* (Khartoum: Khartoum University Press, 1976).

El-Hassan, Ali Mohamed, (ed.), *Growth, Employment and Equity*, (Khartoum: Khartoum University Press, 1977).

Feldstein, Martin S., "Health Sector Planning in Developing Countries," *Economica*, Vol. 37, May, 1970, pp. 139–163.

Fuchs, Victor R., "Some Economic Aspects of Mortality in the United States," mimeograph, (New York: National Bureau of Economic Research, 1965).

Fuchs, Victor M., *Who Shall Live?* (New York: Basic Books, 1975).

Ghez, Gilbert R., "A Theory of Life Cycle Consumption," Ph.D. dissertation, Columbia University, 1970.

Gibson, Robert M., and Charles R. Fisher, "National Health Expenditures, Fiscal Year 1977," *Social Security Bulletin*, July, 1978.

Grossman, Michael, "The Demand for Health: A Theoretical and Empirical Investigation," Ph.D. dissertation, Columbia University, 1970.

Grossman, Michael, "On the Concept of Health Capital and the Demand for Health," *Journal of Political Economy*, Vol. 80, March/April, 1972, pp. 223–255,(a).

Grossman, Michael, *The Demand for Health: A Theoretical and Empirical Investigation*, National Bureau of Economic Research Occasional Paper 119, (New York: Columbia University Press, 1972), (b).

Hagen, Everett E., *The Economics of Development* (Homewood: Richard D. Irwin, 1975).

Idriss, A., P. Lolik, R. A. Khan, and A. Benyoussef, "Sudan: National Health Programme and Primary Health Care, 1977/78–1983/84," *Bulletin of the World Health Organization*, Vol 53, 1976.

International Labour Organization, *Growth, Employment and Equity: A Comprehensive Strategy for the Sudan* (Geneva: International Labour Office, 1976).

Johnson, George E. and Frank P. Stafford, "Social Returns to Quantity and Quality of Schooling," *Journal of Human Resources*, Vol. 8, Spring 1973, pp. 139–155.

Kamarck, Andrew M., *The Tropics and Economic Development* (Baltimore and London: The Johns Hopkins University Press, 1976).

Lancaster, Kelvin, J., "A New Approach to Consumer Theory," *Journal of Political Economy*, Vol. 74, April, 1966, pp. 132–157.

Larmore, Mary Lou, "An Inquiry into an Econometric Production Function for Health in the United States," Ph.D. dissertation, Northwestern University, 1967.

Lefrowitz, Myron J., "Poverty and Health: A Reexamination" *Inquiry*, vol. X, March, 1973, pp. 3–13.

Malenbaum, Wilfred, "Health and Productivity in Poor Areas," in Herbert E. Klarman (ed.), *Empirical Studies in Health Economics* (Baltimore: Johns Hopkins University Press, 1970).

Michael, Robert T., "The Effect of Education on Efficiency in Consumption," Ph.D., dissertation, Columbia University, 1969.

Ministry of Finance, Planning, and National Economy, Southern Region, The Democratic Republic of the Sudan, *Southern Region Budget* (Juba: Ministry of Finance, Planning, and National Economy, Southern Region, 1973).

Ministry of Finance, Planning and National Economy, The Democratic Republic of the Sudan, *Economic Survey*, 1975–76, (Khartoum: Ministry of Finance, Planning and National Economy, 1977).

Ministry of Health, The Democratic Republic of the Sudan, *National Health Programme, 1977–78/1983–84*, (Khartoum: Khartoum University Press, 1975).

Ministry of National Planning, The Democratic Republic of the Sudan, *The Six Year Plan of Economic and Social Development*, (Khartoum: Ministry of National Planning, 1977).

Muth, Richard, "Household Production and Consumer Demand Functions," *Econometrica*, Vol. 34, July, 1966, pp. 699–708.

Newhouse, Joseph P., "Towards a Rational Allocation of Resources in Medical Care," Ph.D. dissertation, Harvard University, 1968.

Ribich, Thomas I. and James L. Murphy, "The Economic Returns to Increased Educational Spending," *Journal of Human Resources*, Vol. 10, Winter 1975, pp. 56–77.

Scrimshaw, Nevin S., "Myths and Realities in International Health Planning," *American Journal of Public Health*, Vol. 64, August, 1974, pp. 792–798.

Sharpston, M. J., "Health and Development," *Journal of Development Studies*, Vol. 9, April, 1973, pp. 455–460.

Sorkin, Alan L., *Health Economics* (Lexington, Toronto and London: D.C. Heath and Company, 1975).

United Nations, Economic Commission for Africa, *African Economic Indicators* (Addis Ababa: United Nations, 1973).

U.S. Bureau of the Census, *Statistical Abstract of the United States:* 1974 (Washington: U.S. Government Printing Office, 1974).

Wen, C. P. and C. W. Hays, "Medical Education in China in the Postcultural Era," *New England Journal of Medicine*, Vol. 19, May 8, 1975, pp. 998–1005.

World Bank, *Health Sector Policy Paper* (Washington: World Bank, 1975).

World Bank, *World Tables, 1976* (Washington and Baltimore: The World Bank and the Johns Hopkins University Press, 1976).

World Health Organization, *Constitution* (Geneva: World Health Organization, 1946).

World Health Organization, *World Health Statistics Annual 1972*, (Geneva: World Health Organization, 1972).

HEALTH EXPENDITURE IN A RACIALLY SEGREGATED SOCIETY:
A CASE STUDY OF SOUTH AFRICA

M. D. McGrath

I. INTRODUCTION

The high degree of racial segregation and the resulting economic segmentation which exists in South Africa raises many interesting questions for economists relating to how the operation of the economy's markets is affected and about the distribution and level of public sector services. This paper concentrates on the structure and distribution of the output of health care and is consequently concerned both with the market and with the public supply of health services.

Health expenditures account for a relatively small proportion of expenditure in most economies. Despite this, their importance in qualitative terms is considerable since good health is a fundamental determinant of the benefits which can be derived from most other forms of consumption expenditure, and improved health is an important factor contributing to economic growth and development. Free or subsidized health services provided by the government must also be considered as part of the "social wage," and in order to obtain a comprehensive description of the distribution of welfare in any society the distribution of the components of the "social wage" should be analyzed in addition to an analysis of the distribution of personal incomes. Little research has, however, been directed toward the analysis of the composition or the distribution of health expenditure in South Africa [O'Reagain (19); Trengrove-Jones (27)].

Section II of the paper describes the sources of supply of health services, and Section III contains an estimate of expenditures on health from 1959 to 1974. In this section an international comparison is made. Universally acceptable norms for determining the optimal level of supply of health services do not exist and comparisons with other countries cannot establish such a norm, but they do provide a useful benchmark for judging the level of the flow of resources relative to more and less developed countries. An analysis of the racial distribution of health services will be attempted in Section IV. Demand and supply factors receive attention in this section, and the regional distribution of health services is also considered. Section V briefly attempts to indicate possible directions which might be taken in the future.

II. THE STRUCTURE OF THE HEALTH CARE SECTOR

The data available for South Africa allow a broad definition of health services to be used. In addition to services which clearly lie within the field, health services which cover preventative measures such as vaccination and immunization, medical examinations of schoolchildren, medical screening for the detection of pathologic lesions such as cancers and tuberculosis, health education, etc. can be included and with two relatively minor exceptions the services which can be covered are the same as those included in Abel-Smith's (2) intercountry studies.[1] In South Africa health services are provided by both the private and public sectors, and in the public sector all levels of government are involved.

The activities of the public sector include the supply of curative services such as general hospitals, outpatient services, mental and infectious disease hospitals, services relating to the prevention of infectious diseases such as tuberculosis and malaria, the provision of medical research, laboratory services, medical training, and the regulation of the private sector. Table 1 shows the proportion of expenditure by the type of service supplied by the public sector for the year 1969–70. Hospitals and nursing services can be seen to be the single most important item of expenditure, and if the subsidies on private sector hospitals and expenditures on mental hospitals are included, they account for over 80 percent of public sector expenditure and 91 percent of provincial expenditure. Provincial government is the largest source of expenditure, accounting for almost three-quarters of the total. The most important item of central government expenditure is on the

Table 1. Expenditure by Type of Service Provided in
the Public Sector, 1969–70
(Percentages)[1]

Service	Level of Government[2]			
	Central	Provincial	Local[3]	All Levels
Administrative	23	2	—	7
	(80)	(20)	(−)	(100)
Mental Hospitals	27	—	—	6
	(100)	(−)	(−)	(100)
Infectious Diseases	42	1	43	11
	(85)	(6)	(9)	(100)
General Hospitals	4	91	57	70
and Nursing	(1)	(97)	(2)	(100)
Subsidies to Hospitals	4	7	—	6
in the Private Sector	(15)	(85)	(−)	(100)
All services	100	100	100	100
	(22)	(75)	(3)	(100)

Notes:
[1] Sources: "Estimate of the Expenditure to be defrayed from Revenue Account during the year ended 31 March 1970." RP 2/1969.
"Part II of the Report of the Controller and Auditor General for the Financial Year 1969/70." RP 51/1970.
"Report of the Provincial Auditor on the Accounts of the Province for the Year ended 31 March, 1970." Province of the Orange Free State, Cape of Good Hope, Transvaal and Natal.
[2] Figures in parentheses show the proportion of the row totals accounted for by each level of government, and those not in parentheses show the proportion of the column totals accounted for by each type of activity.
[3] Local government expenditure is not available in published reports but has been estimated by assuming that the maximum rate of subsidy was paid and using this as a multiplier on all subsidies paid from central government.

control and cure of infectious diseases, and together with local government expenditure accounts for 95 percent of this category. Expenditure on mental hospitals is incurred only by central government. The expenditure of local authorities is divided between the control of infectious diseases and the provision of clinics and nursing services. The year 1969–70 was chosen for this analysis as the Department of Health commenced a takeover of Homeland services shortly afterward, and a similar allocation of expenditure cannot be made for later years.[2] The effects of these changes do not, however, appear to have markedly changed the proportions of total expenditure accounted for by the three tiers of government.

Health services provided in the private sector largely comprise hospital ser-

Table 2. Ownership of Hospitals and Beds, 1975 [1]

Type	Sector					
	Private & Aided Hospitals	Beds	Central Government Hospitals	Beds	Provincial Government Hospitals	Beds
Industrial	71	9,083	—	—	—	—
Infectious [2]	52	9,413	7	4,711	2	165
Mental Disorders	41	13,915	20	19,367	—	—
Old Age & [3] Chronic Sick	53	3,166	—	—	1	189
General [4]	290	18,350	100	28,894	199	50,883
Total	507	53,927	127	52,972	202	51,237

Notes:
[1] Source: H. Engelhardt (ed.), "Hospital and Nursing Year Book of Southern Africa 1976." pp. 19–73. Tomson Publications.
[2] Includes SANTA, State Leper, and other private and aided hospitals specializing only in infectious diseases.
[3] Includes only hospitals specializing in old age nursing. Beds for old patients are also available in some general hospitals.
[d] Includes maternity hospitals. Homeland hospitals and Transkeian hospitals are included under central government.

vices (including the nursing of the aged) and the services of private medical and dental practitioners. An indication of the relative importance of these services is given in Tables 2 and 3.

The ownership of hospitals and beds is shown in Table 2. State hospitals and aided institutions account for almost all the beds for infectious diseases, and the state and private sector share responsibility for the supply of beds for the treatment of mental conditions. Bed accommodation for the aged and chronically sick is provided largely by private and aided institutions, and the supply of general hospital services is shared almost equally by the private sector and provincial

Table 3. Employment of Economically Active Doctors, 1972

	All Doctors	General Practitioners	Specialists
Private Practice	4,716	3,108	1,608
Other Private Institutions	502	474	28
Total Private Employment	5,218	3,582	1,636
Government	2,889	2,167	722
	8,881	6,387	2,494

Note: Interns and doctors in part-time employment are excluded.
Source: "Census of Health Services 1972/73, Medical Practitioners and Dentists." Tables 4.1.1 and 4.3. Department of Statistics, Report No. 02-03-01.

administrations. Private and aided institutions accounted for 61 percent of all hospitals, although they only accounted for 34 percent of the available beds. Approximately 25 percent of the beds supplied by the private sector were purely for profit.[3]

In Table 3 the employment of doctors is analyzed. This table also emphasizes the importance of the private sector in supplying health services. The private sector accounted for the employment of 59 percent of all doctors and the majority of these were in private practice; the proportion of specialists in private sector employment was even larger at 61 percent. In contrast, 90 percent of dentists were in private sector employment.

III. EXPENDITURE

Only current expenditure will be analyzed, as comprehensive data on capital expenditure is more difficult to assemble and the inclusion of investment expenditures in any one year's expenditure can lead to a distorted picture because of the lumpiness of capital expenditures and because their effect is to generate a stream of services in future years. At the same time, current expenditures will understate the amount of resources consumed, since government health costs do not include an allowance for depreciation and interest [Fuchs (9)]. Included in current expenditures are the costs of training medical personnel, although these costs would be more appropriately allocated to capital expenditure.

Cross-section studies of health expenditure have shown that developed countries spend a higher percentage of their Gross National Product (GNP) on health services than do poor countries. Abel-Smith's study around 1960, of data for 17 nations, showed that the proportion of GNP spent on health services ranged from 2.5 to 5.9 percent, with an average increase of $1000 of GNP per capita resulting in an increase of 0.75 percent of the proportion of GNP allocated to health. Publicly provided health expenditures increased at a faster rate with the proportion of GNP spent on health services rising by 0.9 percent for each additional $1000 of per capita GNP [Abel-Smith (1, pp. 40–44)]. Data drawn together a decade later for 16 countries yielded similar results, but in addition this study showed that increases in health expenditure per capita between countries were positively associated with the share of government consumption expenditure in total consumption and negatively associated with the degree of income inequality [Kleiman (12, p. 69)].

Total expenditure in the case of South Africa is shown in Table 4. In 1959–60 health expenditures accounted for 4.2 percent of GNP, but, although GNP grew continuously during the 1960s, the proportion allocated to health services had dropped to 3.8 percent in 1969–70 and had dropped even further to 3.6 percent by the mid-1970s. When the 1959–60 proportion is contrasted with Abel-Smith's data (1, p. 41) for 17 countries at a similar time, the South African percentage was relatively large and was intermediate between the United Kingdom and

Table 4. Health Expenditure in South Africa as a Percentage of Gross National Product at Market Prices, 1959–60 to 1974–75 [1]

	Expenditure in R000,000		Percentage of GNP		
Year	Public Sector	Private Sector [2]	Public Sector	Private Sector	Total
1959/60	96	93	2.1	2.1	4.2
1969/70	242	234	1.9	1.9	3.8
1974/75	515	378	2.1	1.5	3.6

Notes:

[1] *Sources:* Public Expenditure from: "Controller and Auditor General Reports of Central Government and the Provinces" for 1959/60, 1969/70 and 1974/75 and also the Transkei for 1974/75. Private expenditure from: "South African Statistics 1976." Table 21.2; "South African Statistics 1972." Table V-17; and "South African Statistics 1965." Table U-16. Gross National Product at Market Prices from: "A Statistical Presentation of South Africa's Quarterly National Accounts for the period 1960 to 1974." South African Reserve Bank, March 1976; and "South African Reserve Bank Quarterly Bulletin." September 1977.

[2] Private sector expenditure does not include expenditure on industrial hospitals.

France. If South African expenditure had followed the average trend (as indicated by Abel-Smith's cross-section data), the proportion of GNP allocated to health services would have risen by 1974–75 to approximately 5.4 percent of GNP as opposed to the actual 3.6 percent.

The level of public health expenditure may be a better indicator of the quality and quantity of health care available to the typical individual in poor countries, where private health services are available to a relatively small proportion of the population. Abel-Smith's data (1, p. 43) show, for a sample of 25 countries, that the proportion of public health expenditure in the GNP varied between 4.3 percent and 0.5 percent. Even in 1960 this proportion was lower in South Africa's case than for some countries with lower per capita incomes.

Despite the fall in the proportion of health expenditures in GNP, it appears that real expenditures per capita did increase, as is shown in Table 5, although per capita expenditure increased at a slower rate than income. From 1969–70 to 1974–75 real health expenditure per capita and GNP per capita have grown even more slowly, with income per capita increasing at 3.4 percent per annum and health expenditures at 2.5 percent per annum.

In Table 6 the relationship between private health expenditure, public health expenditure, and public consumption expenditure is shown. The results of this table indicate that one reason for the declining proportion of national resources allocated to health is to be found in the slow expansion of public health expenditure. Although public consumption expenditure increased as a proportion of GNP, as would be expected, the proportion of health expenditures in public consumption expenditure decreased noticeably over the 16-year period, and as a result public health expenditure did not rise as a percentage of GNP. The propor-

Table 5. Health Expenditure per Capita in Constant 1970 Prices, 1959–60 to 1974–75 [1]

Year	GNP per Capita [2]	Health Expenditure per Capita [3]
	Rand	Rand
1959/60	356	14
1969/70	596	23
1974/75	705	26
Annual Growth Rate	4.4	3.9

Notes:
[1] Sources: Population data from "South African Statistics 1976." Table 1.6; and "South African Statistics 1968." Table A-9.
[2] Deflated by the national accounting deflator, "South African Statistics 1976," Tables 21.4 and 21.6.
[3] An adequate deflator for all medical services does not exist. The medical index of the consumer price index is unsuitable as it only includes expenditure of White families in major urban areas. No index for medical services in the public sector is available. The national accounting deflator was used here, and the result is more conservative than that obtained using the medical care index of the CPI.

tion of private health expenditure in private consumption expenditure was approximately the same at the beginning and at the end of the period (although it remained at more than 3.0 percent from 1962 through to 1973) and thus the proportion of private health expenditure in GNP fell as the proportion of private consumption in GNP fell. This analysis thus leads to the conclusion that over the 16-year period, a relatively smaller amount of resources has been devoted to health services in South Africa and that by 1974–75, South Africa probably fell well below the average for countries at a similar level of development.

The usefulness of expenditure contrasts between countries is, however, limited because expenditure depends on price levels and wage levels and the price of health service inputs varies enormously between countries. Health services are

Table 6. Percentages of Private and Public Consumption Expenditure Allocated to Health Services, 1959–60 to 1974–75

Year	Private Health as Percent of Private Consumption	Public Health as Percent of Public Consumption	Public Consumption as Percent of GNP
1959/60	2.8	18.7	11.3
1969/70	3.3	17.6	12.0
1974/75	2.8	16.8	14.4

Sources: Consumption expenditure data from "South African Statistics 1976." Table 21.15; and "South African Statistics 1970." Table W-5.

Table 7. Indicators of Physical Quantity of Health Services Supplied by Countries at Different Levels of Development

Country Group by 1973 GNP per Capita	Population per Doctor 1970	Population per Nurse 1970	Population per Hospital Bed 1970
$			
100 and below	21,821	8,243	1,883[1]
101–200	8,879	2,980	1,228[1]
201–375	3,437	1,794	500[1]
376–1000	1,729	1,508	406[1]
1001–2000	1,505	689	153[1]
South Africa	2,500	549	154[2]
2001–5000	762	359	103[1]
5000 & above	764	255	98[1]

Sources:
[1] "World Bank Tables 1976." pp. 496–502, 518–520. Johns Hopkins Press, Baltimore, 1976. In calculating averages for countries with incomes over $1000 per capita, Arab nations have been excluded.
[2] South African health service data from H. MacCarthy (ed.), "Hospital and Nursing Yearbook of Southern Africa 1972." p. 11. H. MacCarthy Publications; "Census of Health Services 1972–73." Table 1.1; and "South African Statistics 1974." Table 4.2.

highly labor-intensive, and in countries where the incomes of medical personnel are high the proportion of GNP spent on health is likely to be above the average. Monopoly factors may also increase the relative price of drugs and other health products in certain nations. As an alternative to a comparison of expenditures, the physical quantity of service supplied can by analyzed. Useful measures of physical capacity are given by numbers of people per hospital bed, per doctor, and per nurse. Such a contrast has an added advantage because data are readily available for a large number of countries, and the results are shown in summary form in Table 7. South Africa falls into the income group $1001–2000 per annum and compares unfavorably with the international average only in terms of population per physician, where the average was more than 60 percent more favorable. Population per nurse statistics were more favorable for South Africa than for the international average, and bed statistics were equal to the average. The impression which this comparison gives is more favorable than the contrast of health expenditure, and in order to gain further insight into the supply of health services, the racial distribution of health services will be considered.

During the period from 1960 to 1975, the number of people per hospital bed in South Africa remained almost constant; this supports the evidence presented earlier which showed a relative decline in the quantity of resources allocated to health services, since the growth rate of the supply of hospital beds was considerably slower than the growth rate of per capita incomes. When, however, the racial distribution of hospital services is examined, a bias in favor of Whites becomes clear—in 1960 there were 100 Whites per bed against 186 Blacks per bed. In 1975 the White ratio had fallen slightly to 96 persons per bed, although the Black ratio remained unchanged.

IV. THE RACIAL DISTRIBUTION

In 1970 Blacks accounted for 83 percent of the population, and Africans accounted for 84 percent of all Black people. At the same time Whites received approximately 72 percent of personal income and Africans 19 percent, resulting in a disparity of 15:1 in the ratio of White to African per capita incomes, and 5.1:1 and 6:1 for the ratios of White to Asian and White to Colored per capita incomes [McGrath (17)]. Rural African incomes are considerably below the average, and the Homeland economies are heavily dependent on the remittances of migrant workers. Even when these remittances are included, 1970 per capita incomes in the Homelands were only R80 (80 rands) per annum, 75 percent of the national average for Africans. Even though African incomes in urban areas are considerably higher, a substantial proportion of the urban families are in poverty; for example, in 1970 approximately 50 percent of African families in the Durban area had incomes below their Poverty Datum Line [Maasdorp (14),

The relatively low incomes of Blacks have two immediate effects on health. The first is the lower level of health which results from poor diet and environment. This is clearly illustrated in Table 8. Infant mortality rates (deaths per 1000 live births) in 1974 were 18.4, 115.5, and 32.0 for Whites, Coloreds, and Asians, respectively. National figures are not available for Africans, but in Durban the corresponding ratios were 5.53, 13.58, and 19.89, while the rate for Africans was 29.98 (3, p. 8). In rural areas African infant mortality rates are probably higher still [Wells (28, p. 1)]. Statistics of the incidence of malnutrition were last published in 1965 when there were 12,062 cases registered among

Table 8. Rate per 100,000 of Population of Notifiable Diseases and Life Expectancy

	Tuberculosis	Typhoid	Meningococcal Infections	Diphtheria	Life Expectancy
White	18.1	1.1	3.8	0.9	65
Coloured	327.7	4.4	12.1	1.9	51
Asian	143.0	5.5	3.0	0.4	60
African	285.2	18.6	5.0	2.3	52

Source: "South African Statistics 1976." Table 4.5–4.7, and J. L. Sadie, "Projection of the South African Population 1970–2020." pp. 41–63. Industrial Development Corporation of South Africa.

Africans, 735 for Coloreds, and 26 among Asians, as against 9 for Whites (24).[4] The conditions listed in Table 8—high infant mortality rates, malnutrition, and low life expectancy rates—are all associated with poverty and are consequently more pronounced among Blacks.

The second direct effect of low Black incomes is on the levels of consumption and on the amount of income available for the purchase of health services. In 1968 per capita consumption expenditure of urban Whites was more than six times that of urban Africans [McGrath (18, p. 161)]. This in itself is a cause of different patterns of disease. An example of the effects of the higher consumption of Whites is given by examining the most common causes of death among Whites and Africans in Durban. In order of importance, these are shown in the accompanying table.

Whites	Africans
Heart diseases	Certain causes of perinatal morbidity and mortality
Neoplasms	Infective parasitic diseases
Diseases of the respiratory system (excluding pulmonary tuberculosis)	Diseases of the circulatory system
	Diseases of the digestive system

The most common causes of death among Whites are the diseases of affluence associated with high levels of consumption, whereas three of the four most common causes of death among Africans are directly linked with low incomes.

Total racial expenditures on health are more unequally distributed than either incomes or total consumption, and in 1975 White expenditure represented approximately 94 percent of private expenditure on health care, causing Whites to receive the major portion of the private health sector's services.[5] Expenditure patterns on medical care in selected urban areas are illustrated in Table 9. The striking features of the table are the importance of medical aid contributions in White and Colored expenditure, the importance of expenditures on doctors and dentists for Whites and Asians, and the large proportion of African expenditure directed to patent medicines. Indeed, the relatively high level of expenditure on patent medicines by Africans in urban areas might be an indication of the difficulties involved in obtaining subsidized medical treatment at hospitals or clinics. When expenditures on hospital services and prescriptions are subsidized, these expenditure data will understate the value of the services actually received.

White and Asian households both spent about 3 percent of their incomes on health care, while the expenditure of Colored and African households was 1.5 and 1.2 percent of income, respectively. This pattern is predictable since higher levels of income and education, as well as a higher average age in the population, have been found to be important determinants of the level of health expenditure, and in South Africa all these factors work to increase the relative level of White

Table 9. Personal Expenditures on Health Care by Race in Selected Areas, 1975

Item	White Rand	%	Colored Rand	%	Asian Rand	%	African Urban Rand	%
Medical Aid & Insurance	100.83	33	13.49	29	19.44	16	2.26	8
Doctors & Dentists	97.11	31	7.15	15	37.23	31	2.55	10
Witch Doctors & Herbalists	—	—	0.01	—	0.24	—	1.33	5
Nurses & Hospitals	28.49	9	4.95	11	11.53	10	1.89	7
Medicines on Prescription	44.08	14	0.47	1	1.19	1	0.11	—
Other Medicines	19.83	6	14.53	31	21.0	18	13.17	49
Dentures & Spectacles etc.	21.10	7	6.22	13	28.33	24	5.66	21
Total Expenditure	311.44	100	44.82	100	119.46	100	26.97	100
Expenditure as a Percentage of Income	3.0		1.5		3.1		1.2	

Sources: White households (in major urban areas): "Survey of Household Expenditure 1975." Department of Statistics, Report No. 11-06-05. Colored households (in the Cape Peninsula): "Income and Expenditure Patterns of Urban Coloured Households in Cape Town." Bureau of Market Research, (B.M.R.) Research Report No. 80.5. Asian households (in Durban): "Income and Expenditure Patterns of Urban Indian Households in Durban." B.M.R. Research Report No. 50.7. African urban households: "African Multiple Households in Johannesburg." B.M.R. Research Report No. 50.3.

and Asian expenditures. The existence of medical insurance also increases the demand for the services which it covers, and in South Africa medical aid benefits apply to 73 percent of the White population (20).

A racial allocation of the distribution of public sector expenditure cannot be made accurately, and even hospital costs and subsidies (which are the biggest items of expenditure) cannot be divided by race. An indication of the distribution of these services can, however, be gained from indexes of the physical quantity supplied, and an indication of the quality of the services can be gained from the analysis of expenditure patterns in racially segregated hospitals.

General hospitals account for 77 percent of all beds provided in the public sector, and Table 10 shows the racial distribution of the services of these hospitals and the racial distribution of services in subsidized hospitals. In 1959 and 1974 beds for Blacks accounted for 68 and 75 percent of total beds. The proportion of hospital services received by Blacks was, however, larger than this, as Blacks accounted for 66 and 81 percent of patient days and 81 and 78 percent of outpatient attendances in 1959 and 1974. The percentage of White beds occupied was low in both years and indicates substantial excess capacity in the supply of White services, while Black beds in both years were intensively used.

Two measures of need might be used to assess the degree of racial equality in the distribution of these public health services. The first approach rests on need,

Table 10. Racial Distribution of Provincial and Homeland Hospital Services and Subsidized Hospital Services, 1959 and 1974

	1974[1]		1959[1]	
	White	Black	White	Black
All Beds	19,128	58,871	15,160	31,918
Patient Days (000's)				
Provincial (& Homeland for 1974)	4,081	18,490[2]	3,469	6,760
Aided	385	98	125	87
Patient Days per 100 of Population	107	98	125	87
Outpatient Attendances (000's)				
Provincial (& Homeland for 1974)	2,992	10,295[2]	1,525	4,885
Aided	99	784	18	1,638
Outpatients per 100 of Population	74	69	51	48
Percentage of Beds Occupied	64	95	69	102

Notes:
[1] *Sources:* "Report of the Director of Hospital Services" for the year 1959 and 1974, Province of Natal, O.F.S., Cape and Transvaal; and for Homelands from the "Hospital and Nursing Year Book of Southern Africa 1976."
[2] The Transkei is not included in 1974.

as shown by the pattern of diseases and the quantity of services available in the private sector. The second approach takes its benchmark simply from the level of services available to the most privileged group (i.e., the White population). In terms of this second standard, the level of public hospital services provided for Blacks in 1975 compared fairly favorably with that provided for Whites. In terms of the first criterion, which is the better guide to need, Black services were deficient. Moreover, the measures used above have abstracted from the quality of services, the distribution of services among regions, and the distribution of services among Black racial groups. Published sources do not allow the last item to be divided any further.

Some indication of the quality of services received can be gained from an analysis of hospital operating costs. In general, costs per patient day vary positively with the proportion of services provided to Whites; for example, in 1969–70 costs per patient day in Natal provincial hospitals predominantly for Whites were R16 per day, whereas in hospitals for Blacks costs were R6 per patient day (26). All the variation in costs between Black and White patients cannot immediately be attributed to a lower quality of service. The composition of the hospital case mix has been shown to influence the level of hospital costs [Feldstein (8, pp. 269–275)], and higher rates of hospital utilization have been shown to result in lower unit costs [Mann (16), pp. 277–280], although this must be associated with a deterioration in the quality of service provided to patients.[6] Additionally, Black professional hospital staff have historically been subject to

Table 11. Contrast of Operation of Two Large Hospitals, 1974 [1,2]

	White Hospital	Black Hospital		
Total Patient Days [3]	290,296	902,217		
Percentage of White In-Patients [4]	68	—		
Percentage of Beds Occupied	64	93		
Ratio of Daily Patients to:				
Doctors	7.2	13.5		
Nurses	0.7	1.7		
	Actual Rand	Actual Rand	Hypothetical Rand [5]	Rand [6]
Operating Costs per Patient Day				
Total	32.10	13.48	24.06	21.89
Salaries, Wages & Allowances	20.85	8.89	16.77	12.91
Provisions	1.69	0.62	1.69	1.69
Medical Supplies & Services	4.55	2.92	4.55	4.55
Other	5.01	1.05	1.05	1.05

Notes:
[1] *Sources:* "Report of the Director of Hospital Services for the year 1974." Province of Natal; "The Provincial Auditor's Report on the Appropriation Accounts for the period 1 April 1974 to 31 March 1975." Province of Natal; and "Estimates of the Expenditure to be defrayed from Revenue during the year ending 31 March 1974." Province of Natal.
[2] The Hospitals are Addington and King Edward VIII, both in Durban.
[3] Patient days are estimated as the sum of inpatient day and one-third of outpatient attendances.
[4] The remainder are Colored.
[5] Provisions and medical supplies at White patient cost, number of patient-days reduced by the ratio daily patients/doctor in Addington to the ratio daily patients/doctor in King Edward.
[6] As for footnote 5, but number of patient-days reduced by the ratio of the percentage of beds occupied in Addington to the percentage of beds occupied in King Edward.

wage discrimination [Knight and McGrath (13, pp. 245–271)], which has lowered the salary and wage bill of Black hospitals relative to White hospitals, and this contributes to the lower cost structure of Black hospitals.

In order to illustrate the effects of these various influences, the operating costs of two large hospitals which render broadly similar services are contrasted in Table 11. In both hospitals, salaries and wages were the largest components of costs per patient day and were about 65 percent of total costs. In the Black hospital salary costs per patient were 43 percent, the cost of provisions 37 percent, and the cost of medical supplies 64 percent of the costs per patient of the same items in the White hospital. Total operating costs per patient in the Black hospital were 42 percent lower than for the White hospital, and the level of bed utilization in both was similar to the national average. In order to examine the effects of the differing quality of service in the Black hospital, its costs were recalculated under two different sets of assumptions.[7] Both sets of assumptions examine the effect on costs in the Black hospital of offering services of the same quality as in the White hospital; although the results are necessarily crude, it appears that the quality of service accounts for between 45 and 57 percent of the differences in these hospitals' costs. The remaining differences are largely attributable to wage differentials between Black and White staff in the two hospitals. There is a considerable variation in the quality of service provided to Blacks and Whites in the two hospitals chosen for this example, and the pattern is likely to be a general one, since this pattern of hospital costs per patient day is found in most provincial hospitals.

Thus the racial segmentation of the labor market, which has depressed Black wages and salaries, lowers the relative cost of hospital services provided to Blacks by the public sector. The racial segregation of hospital services further lowers the cost of treating Black patients because it allows a lower quality of service to be offered.

The regional distribuion of hospital beds indicates a pronounced urban bias for Blacks and a smaller bias for Whites. In 1970 the number of people per bed in the principal urban areas was 92 for Whites and 109 for Blacks, as against a ratio in smaller urban and rural areas of 109 and 191 for Whites and Blacks.[8] The average of persons per bed for Homeland areas was, however, higher at 233 and varied between 527 for Ka Ngwane to 154 for Bophuthatswana (5). The racial distribution of medical personnel for similar regions cannot be obtained, but, in the 13 largest urban areas, the ratio of people to doctors was 969, leaving a ratio of 7612 people per doctor in the rest of the country (see Ref. 23, Tables 1 and 23; also Ref. 6). The average in the Homelands was 23,037 persons per doctor, and this ratio ranged from 7429 in Ka Ngwane to 48,000 in Qwaqwa (5, p. 203). The spatial distribution of the services of doctors therefore appears even more biased toward urban areas than is the distribution of hospital services. These classifications are subject to some error since the distances from which hospitals and doctors draw their patients are unknown. Nevertheless, they give a broad indication of the regional supply.

The picture which appears is one in which White needs are well supplied in both the private and public health care sectors, and, in addition, a large proportion of the White population benefits from medical aid schemes. The Black population consumes a very small proportion of privately supplied medical services, which is explained by their lower income levels. By 1974 the racial proportion of general hospital services supplied by government and aided institutions was almost equal to the racial composition of the population, but the needs of Blacks, as indicated by health standards and the smaller quantity of private services available, require that this proportion should be even higher if racial equity is to be achieved. In addition, the services provided to Blacks are of inferior quality in relation to those supplied to Whites. Moreover in the rural areas and Homelands, where the need for health services is greatest, the supply is at its most deficient level.

V. FUTURE DIRECTIONS

Between 1970 and the end of the century, projections indicate that the South African population (including all the ex-Homelands) will have almost doubled, and growth in the Black population will have contributed most of this increase [Sadie (21)]. Maintaining the standard of health services at their present level will not be too difficult, since this only requires an increase in the supply of beds and doctors of 2.5 percent per annum, and, at this rate of expansion, expenditures on health as a percentage of GNP would probably continue to decline.

There are, however, some forces which may operate to increase the relative amount of resources allocated to health care. The first of these is the high income elasticity of demand of Blacks for health service [Black (4, p. 400)] which, together with rising incomes, will lead to an increase in the demand for medical care in Black urban areas and may lead to the development of a private sector serving these needs. Rising levels of education and the spread of medical aid benefits will add impetus to this development. Secondly, the proportion of expenditure of Whites is not likely to fall, for although the income elasticity of demand of high income groups has been found to be low, changing tastes usually serve to increase the demand for health services as incomes rise.[9] Thus rising incomes are likely to increase the amount of private expenditure on health care. Government has also shown concern over the racial distribution of health services, and this could lead to an increased level of expenditure on Black services [de Beer (7, pp. 431–440)]. Moves to eliminate wage discrimination will also add to current expenditures in the public sector, although this will not in itself increase the supply of services.

Despite government concern, it is unlikely that racial inequalities in the supply of health services will be eliminated by the end of the century. Assuming that the existing standard of services is maintained for Whites, if Blacks are to receive the same level of service by the end of the century the supply of hospital beds will have to increase at an annual rate of 4.8 percent. In order to improve the ratio of beds to doctors to the international average of countries at South Africa's 1973 per capita income, the supply of doctors will have to grow at over 6 percent per annum. If these targets are to be achieved, services in Homelands and rural areas will have to be expanded faster than the national average. For example, the supply of beds in Homelands will have to increase by 7.4 percent per annum, and the supply of doctors will have to grow by 16 percent per annum.

The Department of Health's projections indicate that the supply of doctors to the whole economy will increase annually by only 3.6 percent between 1974 and 1985 [de Beer (7, p. 439)], falling far short of the annual required growth of 6 percent. It is also unlikely that the supply of beds will grow at the rate necessary, and Homelands will probably show the slowest rate of improvement because of the many competing demands for the development of services which will be made on their budgets. If a substantial private sector develops to serve the medical needs of the urban Black population, some resources will be set free for the expansion of services in rural areas, but the growth of formal services to these areas will be retarded since Black doctors and nurses will be attracted away from the public service. Thus it may be difficult to expand the supply of formal services to the poorest areas at a rapid rate and to eliminate the racial inequality which has been discussed, although there is likely to be an improvement in the per capita supply of health services by the year 2000.

The distribution of this increase will have an important influence on the efficiency of the health sector. Several writers have argued that an elaborate hospital-based system of health care does not represent the best solution in countries where there is a predominance of diseases of poverty and a large rural population. The concentration of expenditures on hospital services seems to bring little benefit to people suffering relatively minor or slowly progressive diseases [Sorkin (22, pp. 110–112)]. Simple preventative treatment can eliminate many of the diseases associated with poverty, and, in order to mount effective preventative health campaigns in rural areas, there is a need for a widespread rural health infrastructure [Gish (10, pp. 37–44)]. This cannot be achieved with a hospital based system which leads to a very limited spatial distribution of services.[10]

VI. CONCLUSION

The racial segregation of health facilities and wage discrimination against Black medical personnel in the public sector has enabled South Africa to offer a health service which compares favorably with countries of similar per capita income levels, even though the percentage of the South African GNP allocated to health care now appears considerably lower than average. In many respects this analysis of the distribution of health expenditure has reflected the broader operation of the South African economy. Whites are well rewarded for their output and have used their incomes and their political power to achieve a standard of health service which is high, even by international levels. Blacks are poorly cared for by comparison in a society which has prevented the achievement of their full economic potential. Services to Blacks in the areas which form the labor pools of the economy are the most inadequate, although the urban industrial labor force is provided for more adequately. Even in the urban areas, the racial segmentation of the supply of health services has allowed a lower quality of services to be supplied to Blacks. Influx control and the migrant labor system ration the demands which can be made on the better urban facilities.

This paper indicates clearly that the amount of resources allocated to health care must be increased in order to eliminate racial inequalities in the distribution of these services and thereby improve the quality of life for Blacks, but these increases should be channeled into supplying services which will benefit the largest number of people for the lowest cost possible. The achievement of this goal will require a change in the structure of public expenditures, with a change of emphasis from curative to preventative services and a relative increase in the supply of services to rural areas.

ACKNOWLEDGMENTS

The author is grateful to Jill Nattrass, Gavin Maasdorp, and Trevor Jones for their comments.

NOTES

1. The two differences are that military health services have been omitted and expenditure on mental patients detained in institutions but not receiving remedial treatment is included, whereas it was excluded by Abel-Smith.
2. By 1975 the central government was responsible for the maintenance expenditures of 71 Homeland hospitals, whereas in 1969 it had administered only 3. Total expenditure and its division between level of government was estimated for 1974–75, and when the health outlays of Homeland governments (including the Transkei) are included with the central government's expenditures, the proportions are 24, 73, and 3 percent for central, provincial, and local government, respectively.
3. This was 13,458 beds in 1974 (20).
4. The true rate has been argued to be about four times higher (28).
5. White expenditure is estimated from the "Survey of Household Expenditure" (25). It has been assumed that the average per capita expenditure of urban White households (in major urban areas) applies to the whole White population. In 1972, 86 percent of doctors in private practice were White, and custom and legislation would have ensured that Whites received the major benefit of these services. Whites would also have occupied the majority of the beds available purely for profit, which accounted for 25 percent of private and aided beds in 1974.
6. This result is inevitable when large components of cost such as salaries and wages are semifixed and where the marginal cost of other activities such as feeding or medicines is probably constant (or possibly falling).
7. The assumptions are stated in the footnotes to Table 11.
8. Calculated from (15) and (23) Tables 1.23 and 1.24.
9. Klaarman reports that in the United States the proportion of income spent on health has either risen or remained constant over time despite low income elasticities (11) pp. 31–36. In South Africa expenditure of White families rose from 2.9 percent of income in 1955 to 3.0 percent in 1975.
10. There is no evidence for South Africa, but Gish (10) p. 40 cites evidence showing that the catchment area of Tanzanian hospitals is extremely limited: over 91 percent of patients at the national referral hospital in the capital came from a radius of 15 km.

REFERENCES

1. Abel-Smith, B. (1967) "An International Study of Health Expenditure," Geneva: *World Health Organization* pp. 40–44.
2. _____. (1977) "Paying for Health Services," Geneva: *World Health Organization*.
3. "Annual Report of the City Medical Officer of Health, Year Ended 31 December 1975," City of Durban, p. 8.
4. Black, P. A. (1977) *South African Journal of Economics* 45:400.
5. "Black Development in South Africa," Pretoria: B.E.N.B.O. p. 203.

6. "Census of Health Services 1972/1973, Medical Practitioners and Dentists," Department of Statistics Report No. 02-03-01, Pretoria: Government Printer.
7. de Beer, J. (1976) *South African Medical Journal* 50:431–440.
8. Feldstein, M. S. (1973) in *Health Economics*, (eds.) M. H. Cooper *et al.*, Penguin, Harmondsworth pp. 269–275.
9. Fuchs, V. R. (1973) in *Health Economics*, (eds.) M. H. Cooper *et al.*, Penguin, Harmondsworth.
10. Gish, O. (1973) *World Development* 1:37–44.
11. Klaarman, H. E. (1965) *The Economics of Health*, New York: Columbia University Press.
12. Kleiman, E. in *The Economics of Health and Medical Care*, (ed.) M. Perlman, New York: Macmillan p. 69.
13. Knight, J. B. and M. D. McGrath (1977) *Oxford Bulletin of Economics and Statistics* 39:245–271.
14. Maasdorp, G. *et al.* (eds.) (1978) *From Shantytown to Township*, Cape Town: Juta pp. 109, 110.
15. MacCarthy, H. (ed.) *Hospital and Nursing Yearbook of Southern Africa 1970*, Johannesburg: H. MacCarthy Publications.
16. Mann, J. K. *et al.* (1973) in *Health Economics*, (eds.), M. H. Cooper *et al.*, Penguin, Harmondsworth pp. 277–280.
17. McGrath, M. D. (1977) "Racial Income Distribution in South Africa," Black/White Income Gap Project Report, University of Natal.
18. _____. (1978) in *Change, Reform and Economic Growth in South Africa*, (eds.), L. Schlemmer *et al.*, Johannesburg: Ravan Press, p. 161.
19. O'Reagain, Mary (1970) "The Hospital Services of Natal," Durban: University of Natal.
20. "Report of the Secretary for Health for the Year Ended 31 December 1975," Annexure 7, RP26/76, Pretoria: Government Printer.
21. Sadie, J. L. (1970) "Projections of the South African Population 1970–2020," Industrial Development Corporation of South Africa, Johannesburg.
22. Sorkin, A. L. (1976) *Health Economics in Developing Countries*, Lexington: Lexington Books, pp. 110–112.
23. "South African Statistics 1974," Tables 1.23 and 1.24, Pretoria: Government Printer.
24. "Statistical Year Book 1966," Table D16, Pretoria: Government Printer.
25. "Survey of Household Expenditure, 1975," Department of Statistics Report No. 11-06-05, Pretoria: Government Printer.
26. "The Provincial Auditor's Report on the Appropriation Accounts and the Finance Accounts for the Period 1 April 1969 to 31 March 1970," Province of Natal.
27. Trengrove-Jones, S. (1977) "A Study of Health and Health Services in South Africa," Unpublished M.Com. dissertation, University of Natal.
28. Wells, L. G. (1974) *Health, Healing and Society*, Johannesburg: Ravan Press, p. 1.

HEALTH MANPOWER IN RELATION TO SOCIOECONOMIC DEVELOPMENT AND HEALTH STATUS

T. Fülöp and W. A. Reinke

> To look forward with vision, it is wise to glance backward with perception—not to be bound by history, nor to blame ourselves or our predecessors, but to learn lessons as a springboard to the future.
>
> H. Mahler, Director-General, WHO, 1978

I. INTRODUCTION

Since its inception 30 years ago, the World Health Organization (WHO) has engaged in continuing cooperation with its member states in their efforts to improve the health of their populations by addressing, *inter alia*, both qualitative and quantitative issues of health manpower development (HMD). While achievements in this field, as well as in many other fields of the organization's work, may in some instances be quite obvious, there has been little systematic effort to document the relationship between enunciated policy, objectives, program setting and implementation, and ultimate outcome at country level. There seems, therefore, to be a need to carry out such an analysis, with the expectation that careful delineation of the forces which led to success or resulted in failure to achieve major objectives in the past may be helpful in planning for the future.

The methods that have been selected for the analysis included document analysis, literature research, expert opinion study, country case studies, and statistical analysis. The latter part of the study was intended to include the analysis of data pertaining to the 152 member states of WHO (in 1979). Finally, only 131 were available for analysis. The present publication highlights some of the results of the statistical analysis.

The fundamental issues that had to be studied were the interrelationships between health services/health manpower development and health improvements, on the one hand, and socioeconomic development and health services/health manpower as well as health status development on the other. Then the further main issue was the extent to which WHO/HMD programs have promoted health services/health manpower development and through it health status improvements in member states. Some of these issues have already been intensively studied with mixed results because of the lack of clarity in the relationships of interest, number of interdependent factors which play a more or less important role, and lack of data to measure whatever relationships are hypothesized. An extensive bibliography compiled by Johns Hopkins University (5) served only to catalog investigative approaches used and unresolved issues. More recently Barlow (3) has presented a general model of health and development which depicts schematically the possible relationships linking income, education, fertility, nutrition, and health. Within this framework, interest herein centers on the way in which health is affected by the other four factors. Barlow further classifies processes of analysis according to whether investigation is at the macro or micro level and whether a single-equation model or a set of simultaneous equations are employed. Of the five studies cited at the macro level, all were of the single-equation type. Two of the studies (12, 13) involved health, education, and income only, while the other three (4, 8, 14) included the nutrition factor as well.

Conclusions reached from large-scale studies vary substantially. Arriago and Davis (1), Preston (14), and Stolnitz (18) are persuaded that public health activities deserve major credit for measurable improvements in health status. Fuchs (7), Fülöp (6), Krishnan (12), Shchepin (15, 16), Smucker (17), and Venediktov (19), on the other hand, stress the importance of socioeconomic improvements. Simple correlation analysis of data for 68 countries in an earlier WHO study (9) yielded high associations between life expectancy and both per capita income and health resource factors such as medical density and bed/population ratios. However, the latter investigators cautioned about the quality of the data and apparent differences in relationships between richer and poorer countries. Examining evidence from a historical perspective, Grosse (10) and Preston (14) have suggested that health conditions, e.g., environmental sanitation, may be increasingly important determinants of life expectancy in place of formerly dominant socioeconomic factors.

In view of the pattern of ambiguous and conflicting findings, it is understandable that some studies have been limited to selected populations for which data of uniform quality are available on a few variables relevant to rather narrowly defined relationships. While these studies may gain advantages in validity, they suffer from problems of generalizability.

Krishnan's study (12) in India is a noteworthy case in point inasmuch as the model tested closely parallels that employed here. Changes in crude death rates between 1951 and 1961 were investigated as a function of per capita income, literacy rate, doctors per capita, hospital beds per capita, and per capita state expenditures on health services. Since the study was confined to 11 Indian states, however, its statistical meaning and general value have been questioned (3).

Analysis of the relationship between education and health has been hampered by lack of meaningful data in many countries. Auster *et al.* (2) have conducted one of the more definitive studies establishing such a relationship, but attention was limited to statewide indicators in the United States.

Examinations of the relationship between fertility and health are likewise rare. One example is provided by Heller and Drake (11), who compared children's morbidity experience and their weight/height ratios with birth order and the number of young children in the family. While the analysis dealt with an important issue, its scope was extremely narrow.

In general, review of the literature underscores a continuing dilemma. Large-scale comprehensive studies run the risk of having to deal with nonspecific indicators of variable quality from heterogeneous populations. On the other hand, more circumscribed investigations are likely to produce results of questionable general validity.

II. PURPOSE OF THE STATISTICAL ANALYSIS

A quantitative analysis, in addition to the qualitative one, to attempt to clarify basic relationships at national level between socioeconomic factors, health sector (health resource) variables, and health status was in order if we wanted to study the extent to which WHO programs in HMD have enhanced the development of health services and, through them, promoted an improvement of health status in member states. Such a study, broad in purpose, would then be limited in implementation by the availability of valid and reliable data concerning the majority of member states.

The study also tries to analyze changes in the above relationships over time. Thus the expectation was that careful analysis of the basic relationships and their dynamics would provide a framework for assessing the effects of specific interventions, e.g., HMD programs, national, WHO, and other. The basic analyses, some of the results of which are published here, are intended to establish points of reference, departures from which could be judged, by methods of qualitative analysis, in relation to programmatic efforts which had taken place aiming at change.

III. PREPARATION FOR ANALYSIS

The relationships between the three basic components of the study are postulated in the following, very much simplified, way:

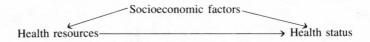

This does not take into account that those relationships are, in fact, interrelationships.

These relationships could be described as meaning that the countries with more favorable socioeconomic conditions, including relatively high gross national product (GNP) per capita, are likely to allocate more substantial material and personnel resources to the health sector. It is assumed, though not proved, that these resources will contribute to improve measurable indices of health status, e.g., life expectancy. These indices are, of course, also influenced by socioeconomic factors both directly and indirectly, *inter alia*, through their effect on health resources development.

This model of relationships and the analytical purpose inferred from it led to a search for indicators of the variables discussed above. Those indicators, we decided, were to meet three criteria.

First, each indicator should broadly represent one of the three categories of variables depicted in the model above. Second, each indicator should be generally available to permit comprehensive analysis of as many of the 152 WHO member states as possible. Third, where appropriate, changes in the indices over time should be available.

With the analytical purpose and these criteria in mind, a list of more than 30 potential measures was assembled for consideration. In spite of thorough review of reports of different international agencies, some of the candidate measures were found for only a limited number of countries, if at all. In other cases in which indicators were intended to measure similar things, a study of the data was conducted to select the most accurate and consistent measures. For example, instances of obvious underreporting of infant mortality were so common that World Bank estimates of life expectancy were used instead. As a result of the search, the following 12 variables[1] were finally selected for analysis.

Socioeconomic variables
- GNP—per capita gross national product
- PE—percent of primary school age children in primary education
- TOFE—ratio of percent of total children in primary education to percent of female children in primary education
- PUTE—pupil/teacher ratio in primary education

Health resource variables
- HLTH—health employees per 1,000 population
- MD—absolute number of physicians
- DR—physicians per 10,000 population
- NMW—nurses, midwives per physician
- BED—hospital beds per physician

Health status variables
- CBR—crude birth rate
- CHL—percent of population aged 0–14
- LE—life expectancy at birth

The first four variables are measures of *socioeconomic status,* of which three have to do with educational opportunities. In addition to the overall measure of participation in education (PE), relative opportunity for females (TOFE) was noted. Availability of teachers at the educational base (PUTE) was intended as a very crude measure of the quality of education.

Five *health resource* indicators were employed, four of which related to manpower, a matter of particular concern to this study. The number of health employees per 1000 population (HLTH) was used as an overall indicator of manpower availability. As an important category of manpower, physicians were viewed in two ways, both in absolute numbers (MD) and relative to population size (DR). Nurses and midwives were related to number of physicians (NMW) rather than to the general population—for two reasons. First, countries with a relative abundance of one category of health manpower tend to have above average personnel/population ratios in other categories as well. To put all manpower indices on the same population base, therefore, was likely to introduce confusing correlations and redundancy into the analysis. The second reason for relating nurses and midwives to physicians was an effort to obtain a measure of manpower balance. For similar reasons, numbers of hospital beds (BED) were related to numbers of physicians rather than to total population. In effect, the BED variable was intended to provide a very crude indication of the health infrastructure in which personnel were employed, just as PUTE gave a semiqualitative indication of the educational system.

The last three variables in the list are "outcome" measures, indices of *health status*. The crude birth rate (CBR) and the relative age of the population (CHL) are frequently used indicators of societal well-being. Life expectancy (LE) is a more direct measure of this condition.

Periodic measurements were available for 6 of the 12 variables. Considering the interest in relating study findings to WHO programs throughout the organization's history, the two dates 1950 and 1975 were chosen for analysis. In the case of PE, for example, the initial (1950) statistic was labelled IPE, while "current" (1975) status was referenced as CPE. In addition, the rate of change between the two years was calculated. In the case of PE the change variable was defined as

$$\text{DPE} = \left(\frac{\text{CPE} - \text{IPE}}{\text{IPE}}\right) \times 100.$$

Other change variables were defined similarly. Where data were not available for the exact years 1950 and 1975, measurements were recorded for the nearest available year.

To illustrate the procedure, it was found, e.g., for country X, that

$$\text{IPE} = 46$$
$$\text{CPE} = 77$$
$$\text{DPE} = \left(\frac{77 - 46}{46}\right) \times 100 = 67,$$

indicating that the percentage of children in primary education increased by 67 percent in the interval between 1950 and 1975.

In effect, 6 of the 12 basic variables generated three variables (C, I, D) each, so that a total of 24 variables was available for analysis.

Twenty-one of the 152 member states had 5 or more missing observations among the 24 variables. These states were excluded, so that the final data set consisted of 24 observations of 131 member states. Of the

$$131 \times 24 = 3144$$

required data points, 38 observations were missing and had to be estimated. In all cases reasonable approximations could be made, and in any event the number of estimates was considered small enough (1.2 percent) not to distort the analysis.

IV. ANALYTICAL MODEL

The initial analytical model is shown schematically in Figure 1. At the first stage socioeconomic factors were related to health resources as the dependent variables. The second-stage analysis related socioeconomic factors and health resources to health status variables.

The two stages relate to the rows of Figure 1. At each stage three sets of analyses were conducted in accordance with the columnar arrangement of the table. The first set dealt with initial conditions. The second concerned current conditions. The third set related initial states and measures of change in the independent variables to current status of the dependent variables. Thus the third set tested the presence of a lagged effect.

All the foregoing analyses are contained within the broken-line boundary of Figure 1. The hope was that a synthesis of results obtained would be meaningful in itself and could then be related descriptively to HMD activities.

Figure 1. Initial analytical format

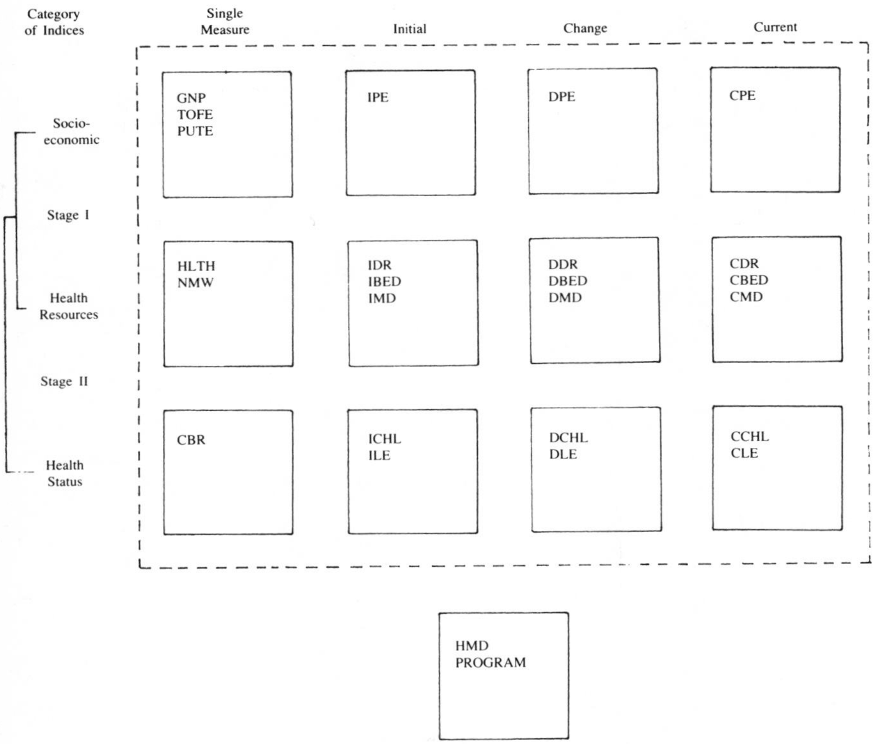

	Single Measure	Initial	Change	Current
Socio-economic	GNP TOFE PUTE	IPE	DPE	CPE
Health Resources	HLTH NMW	IDR IBED IMD	DDR DBED DMD	CDR CBED CMD
Health Status	CBR	ICHL ILE	DCHL DLE	CCHL CLE

Category of Indices — Stage I (Socio-economic, Health Resources), Stage II (Health Status)

HMD PROGRAM

Multiple linear regression analyses were conducted according to the foregoing format. Findings indicated that certain of the variables were generally of little or no consequence. In particular, the only change variable which showed promise related to the number of hospital beds per physician.

On the other hand, it became apparent that some of the significant relationships could be further improved through logarithmic transformations. The pruning and refinement of variables produced the modified format of Figure 2 which served as the basis for the final analysis reported herein.

V. FINDINGS FROM REGRESSION ANALYSIS

A. Resource Predictions

The ability of socioeconomic [abbreviated as exog(enous)] factors to explain health resources variation is displayed in Table 1, using standardized beta coefficients as indicators of relative importance.[2] We see that predictability is good

Figure 2. Final analytical format

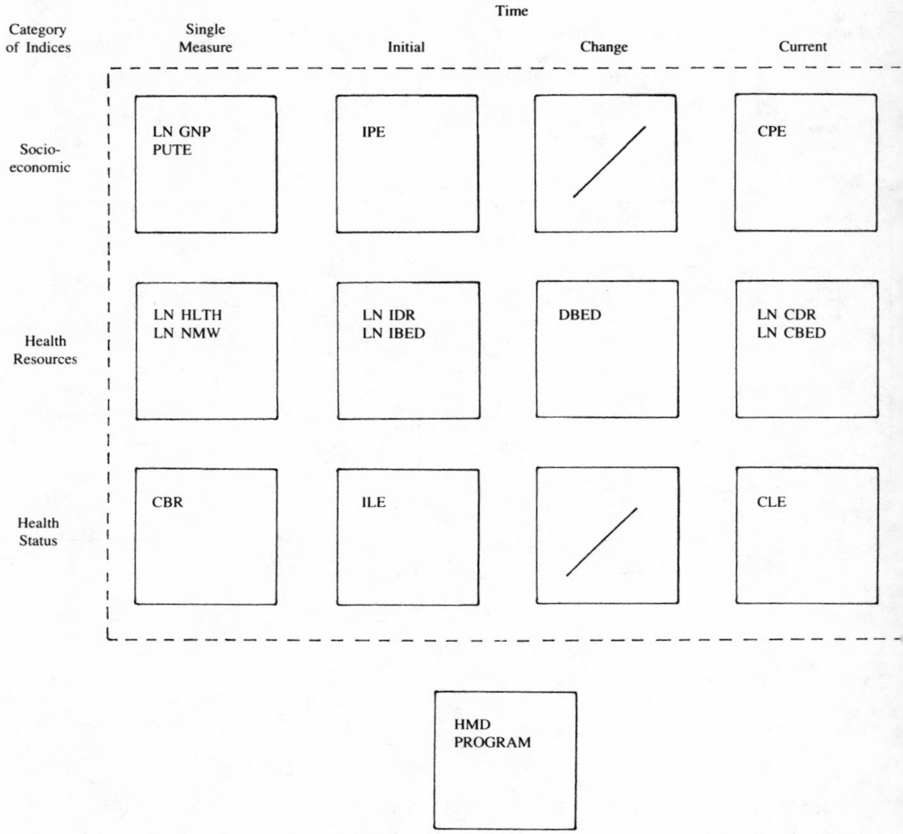

Category of Indices	Single Measure	Initial	Change	Current
Socio-economic	LN GNP PUTE	IPE	/	CPE
Health Resources	LN HLTH LN NMW	LN IDR LN IBED	DBED	LN CDR LN CBED
Health Status	CBR	ILE	/	CLE

HMD PROGRAM

with respect to health workers per 1000 population and physicians per 10,000 population, regardless of the time frame used. Coefficients of determination (R^2) indicate that 67–84 percent of resource variation is explained. Per capita GNP is the principal explanatory factor, although percent of children in primary education is also quite significant. Pupil/teacher ratios are generally less important, but they do provide some evidence that relatively large numbers of teachers and large numbers of health workers tend to go together.

Socioeconomic factors are less successful in predicting numbers of nurses and midwives and hospital beds per physician, producing R^2 values between .06 and .32. The only relationship that is uniformly significant is that between the pupil/teacher ratio and the bed/physician ratio.

Table 1. Summary of Regression Analysis Findings: Resource Variables

	Beta Coefficients			
Function	LN GNP	PUTE	PE	R^2
HLTH				
LN HLTH = f(EXOG$_o$)	.50***	−.17**	.26***	.69
LN HLTH = f(EXOG$_t$)	.58***	−.21**	.17*	.67
DR				
LN IDR = f(EXOG$_o$)	.40***	−.07	.51***	.79
LN CDR = f(EXOG$_o$)	.51***	−.15***	.37***	.84
LN CDR = f(EXOG$_t$)	.59***	−.21***	.27***	.83
NMW				
LN NMW = f(EXOG$_o$)	−.00	.10	−.25	.07
LN NMW = f(EXOG$_t$)	−.09	.13	−.13	.06
BED				
LN IBED = f(EXOG$_o$)	−.02	.29**	−.25*	.21
LN CBED = f(EXOG$_o$)	−.20	.27**	−.20	.32
LN CBED = f(EXOG$_t$)	−.26*	.30**	−.13	.31

Notes:
*p < .05
**p < .01
***p < .001

B. Health Status Predictions

From Table 2 it can be seen that socioeconomic factors, notably per capita GNP, can be used to explain more than 60 percent of the variation in crude birth rate. All the regression coefficients are in the expected direction.

Adding resource variables to the analysis further raises R^2 to .71. More striking, the physician and bed variables are the dominant explanatory factors, replacing per capita GNP in importance. However, regression analyses are notoriously weak in disentangling the effects of correlated independent variables, and the above results must be treated with great caution.

Turning to the Table 2 results on life expectancy, we find that each of the socioeconomic factors is highly significant, and together they explain as much as four-fifths of the variation in life expectancy. Inclusion of the resource variables produces little increase in values of R^2, but once again the doctor/population ratio becomes the outstanding explanatory factor in place of the socioeconomic variables. Roughly similar results are obtained when the crude birth rate enters

Table 2. Summary of Regression Analysis Findings: Health Status Findings

				Beta Coefficients					
Function	LN GNP	PUTE	PE	LN HLTH	LN DR	LN NMW	LN BED	DBED	CBR	R_2
CBR										
CBR = f(EXOG$_o$)	−.38***	.18**	−.38***							.67
CBR = f(EXOG$_t$)	−.60***	.20**	−.08							.61
CBR = f(EXOG$_o$, RES$_o$)	−.13	.16*	−.14	−.10	−.51***	−.12*	−.09	.00		.71
CBR = f(EXOG$_t$, RES$_t$)	−.08	.10	.15*	.04	−1.10***	−.08	−.36***			.71
LE										
ILE = f(EXOG$_o$)	.31***	−.20***	.51***							.82
CLE = f(EXOG$_o$)	.26***	−.19***	.58***							.83
CLE = f(EXOG$_t$)	.53***	−.25***	.23***							.71
CLE = f(EXOG$_o$, RES$_o$)	.02	−.14***	.31***	−.00	.51***	.02	−.04	−.11**		.88
CLE = f(EXOG$_t$, RES$_t$)	.05	−.10	.00	−.08	.93***	−.01	.11			.81
CLE = f(EXOG$_o$, CBR)	.14*	−.13**	.46***						−.31***	.86
CLE = f(EXOG$_t$, CBR)	.24***	−.15**	.19***						−.48***	.80
CLE = f(EXOG$_o$, RES$_o$, CBR)	−.01	−.10*	.28***	−.02	.38***	−.01	−.06	−.11***	−.25***	.90
CLE = f(EXOG$_t$, RES$_t$, CBR)	.02	−.06	.06	−.06	.50*	−.04	−.03		−.39***	.85

Notes:
*p < .05
**p < .01
***p < .001

the equations in place of health resource variables. However, the crude birth rate does not appear to be as dominant a force as the doctor/population ratio. Inclusion of both health resources and crude birth rate in the analysis yields R^2 as high as .90. The persistence of the pattern of shift in importance away from socioeconomic variables tends to increase somewhat the credibility of the shift.

As noted earlier, the set of R^2 values associated with a particular dependent variable tend to be rather similar. It is interesting to observe, however, that without exception the highest R^2 in a set is that which relates current status of the dependent variable to initial state of the independent variables. Thus there is some evidence in support of a lagged response. Further investigations below are limited to equations within each set which incorporate these lags.

C. Synthesis of Findings

Of the 23 regression equations derived for 6 dependent variables (Tables 1 and 2), attention ultimately focused on the 6 given in Table 3. Since these are prediction equations, the original (unstandardized) regression coefficients are listed, along with their t values. The table also presents the simple correlation matrix in order to (1) permit comparison of two-way associations with those derived after taking other factors into account; and (2) display areas of high correlation as a signal of skepticism concerning the corresponding partial regression coefficients.

To illustrate the comparison of two-way and multivariate associations, consider the life expectancy dependent variable (CLE). IPE, LN IDR, and CBR all show high simple correlations with CLE, and the associations are retained in multiple regression equation (8). Independent variables LN HLTH and CBR also show a high simple association with CLE, but they lose importance in the multiple relationship. Note, however, that the latter independent variables are highly correlated with the former. DBED presents a more surprising phenomenon. Its simple correlation with CLE is slight, and it is not closely associated with any of the other independent variables. Yet after adjustment for them, the importance of DBED shows forth in equation (8).

Taken together, the six regression equations of Table 3 produce the rather straightforward set of relationships diagramed in Figure 3. Socioeconomic factors show a strong association with two of the health manpower indicators, especially with the number of physicians per 10,000 population. These socioeconomic factors are also associated with crude birth rate and life expectancy. When the health resource conditions (DR and BED) are combined with the socioeconomic factors, further increases in correlations with crude birth rate and life expectancy are observed. Similarly, crude birth rate combined with socioeconomic variables explains somewhat more of the variation in life expectancy than socioeconomic factors alone. Finally, inclusion of both crude birth rate and the resource variables DR and BED, along with socioeconomic variables, yields the most explained variation in life expectancy. We emphasize again,

Table 3. Major Regression Equations and Simple Correlation Matrix
(t Values in Parentheses under Partial Regression Coefficients)

(1) LN CDR (R^2=.84)	=	−3.165	+ 0.571 LN GNP (8.65)	− 0.020 PUTE (3.52)	+ 0.015 IPE (6.95)
(2) LN CBED (R^2=.32)	=	2.803	− 0.144 LN GNP (1.60)	+ 0.024 PUTE (3.00)	− 0.006 IPE (1.85)
(3) LN HLTH (R^2=.69)	=	−0.116	+ 0.457 LN GNP (5.99)	− 0.019 PUTE (2.79)	+ 0.009 IPE (3.50)
(4) LN NMW (R^2=.07)	=	1.059	− 0.002 LN GNP (0.00)	+ 0.009 PUTE (0.91)	− 0.007 IPE (1.92)
(5) CBR (R^2=.71)	=	51.615	− 1.183 LN GNP (1.26)	+ 0.178 PUTE (2.55)	− 0.050 IPE (1.56)
			− 4.366 LN IDR (4.02)	− 1.227 (1.27)	LN IBED + 0.001 DBED − 0.986 LN HLTH (0.06) (1.09)
			− 1.467 LN NMW (2.04)		
(6) CLE (R^2=.90)	=	64.741	− 0.078 LN GNP (0.13)	− 0.116 PUTE (2.66)	+ 0.097 IPE (4.99)
	+	3.275 LN IDR (4.66)		− 0.890 LN IBED (1.51)	− 0.027 DBED − 0.251 LN HLTH (3.44) (0.45)
	−	0.097 LN NMW (0.22)		− 0.252 CBR (4.57)	

Correlation Matrix

Resource Dependent Variables

		Socioeconomic Factors			Resource Dependent Variables			
		PUTE	IPE	LN CDR	LN CBED	LN HLTH	LN NMW
Socioeconomic	LN GNP	−.61	.76	.88	−.52	.80	−.25
	PUTE		−.48	−.64	.49	−.60	.22
	IPE			.82	−.48	.72	−.30
Resource	LN CDR				−.72	.84	−.41
	LN CBED					−.44	.54
	LN HLTH						−.16

Resource Independent Variables

		LN IDR	LN IBED	DBED	LN HLTH	LN NMW
Socioeconomic	LN GNP	.83	−.38	−.31	.80	−.25
	PUTE	−.56	.42	.19	−.60	.22
	IPE	.84	−.40	−.24	.72	−.30
Resource	LN IDR		−.58	−.16	.78	−.43
	LN IBED			−.14	−.34	.57
	DBED				−.21	.04
	LN HLTH					−.16
	LN NMW					

Health Status

		CBR	CLE
Socioeconomic	LN GNP	−.77	.81
	PUTE	.59	−.63
	IPE	−.75	.87
Resource	LN IDR	−.80	.91
	LN IBED	.34	−.50
	DBED	.22	−.30
	LN HLTH	−.75	.76
	LN NMW	.17	−.35
Health Status	CBR		−.85

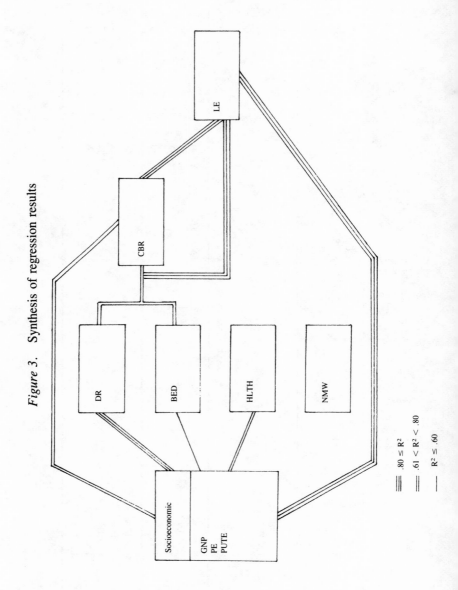

Figure 3. Synthesis of regression results

Table 4. Implications of Regression Coefficients Relating to CLE

Variable	Beta	Regression Coefficient	Mean	Needed to Increase Expected Life Expectancy, by One Year	
				Change	% of Mean
% Children in Primary Education (IPE)	.28	.0974	72.1	10.3	14.3
Physicians per 10,000 Population (IDR)	.38	3.2747	1.64[1]	.58[1]	35.4
Crude Birth Rate (CBR)	−.25	−.2516	37.0	−4.0	−10.8

Note:
[1] Numbers converted from logarithms.

however, that the small incremental improvement in R^2 relative to life expectancy is not as striking as is the shift in significance from the socioeconomic factors to selected health resource variables as these are introduced into the equations. There is some evidence, therefore, that socioeconomic conditions affect life expectancy also through their influence on the health infrastructure, notably the number of physicians and hospital beds.

In the analytical sequence that has been followed, the ultimate outcome variable is life expectancy, and its analysis produced the largest R^2 (.90). For this especially noteworthy case, it is of interest to examine more carefully the implications of selected regression coefficients, emphasizing again that they must be treated with considerable skepticism. To make the investigation meaningful it is necessary to reconvert LN IDR to original rates, i.e., numbers of physicians per 10,000 population. The three independent variables of special interest are IPE, LN IDR, and CBR. Information concerning them is summarized in Table 4.

The regression coefficient for IPE suggests that each 1 percent increase in school enrollment is expected to increase life expectancy by 0.0974 year. Thus, an increase of 10.3 percent in enrollment would be required to add a full year to life expectancy. Assuming a country of average status, i.e., approximately 72 percent of eligible children in primary education, the required increase would be 14.3 percent beyond the present level. Corresponding calculations, with similar interpretations, are provided for the other variables of interest.

Although the physician/population ratio was found to be highly significant statistically, the practical implication is that an increase in the ratio by 35 percent is necessary on the average to achieve a single year of improvement in life expectancy. This is not likely to happen holding other factors constant, notably GNP. In practice, as well as statistically, it is difficult to separate socioeconomic factors from the impact of health resources development. By comparison, Table

4 suggests that a 1-year improvement in life expectancy can be achieved through a reduction in crude birth rate by 4 per 1000, or about 11 percent of the average.

VI. ANALYSIS OF RESIDUALS

From the foregoing analyses it is clear that, as it is well known, direct comparisons of health resource levels between countries can be misleading because of different local conditions. It would be preferable to make the comparisons after adjusting for the differences. The regression results are encouraging in this regard. The six equations of Table 3 succeeded in accounting for most of the national differences, even if little can, of course, be inferred concerning cause and effect. They provide individualized expectations of health resource and health status to be compared with actual conditions. Residual findings (actual minus expected) can then be examined to determine whether discrepancies are purely random or whether systematic patterns exist which might be related to special needs or interventions, e.g., WHO health manpower development programs, not factored into the basic equations, because of inherent difficulties of meaningful quantification of the latter.

A. Residuals by WHO Region

To begin with, residuals for each country on each of the six dependent variables were compiled by WHO region. The resulting average residuals are recorded in Table 5. The corresponding standard error of the calculated average was also determined in each case, and t statistics were developed. The t values are also shown in the table, along with indications of significance. Residuals were determined from the standardized form of the regression equations. Thus, for example, a residual of .5 is a measure of relative discrepancy which has essentially the same meaning when applied to any variable. The t values, of course, are also subject to uniform interpretation.

No significant regionwide departures from expectation are found with regard to total health manpower. Decided imbalances are noted within categories, however. The African region as a whole has a severe shortage of physicians, even after adjustment for the low socioeconomic status of many countries in the region. A corresponding shortage of nurses and midwives is not observed. On the contrary, there is a relative abundance in comparison with physicians. The same is true in even larger degree for hospital beds. The reverse pattern is found in the American and eastern Mediterranean regions, although the situation is not as striking as in Africa. The only other health resource finding of significance occurs in the Southeast Asia region. There the number of hospital beds does not match the quantity of doctors, which is more than up to expectation.

Separate global analysis of residuals reveals that those for NMW and CBED are positively correlated and both are negatively associated with CDR residuals. It seems that the number of hospital beds is statistically more closely associated

Table 5. Summary of Average Residuals by WHO Region
(Measured as Deviations from Expected Values in Standardized Form
of Regression Equations)

Region	Numbers of Countries	Dependent Variable					
		LN HLTH	LN CDR	LN NMW	LN CBED	CBR	CLE
		a. Mean Residuals					
African	41	.06	−.20	.37	.57	.01	−.09
American	26	−.11	.18	−.56	−.45	.24	.07
Eastern Mediterranean	17	−.17	.12	−.41	−.59	.36	.02
European	29	.15	.06	.03	.17	−.39	.01
South-East Asia	8	.10	.17	.04	−.75	−.08	.01
Western Pacific	10	−.19	−.16	.50	−.05	−.08	.14
	131						
		b. t-Values					
African		0.69	−3.56***	2.68*	5.42***	0.07	−1.22
American		−0.98	2.58*	−3.20**	−3.38**	2.71*	1.18
Eastern Mediterranean		−1.27	1.36	−1.92	−3.63**	3.24**	0.20
European		1.43	0.88	0.20	1.34	−4.52***	0.13
South-East Asia		0.54	1.32	0.14	−3.16*	−0.50	0.09
Western Pacific		−1.08	−1.42	1.78	−0.26	−0.58	1.43

Notes:
*p < .05
**p < .01
***p < .001

with the number of nurses than with the number of doctors. These findings, coupled with the nonsignificance of HLTH residuals in Table 5, suggest that the ratio of health workers per 1000 population tends to follow a standard pattern in relation to development, but that the composition of the manpower pool tends to be more individualized.

Crude birth rate residuals show distinct regional patterns. Rates in the American and eastern Mediterranean regions were substantially higher than expected, whereas those in Europe were below expectations. Clearly, cultural factors are important in fertility, and these were not included in the regression equations.

The equation involving life expectancy accounted for as much as 90 percent of its variation, and Table 5 reveals no systematic pattern for the small residual. It is noteworthy, however, that residuals from earlier equations (not shown here) involving only socioeconomic factors showed significantly higher than expected life expectancies in the American and European regions, while residuals in the African and eastern Mediterranean regions were significantly negative. Only when health resources availability came into the equations did these systematic patterns dissipate. This provides one further bit of evidence of the differential effects of health resources and socioeconomic factors in spite of their joint correlation.

B. Individual Residuals

While examination of residuals by WHO region has been rewarding, individual discrepancies of note can easily become buried in averages. Because of the large number of individual cases,

$$131 \text{ countries} \times 6 \text{ dependent variables} = 786,$$

however, it is necessary to investigate them purposefully. Two approaches have been used. First, the 10 most extreme positive and negative residuals for each variable have been selected out and highlighted in Table 6. Second, in view of the overriding concern for present purposes with health manpower programs, all 131 countries have been listed in Table 7, within region according to magnitude of LN HLTH residual. Corresponding residuals for the CDR variable have been recorded in the adjacent column.

In many respects the array of countries conforms to the regional findings cited above. In addition, the lists contain countries which may be found to be exceptional in other regards. The Southeast Asia region in particular is seen to be a region of wide individual differences which tend to cancel each other in overall averages.

We were especially interested in investigating exceptional cases in relation to particular health manpower development initiatives. One quantitative measure, the number of WHO fellowship awards by country, was available for statistical examination in this regard. Fellowships indices were created by country to reflect population size and number of years of membership in WHO. The indices were then compared by means of t tests for exceptional cases in Table 5 regarding ratios of total health manpower and physicians to population. While averages were in the expected direction, variability in the indices was so large that results were not statistically significant. This suggests that reliance must be placed, as originally planned, upon more qualitative approaches, including document analysis, expert opinion study, and detailed country case studies.

VII. CONCLUSIONS

An in-depth statistical analysis of interdependence of some indices characterizing national socioeconomic conditions, material and personnel health resources, and health status was carried out, using the available data of 131 WHO member states, out of a total of 152. The aim of this study was to complement an analytical review of the World Health Organization's HMD program for the years 1948–78.

Health Resources and Health Status

Table 6. Individual Country Extreme Residuals

Most Negative	Most Positive	Most Negative	Most Positive
LN HLTH		**LN NMW**	
Malaysia	Burma	Spain	Cameroon
Guyana	Mongolia	Colombia	Nigeria
Panama	Sri Lanka	Algeria	Sweden
Australia	Congo	Nepal	Finland
Nepal	USSR	Ecuador	Indonesia
Somalia	Albania	Pakistan	Gabon
Switzerland	Swaziland	Mexico	Trinidad + Tobago
GDR	Sweden	Belgium	Laos
Cent. Afr. Emp.	Gambia	Iran	Guyana
Bangladesh	Pakistan	Uruguay	Cent. Afr. Emp.
LN CDR		**LN CBED**	
Thailand	Mongolia	Pakistan	Zaire
Libya	Pakistan	Mexico	Gabon
Malawi	India	Paraguay	Guyana
Gabon	Burma	Afghanistan	Malawi
Malaysia	Nicaragua	Jordan	Uganda
Luxembourg	Argentina	Syria	Cameroon
Uganda	USSR	Nicaragua	Burundi
France	Bulgaria	Peru	Lesotho
Indonesia	Bolivia	Saudi Arabia	Luxembourg
Papua New Guinea	Albania	Venezuela	Zambia
CBR		**CLE**	
Nepal	Kuwait	Gabon	Jamaica
Chad	So. Africa	Bolivia	Albania
Belgium	Libya	So. Africa	Guyana
Haiti	Iraq	Guinea-Bissau	Comoros
Spain	Nicaragua	Congo	El Salvador
Bulgaria	Honduras	Cameroon	Sri Lanka
Mozambique	Dominican Republic	Mauritania	Sudan
Greece	Lebanon	Gambia	Costa Rica
Afghanistan	Mongolia	Argentina	Morocco
Poland	Sawziland	Senegal	Mexico

The analysis was limited by the lack of reliable data on many variables which had been planned for inclusion in the study, on the one hand, and by the inherent difficulties of meaningful quantification of HMD programs in general and especially of WHO's HMD program, on the other.

Twenty-four variables were used for each of the 131 member states. The statistical analysis (computer operation) of the 3144 data allowed some interesting conclusions, such as the following:

Table 7. Country Residuals in Order by Region and Magnitude of LN HLTH Residual (Standardized)[a]

AFRICAN REGION

Country	LN HLTH	LN CDR
Cent. Afr. Rep.	−.78	−.04
Algeria	−.70	−.27
Cape Verde	−.62	−.11
Burundi	−.55	−.48
Niger	−.49	−.42
Uganda	−.40	−.63
Lesotho	−.39	−.53
Guinea-Bissau	−.36	.20
Rwanda	−.20	−.44
Ghana	−.17	−.51
Ethiopia	−.15	−.58
Togo	−.15	−.13
Nigeria	−.11	−.30
Zaire	−.08	−.41
Malawi	−.06	−.91
Gabon	−.04	−.77
Upper Volta	−.04	−.24
Cameroon	0	−.40
Ivory Coast	.02	−.43
Tanzania	.08	.07
Kenya	.09	−.34
Mali	.09	−.31
Chad	.12	−.06
Madagascar	.13	.38
Mozambique	.15	.26
Angola	.17	.05
Zambia	.17	−.06

AMERICAN REGION

Country	LN HLTH	LN CDR
Guyana	−1.88	−.44
Panama	−1.51	.11
Cuba	−.72	.26
Brazil	−.41	.03
Haiti	−.32	.17
Nicaragua	−.27	.70
Uruguay	−.22	.21
Trin. & Tobago	−.18	−.41
Mexico	−.12	.38
Bolivia	−.04	.55
El Salvador	.04	.12
Venezuela	.07	.19
Argentina	.08	.63
Ecuador	.09	.34
Dom. Rep.	.10	.44
Paraguay	.10	.51
Canada	.11	−.26
Chile	.14	−.11
Honduras	.14	.36
Colombia	.16	.41
Costa Rica	.16	.14
Peru	.18	.40
Guatemala	.23	.24
Jamaica	.31	−.03
USA	.38	−.32
Barbados	.64	.16

EUROPEAN REGION

Country	LN HLTH	LN CDR
Switzerland	−.93	.10
GDR	−.92	−.01
Belgium	−.68	.06
Italy	−.54	.19
Morocco	−.43	−.54
Spain	−.33	.23
GFR	−.15	−.31
France	−.12	−.61
Austria	−.06	.08
Netherlands	−.05	−.03
Greece	.01	.48
Turkey	.01	.37
Norway	.02	−.32
Luxembourg	.08	−.65
Portugal	.21	.10
Yugoslavia	.21	.27
Denmark	.23	−.30
Poland	.32	.17
Finland	.50	−.10
Ireland	.51	−.06
UK	.52	.05
Bulgaria	.63	.56
Hungary	.65	.38
Czechoslovakia	.68	.41
Rumania	.70	.30
Iceland	.72	−.09
Sweden	.74	−.18

348

Mauritania	.18	−.10
Mauritius	.26	.13
Benin	.31	−.21
Guinea	.36	−.23
Liberia	.36	−.04
Senegal	.40	−.06
Sierra Leone	.45	−.05
Botswana	.51	.07
So. Africa	.58	.01
Comoros	.72	.27
Gambia	.73	−.06
Swaziland	.85	−.04
Congo	.90	.06

Albania	.85	.52
USSR	.87	.62

EASTERN MEDITERRANEAN REGION

Somalia	−1.08	.34
Afghanistan	−.76	.03
Lebanon	−.51	.06
Jordan	−.50	.31
Saudi Arabia	−.50	.04
Kuwait	−.49	−.91
Syria	−.44	.21
Iraq	−.33	−.09
Israel	−.29	.43
Dem. Yemen	−.19	−.55
Libya	.02	.13
Egypt	.05	.34
Tunisia	.06	.02
Iran	.25	.15
Cyprus	.42	.27
Sudan	.67	.19
Pakistan	.73	1.05

SOUTH-EAST ASIA REGION

Nepal	−1.11	−.32
Bangladesh	−.78	.27
Thailand	−.65	−.95
Indonesia	−.43	−.61
India	.43	.89
Sri Lanka	1.00	.20
Mongolia	1.17	1.12
Burma	1.19	.75

WESTERN PACIFIC REGION

Malaysia	−1.95	−.67
Australia	−1.13	−.13
Korea	−.46	.02
Singapore	−.08	−.32
Philippines	−.05	.02
Japan	−.03	−.19
Fiji	.16	.16
Papua New Guinea	.42	−.59
New Zealand	.56	−.08
Laos	.70	.17

[a]Omitted from analysis
African Region
  Sao Tome & Principe
  Seychelles
  Namibia (associate members)
  Southern Rhodesia (associate members)
American Region
  Bahamas
  Grenada
  Surinam
Eastern Mediterranean Region
  Bahrain
  Djibouti
  Oman
  Qatar
  United Arab Emirates
  Yemen
European Region
  Byelorussian SSR
  Malta
  Monaco
  Ukranian SSR
South-East Asia Region
  DPR Korea
  Maldives
Western Pacific Region
  China
  Dem. Kampuchea
  Samoa
  Tonga
  Vietnam

1. In the relationship of socioeconomic and health resource data, the variations in the total number of health workers and in the number of physicians are statistically explained to 67–84 percent by the socioeconomic independent variables, mainly by the GNP per capita and to a lesser extent by the percentage of primary-school-age children in primary education;
2. In health status predictions, socioeconomic factors also explain more than 60 percent of the variation in crude birth rate and over 80 percent of life expectancy differences;
3. Inclusion of resource variables in the analysis appears to cause them to replace in importance the socioeconomic ones in explaining variations in health status indices;
4. There seems to be a "lagged" response—i.e., the highest correlations are invariably observable between the current status of the dependent variable and the initial state of the independent variable (e.g., between present life expectancy and initial resource variables);
5. Analysis of the regression coefficients shows that, statistically, a 35 percent increase in the physician/population ratio would "produce" a full year of improvement in life expectancy;
6. The study of residuals (*actual* figures in health resources and health status *minus* those *expected* on the basis of the socioeconomic situation) shows that the human health resource variables tend to follow a standard pattern in relation to socioeconomic development at the level of the different WHO regions—i.e., in the majority (78 percent) of cases, with the notable exception of Africa, they are at the expected level—but the analysis of the cases of individual member states reveals rather striking deviations from what is expected.

The analysis has proved that the statistical approach of the problem in question, however useful it may be, can only be a complementary one to other, qualitative ones. It could, however, reveal some interesting statistical relationships as described above and detect those countries which show deviation in their health personnel resources (and health status) indices markedly more or less than what would have been expected on the basis of their socioeconomic situation characterized by some available indicators. The explanation for these phenomena could and should be sought by other methods such as document analysis, expert opinion study, country case studies, etc., and in the case of the analytical review of WHO's HMD program this is now being done.

VIII. SUMMARY

The analysis that has been reported was meant to serve two purposes. First, it has been the latest in a series of efforts to clarify the relative importance of the health system itself and of socioeconomic factors to the health of populations. Second, the study has quantified national and regional deviations from general patterns as a basis for selective investigation of the effects of programmatic interventions nationally and through WHO.

With respect to the clarification of relationships, the analysis was unusually comprehensive in that it included 131 WHO member states. As in a number of other studies, socioeconomic factors were found to account for much of the national variation in life expectancy. Inclusion of health resources variables provided an additional, special lagged effect which succeeded in ultimately accounting for 90 percent of life expectancy variation. Potentially more important, evidence was obtained that socioeconomic factors may operate to a certain extent through their facilitation of health resources development. It appears, therefore, that though socioeconomic factors are necessary conditions for health improvement, they are not sufficient in the absence of corresponding development of a viable health services infrastructure.

Residual deviations from the general pattern varied systematically by WHO region in 30 percent of the cases. Most notably, in the African region the number of physicians, but not other health workers or hospital beds, is well below even the modest level expected on the basis of socioeconomic situation in the region. Individual countries do not necessarily conform to regional characteristics, however. The country residuals presented in the paper represent, in effect, a new set of variables, adjusted for general socioeconomic and health resources effects, to be investigated further in relation to conditions peculiar to individual countries.

NOTES

1. Main sources of data: Volumes of World Health Statistics Annual, World Health Organization, Geneva, 1951–1979; World Atlas of the Child, The World Bank, Washington, D.C., 1979; World Development Report, 1979, The World Bank, Washington, D.C., August 1979.

2. The usual regression coefficients have value in estimating the amount of change in the dependent variable produced by unitary changes in each of the independent variables. Because of differences in the magnitude of variability exhibited by the latter, however, the meaning of "unit of change" is not uniform. For example, a $1 increase in per capita GNP has different connotations of practicability from a 1 percent increase in primary school enrollment. To facilitate comparability, beta coefficients measure effects in standard deviation units. This conversion procedure also produces differences between "observed" and "expected" values of the dependent variables which are similarly comparable. To further facilitate comparability, the original regression coefficients have been compared with their standard errors to produce "t" statistics. Levels of significance of the t values are indicated in Tables 1 and 2 by asterisks.

REFERENCES

1. Arriaga, E. E. and K. Davis (1969) "The Patterns of Mortality Change in Latin America," *Demography* 6(3):223–242.
2. Auster, R., I. Leveson, and D. Sarachek (1969) *Journal of Human Resources* 4:411–326.
3. Barlow, R. (1979) "Health and Economic Development: A Theoretical and Empirical Review." in Sirageldin, Ismail (ed.), *Research in Human Capital and Development* 1:45–75.

4. Correa, H. (1975) "Population, Health, Nutrition, and Development," Lexington, Mass.: D. C. Heath.
5. Department of International Health. (1972) "Health and Development: An Annotated, Indexed Bibliography," Johns Hopkins University, Baltimore.
6. Fulop, T. (1978) " Egészégugyi Szervezéstan," Medicina, Budapest.
7. Fuchs, V. R. (1979) *Milbank Memorial Fund Quarterly* 7:153–182.
8. Galenson, W., and G. Pyatt (1964) "The Quality of Labor and Economic Development in Certain Countries," Geneva: International Labour Office.
9. Gilliand, P., and R. Galland (1977) *World Health Statistics Reports* 30:227–238.
10. Grosse, R. N., and B. H. Perry (1979) "Correlates of Life Expectancy in Less Developed Countries," University of Michigan (processed).
11. Heller, P. S., and W. D. Drake (1976) "Malnutrition, Child Morbidity, and the Family Decision Process," Discussion Paper No. 58, Center for Research on Economic Development, University of Michigan, Ann Arbor.
12. Krishnan, P. (1975) *Social Science and Medicine* 9:475–479.
13. Malenbaum, W. (1970) In *Empirical Studies in Health Economics,* (ed.), H. E. Klarman, pp. 31–54, Baltimore: The Johns Hopkins University Press.
14. Preston, S. H. (1976) "Mortality Patterns in National Populations," New York: Academic Press.
15. Shchepin, O. P. (1978) "Aktualnie Problemi Zarubezhnovo Zdravoohraneni a," Tsoliuv, Moscow.
16. Shchepin, O. P. (1976) "Problemi Zdravoohraneni a Razviva ushchihs a Stran," Meditsina, Moscow.
17. Smucker, C. M. (1975) "Socio-Economic and Demographic Correlates of Infant and Child Mortality in India," Unpublished doctoral dissertation, University of Michigan.
18. Stolnitz, G. J. (1975) "The Population Debate: Dimensions and Perspectives, Papers of the World Population Conference, Bucharest, 1974," *Population Studies* I(57).
19. Venediktov, D. D. (1976) "Mezhdunarodnie Problemi Zdravoohraneni a," Meditsina, Moscow.

SUMMARY AND DISCUSSION OF PART III

Alan L. Sorkin

McGrath's paper is a pioneering effort to analyze expenditures on health services in a society where racial segregation is required by law. Although information on the distribution of health resources by race is limited, some important trends clearly emerge.

South Africa is one of the few developed countries in the world that has experienced a decline in the proportion of Gross National Product allocated to health services. Thus, from 1960 to 1975, the percentage of GNP spent on health services declined from 4.2 to 3.6 percent. The latter figure is considerably below the average for nations at a similar level of development (1, p. 40).

Because of the slow growth in health expenditures, the Black population has achieved little gain in availability of health services. As McGrath indicates, over 94 percent of private expenditures on health care are obtained by the one-sixth of the population that is White. The Black majority is forced to rely almost entirely on inadequate public facilities and services. The only health items widely purchased by Blacks are patent medicines—products of dubious therapeutic value.

It should be noted that the statistics on private health expenditure developed by McGrath exclude rural Blacks, including the large number living in the Homelands. In the rural portions of South Africa indigenous practictioners supply a large share of the personal health services which Blacks receive. Although the proportion of income spent by South African Blacks on the services of indigenous practitioners is unknown, Myrdal found that in the poorest rural communities of India such expenditures accounted for as much as 2–3 percent of personal income (6, p. 1579).

McGrath's statistics on hospital expenditures clearly show the effects of racial discrimination in the provision of hospital services. However, recent studies of publicly supported health programs indicate similar racial differentials in the utilization of hospital facilities in the American South. For example, in regard to the Medicaid program, Southern Whites received 1.79 times as much in Medicaid payments per recipient as Southern Blacks. Regarding Medicare, Whites received 55 percent more for inpatient hospital care, 95 percent more for physician services, and almost two-and-one-half times as much for extended care services as compared to Blacks (2, pp. 51, 53).

Although patently immoral, to the extent that the South African government views health expenditures as a factor accelerating development, the current discriminatory pattern may seem highly rational. Most South African Blacks are either excluded from the modern sector of the economy or are restricted to menial positions. Thus the South African government may deem it inappropriate to spend money on groups whose economic potential is very limited. Until apartheid is ended, most Blacks in South Africa will continue to receive far less than the minimum adequate level of health services.

Grosse and Perry investigate the relationship between life expectancy in developing countries and a multitude of independent variables. The latter can be grouped into five categories: economic, social, health personnel and facilities, health expenditures, and sanitation. The study is important partly because of the extensive data base—approximately 90 countries and 200 variables. The study is also important because it seems to confirm what other investigators using less extensive amounts of information concluded, namely that economic factors have become progressively less important over time in explaining differences in life expectancy or mortality between countries.

One important consideration in this paper that requires further study is the reliability of the data. Most of the countries included in this study have fairly primitive data collection systems. In fact, for many of the variables considered, one could question the reliability of such information in countries with highly sophisticated data collection systems.

Some pieces of information seem particularly in need of checking. Appendix 1 indicates that in 1974 the infant mortality rates in Vietnam, Cambodia, and Laos were 100.0, 120.7, and 123.0, respectively. Since these countries were in a theater of war in 1974, one wonders whether these rates are even roughly correct.

While it is possible that regression results obtained with "perfect data" would lead to the same conclusions, this may not be the case. In any event, the conclusions should be accepted cautiously.

While the models developed by Grosse and Perry explain most of the variance in life expectancy in 1960–65 and 1970–75, they acount for only a small proportion of the variance in the *change in life expectancy* between developing countries from 1950 to 1970. Clearly factors other than those considered by the authors are important.

One wonders why economic factors are of decreasing importance in explaining cross-sectional differences in life expectancy. Is it because once national income reaches a certain threshold level it ceases to be of much importance? Does this result occur because health interventions were less effective in improving health status in earlier times than in the more recent period?

Finally, one can question the policy implications of macroeconomic studies of this type. The fact that sanitation is positively associated with life expectancy in a multivariate model does not necessarily imply that sanitation projects within particular countries will yield the same result. It seems that these macroeconomic studies can yield suggestive information but that this data would need to be supplemented by micro-level studies in order to determine whether the program should be part of national policy.

As indicated by Vogel and Greenspan, Sudan has a number of health problems which occur in other developing countries. These include a severe shortage and maldistribution of health professionals, a lack of appropriate sanitation facilities, and a very high morbidity rate from infectious diseases. However, the country has some health problems that are fairly unique, such as the difficulty of providing health care to a significant nomadic population and a situation regarding potential conflict between health programs and programs for agricultural development.

It is clear that in a country, such as Sudan, in which agricultural acreage can only be expanded as a result of irrigation, agricultural experts will likely emphasize such schemes as part of an overall development plan. However, irrigation projects can facilitate the spread of certain waterborne diseases such as malaria and schistosomiasis (3, p. 1011). Given this possibility, what is clearly needed is an interface between those responsible for health planning and the development planners. One possible result of cooperation between these experts is an understanding by the economists that the expected benefits of an irrigation project should be reduced if it is likely to result in an increase in illness. This is because the latter imposes a burden on society with calculable economic costs. If these disease-related costs are sufficiently high, it is possible that in some cases the project would not be undertaken.

Vogel and Greenspan indicate the high rate of absenteeism at harvest time resulting from malaria. This makes it difficult for farmers to get the cotton harvested on time. Similar health-related labor shortages have been noted in other developing countries.[1] This implies that in the rural areas of Sudan (and elsewhere) there is less disguised unemployment than economists commonly believe. For if disguised unemployment were widespread, labor shortages would not occur even with high rates of absenteeism.

The lack of sanitation facilities in rural Sudan is one of the major causes of schistosomiasis. Thus people use the irrigation ditches as latrines and subsequently become infected with the disease. Although a health education program could possibly improve sanitary habits, the experience of other developing countries with community and individually owned latrines leaves one skeptical about a short-term solution to this problem.

Vogel and Greenspan discuss at considerable length the health program measures and objectives summarized in Table 13 of their paper. What is missing from the discussion is the way in which these objectives were determined. It is clear that a number of items of importance to the health of the Sudanese population have been enumerated. However, although from no fault of the authors in this type of analysis, one is uncertain as to whether the objectives reflect the desires of a cross section of the Sudanese population at large or only represent the views of the planners. Were these objectives developed solely in Khartoum, or did persons from other regions of the country provide input?

There are two major themes that permeate the Malenbaum paper. First, continuing a theme he developed in two earlier papers (4, 5), is a discussion of the importance of health inputs in changing the attitude and motivation of the rural poor toward innovation and development. It is postulated that the *modification in attitudes* associated with health service inputs is the way in which health programs accelerate development.

The second theme of the paper is that there exists in the world today a serious and even worsening maldistribution of income and social services, including health services. Although one may question Malenbaum's statement that three-fourths or more of all health services in poor lands were provided to the top 20–25 percent of the population, in terms of income, it is nevertheless true that those in poverty, and particularly the rural poor, have minimal access to modern health care servies in most developing countries.

The most important suggestion made in the paper is for the development of an accounting framework which would, for example, permit the delineation of health service usage by income level and geographic area. Such utilization could also be compared with the individual's or family's tax burden. This information would generally reveal an imbalance in health services between urban and rural dwellers and between the poor and nonpoor. The data would also clearly indicate an overemphasis on curative as opposed to preventive health services. Such a framework could also provide an impetus for policies which would increase access to health services for the rural poor, including a variety of organizational arrangements to accomplish this goal.

Malenbaum seems to question the relevance of the theory of demographic

Summary and Discussion of Part III

transition to the experience of developing countries. While it is true that in the 1940s and 1950s there was little change in birth rates in poor nations, there has been considerable decline in the 1960s and 1970s, particularly in countries like Sri Lanka which have egalitarian policies in regard to the distribution of social services (8).

One possible explanation of such a decline in birth rates is the child survival hypothesis. Thus, as the community perceives a decline in the infant and child mortality rate, the birth rate declines since fewer children are required to reach the target family size that the parents deem appropriate to insure their support in old age.[2]

The demographic transition has occurred in developing countries which have made limited economic progress. This contrasts with the historical experience of northwestern Europe and the United States, where economic growth preceded the decline in mortality and fertility.

Malenbaum recognizes the importance of micro-level case study research in providing links between health, attitude change, and development. These studies are particularly important in the planning context where one is attempting to engage in decentralized or localized planning as opposed to centralized decision making.

The Fülöp and Reinke paper bears some similarities to the Grosse and Perry paper. Both examine correlates of life expectancy, and both are sensitive to the importance of socioeconomic factors and health variables in explaining life expectancy. While Fülöp and Reinke focus somewhat more attention on the problem of data quality than do Grosse and Perry, there is room for more detailed examination. Both papers have raised various policy issues that deserve future attention.

Fülöp and Reinke develop a clear analytical framework for their analysis and develop and present their variables in a concise and imaginative manner. One very important part of the analysis which is a most useful aid to the reader is the information presented in Table 4. Fülöp and Reinke emphasize at several previous points in the paper that health resource variables are highly significant statistically and that when included in regression equations with socioeconomic variables the latter become less significant. However, as Table 4 indicates, an increase in the physician/population ratio of 35 percent is necessary on average to achieve a gain of 1 year in life expectancy. This is highly unlikely to occur, holding other variables constant, particularly GNP. Thus, while a variable may be important in the sense of being highly significant in statistical terms it may be of minor importance in terms of policy.

A major contribution of the Fülöp and Reinke paper is to focus attention on the analysis of residuals. For example, the African region is particularly short of physician manpower, even after adjustment for the low socioeconomic status of the region, which includes a number of the poorest countries in the world. The residuals, vis-à-vis particular countries, provide an additional set of variables which can be systematically analyzed in relation to conditions prevailing in those countries.

An important conclusion of the Fülöp and Reinke paper is the suggestion that socioeconomic variables exert their influence by facilitating health resource development. Thus external factors such as income and educational attainment are necessary conditions for health improvement but are not sufficient given the lack of an adequate health service infrastructure.

Finally, Fülöp and Reinke indicate the importance of supplementing statistical results with *qualitative* information such as national case studies or expert opinion. In an age where computer technology permits one to obtain extensive results in the absence of theoretical foundations, such advice is sound indeed.

NOTES

1. Thus, for example in India the labor shortage in Punjab State at harvest time is sufficiently great that workers must be imported from Rajastan.
2. See Taylor *et al.* (7) for some supporting evidence regarding the child survival hypothesis.

REFERENCES

1. Abel-Smith, B. (1967) *An International Study of Health Expenditure and Its Relevance for Health Planning*, Public Health Papers, No. 32, Geneva, Switzerland: World Health Organization.
2. Davis, K. (1975) *National Health Insurance: Benefits, Costs and Consequences*, Washington, D.C.: The Brookings Institution.
3. Lanoix, J. N. (1958) "Relation Between Engineering and Belharziasis," *Bulletin of the World Health Organization* 18.
4. Malenbaum, W. (1972) "Health and Expansion in Poor Lands," *International Journal of Health Services* 3(2).
5. Malenbaum, W. (1970) "Health and Productivity in Poor Areas," in *Empirical Studies in Health Economics*, (ed.) Herbert E. Klarman, Baltimore: The Johns Hopkins Press, pp. 31–54.
6. Myrdal, G. (1968) *Asian Drama: An Inquiry into the Poverty of Nations*, Volume 3, New York: Pantheon.
7. Taylor, C. E.; J. S. Newman; and N. V. Kelly. (1976) "The Child Survival Hypotheses," *Population Studies* 30(2):263–278.
8. Watson, W. and R. Lapham (1975) "Family Planning Programs: World Review, 1975," *Studies in Family Planning* 6(8):219, Table 6.

Author Index

Abel-Smith, B., 29, 56, 312, 315, 316, 327, 358
Adelman, Irma, 306, 307
Ahmad, P. I., 27
Ali, Abdel Gadir Ali, 307
American Dental Association, 89, 107
Annual Report of the City Medical Officer of Health, South Africa, 327
Anthony, C. R., 269
Arnold, C. B., 204, 207
Arriaga, Eduardo E., 220, 253, 330, 351
Arrow, K. J., 41, 56
Auster, Richard D., 306, 307, 351

Babigian, H. M., 28
Baker, T. D., 31, 56
Banerji, D., 72, 74
Bank of Sudan, 307, 308
Barlow, R., 42, 50, 56, 351
Bartel, A., 184, 185, 204, 206
Becker, Gary S., 79, 80, 103, 107, 131, 178, 180, 205, 206, 272, 308
Bengoa, J., 268

Ben-Porath, Yoram, 79, 107, 142, 143, 178, 180
Benyoussef, A., 308
Berg, R. L., 27, 268, 269
Berkanovic, E., 205, 206
Birch, Herbert G., 90, 107
Black, P. A., 325, 327
Black Development in South Africa, 327
Blaug, Mark, 306, 308
Bloomfield, J. M., 78, 107
Breslow, L., 78, 107
Brook, R. H., 27, 28
Bross, I., 204, 206
Bryan, R. B., 124
Bush, J. W., 30, 43, 56
Butler, N. R., 78, 108

Call, D. L., 268, 269
Cardus, D., 27
Carnegie Council on Children, 78, 108
Census of Health Services, South Africa, 314, 328
Chase, A., 124
Cheng, M. K., 27
Chernichovsky, Dov, 112, 124

Chlang, C. L., 27
Christakis, George, et al., 123, 124
Christian, B., 50, 56
Coate, Douglas, 112, 124
Coelho, G. V., 27
Colle, A., 101, 103, 107, 178, 180
Conly, G. N., 268, 269
Constitution of the World Health Organization, 27
Consumer Reports, 105, 107
Cooper, C. M., 74
Cooper, M. H., 328
Corbit, J. D., 205, 207
Cornfield, J., 207
Correa, H., 351
Cravioto, J., 268
Cvjetanovic, B., 50, 56

Daniel, Jr., W. A., 87, 107
Dasgupta, B., 74
Davie, R., 78, 108
Davies, G. W., 50, 56
Davies-Avery, A., 27, 28
Davis, Kingsley, 220, 253, 330, 351, 358
de Beer, J., 325, 328
Department of International Health, 352
DeTray, D., 131, 132, 178, 180
Development Assistance Committee, 257, 258, 259, 266, 268, 269
Diamond, P. A., 204, 206
Diethelm, P., 50, 56
Division of Dental Health of the National Institutes of Health, 89, 107
Djukanovic, V., 269
Donald, C. A., 27
Dougherty, Christopher, 306, 308
Douglas, J. W. B., 77, 107
Drake, W. D., 331, 352
Dubos, R., 27

Dunn, H., 27
Dutton, Diana B., 77, 104, 107

Easterlin, R. A., 269
Edozien, J. C., 124
Edwards, Linda N., 79, 100, 102, 103, 107, 180
El-Hassan, Ali Mohamed, 308
Engelhardt, H., 314

Fanshel, S., 30, 43, 56
Feldstein, Martin S., 305, 307, 308, 322, 328
Fielding, J. E., 204, 206
Fisher, Charles R., 281, 308
Fisher, D., 50, 56
Fried, M., 28
Friedlander, Lindy J., 79, 103, 106, 109, 205, 207
Friedman, B., 178, 180
Fuchs, Victor R., 78, 107, 127, 180, 204, 205, 206, 212, 214, 221, 253, 306, 308, 315, 328, 330, 352
Fülöp, T., 330, 352

Galenson, W., 352
Galland, R., 352
Gallard, Rene, 253
Geijerstam, G., 168, 180
Ghez, Gilbert R., 308
Gibson, Robert M., 281, 308
Gilliand, Pierre, 253, 352
Gish, O., 326, 327, 328
Goldberger, A., 206
Goldman, Fred, 103, 108, 178, 180
Goode, Richard, 306
Grab, B., 50, 56
Grosse, Robert, 242, 331, 352
Grossman, Michael, 78, 79, 100, 101, 102, 103, 107, 108, 129, 138, 139, 178, 180,

Author Index

205, 206, 212, 214, 272, 274, 277, 306, 308
Gussow, Joan Dye, 90, 107

Hagen, Everett E., 308
Haggerty, Robert J., 78, 88, 108
Hall, H. S., 28
Hamilton, J., 204, 206
Harbison, F. H., 269
Harris, J., 206
Hauser, P. M., 207
Hays, C. W., 307, 309
Health Services and Mental Health Administration DHEW, 124
Heller, Peter (Introduction, please add page number), 331, 352
Heston, Alan W., 253
Higgins, G., 204, 206
Hofer, B., 28
Hu, Teh-Wei, 77, 108

Idriss, A. P., 308
Inman, Robert P., 77, 108, 138, 178, 181, 205, 206
Institute of Medicine, 128, 181
International Conference on Primary Health Care, 268, 269
International Labour Organization, 304, 307, 308
Ishikawa, 205, 207

Jablon, S., 184, 207
Jaffe, F. S., 181
Jazairi, N. T., 28
Jelliffe, D. B., 125
Johnson, B., 74
Johnson, George E., 308
Johnston, B. F., 269
Johnston, J., 179, 181
Johnston, S. A., 27

Kamarck, Andrew M., 307, 308
Kannel, W., 207

Kellmer-Pringle, M. L., 78, 108
Kelly, James E., 87, 108
Kelly, N. V., 358
Keniston, Kenneth, 78, 108
Kessner, David M., 78, 108
Khan, R. A., 308
Kitagawa, E. M., 207
Klarman, Herbert E., 31, 36, 56, 276, 327, 328
Kleiman, E., 315, 328
Klein, B., 78, 107
Kleinman, Joel C., 101, 108
Knight, J. B., 323, 328
Knight, S. R., 28
Knodel, T., 269
Kolker, A., 27
Koranyi, E. K., 28
Krasner, M., 181
Kravis, Irving R., 253
Kirshnan, P., 253, 330, 331, 352
Kristen, M. M., 204, 207

Lancaster, Kelvin J., 43, 44, 56, 272, 308
Lanoix, J. N., 358
Lapham, R., 358
Larmore, Mary Lou, 306, 308
Lave, L., 186, 207
Lefrowitz, Myron J., 308
Leibenstein, H., 269
Leibowitz, A., 178, 180
Leveson, Irving, 306, 307, 351
Lewis, H. F., 79, 107, 131, 178, 180
Lewit, Eugene M., 102, 108
Levy, E., 28
Lolik, P., 308
Luft, H., 184, 207

Maasdorp, G., 319, 328
MacCarthy, H., 318, 328
Mach, E. P., 269
Mahler, H., 329

AUTHOR INDEX

Malenbaum, Wilfred, 268, 269, 276, 277, 306, 308, 352, 358
Manheim, Lawrence, 78, 108
Mann, J. K., 322, 328
Mannheim, L., 205, 207
Manning, Jr., Willard G., 96, 97, 106, 108
Marmor, Theodore R., 78, 108
Marschak, T. A., 206, 207
McClure, F. J., 104, 108
McGrath, M. D., 319, 320, 323, 328
Michael, Robert T., 103, 108, 125, 138, 178, 181, 272, 308
Ministry of Finance, Planning and National Economy, The Sudan, 291, 303, 309
Ministry of Finance, Planning and National Economy, Southern Region, The Sudan, 294, 307, 308
Ministry of Health, The Sudan, 282, 293, 298, 300, 302, 303, 305, 306, 307, 309
Ministry of National Planning, The Sudan, 286, 288, 295, 307, 309
Mirrlees, J. A., 204, 206
Morris, J. N., 59
Morris, Portia M., 123, 125
Morrison, D. F., 206, 207
Muller, C., 149, 181
Murphy, James L., 306, 309
Mushkin, Selma, 31, 56, 107
Muth, Richard, 272, 309
Myrdal, G., 354, 358

National Academy of Sciences, 175, 181
National Center for Health Statistics, 88, 103, 108, 125
National Heart, Lung, Blood Institute, 88, 108
Newberger, Carolyn Moore, 78, 109
Newberger, Eli H., 78, 109
Newhouse, Joseph P., 79, 103, 106, 107, 109, 205, 207, 306, 309
Newman, J. S., 358
Newman, P., 48, 56
Neyman, J., 185, 207
Nortman, D., 181

O'Hara, D. J., 130, 131, 181
O'Reagain, Mary, 312, 328
Otto, R., 74
Owen, George M., 123, 125

Parsons, T., 28
Pauly, Mark V., 109
Pederson, A. M., 28
Peltzman, S., 207
Perlman, M., 328
Perry, B. H., 352
Phelps, Charles E., 96, 97, 108
Piot, M., 50, 56
Pless, Ivan B., 78, 88, 108
Pollak, R., 205, 206, 207
Preston, Samuel H., 220, 221, 253, 330, 331, 352
Prothero, R. Mansell, 56
Psacharopoulos, George, 306, 308
Purola, T., 28
Pyatt, G., 352

Rabkin, J. G., 207
Ray, D., 50, 56
Raymond, L., 28
Report of the Preparatory Scientific Working Group on Social and Economic Research, 268, 269
Report of the Secretary for Health, South Africa, 328
Report on Socio-Economic Research, 268, 269
Report on Socio-Economic Research in the Special Tropical Disease Program, 258, 269
Ribich, Thomas I., 306, 309

Author Index

Rice, D., 183, 207
Richmond, Julius, B., 78, 109
Rogers, W. H., 27
Roghmann, Klaus J., 78, 88, 108
Rosett, Richard, 108
Ross, S., 178, 181
Russell, A. L., 87, 93, 109

Sadie, J. L., 319, 325, 328
Sanchez, Marcus J., 87, 108
Sarachek, Deborah, 306, 307, 351
Scheffler, Richard M., 107
Schelling, T., 35, 56
Scientific Working Group on Epidemiology, 268, 269
Scrimshaw, Nevin S., 268, 269, 309
Seskin, E., 186, 207
Sharpston, M. J., 309
Shavell, S., 204, 206, 207
Shchepin, O. P., 330, 352
Shuval, Hillel, 242
Silber, Ralph S., 77, 107
Silverman, William, 97, 106, 109
Sims, Laura S., 123, 125
Singer, H., 74
Sirageldin, I., 351
Sivard, Ruth Leger, 252
Sloan, Frank, 221, 253
Smucker, Celeste M., 220, 253, 330, 352
Solomon, R., 205, 207
Sorkin, A. L., 268, 269, 306, 309, 326, 328
South African Reserve Bank, 316
South African Statistics, 317, 319, 328
Stafford, Frank P., 308
Statistical Year Book, South Africa, 328
Sterling, T., 204, 207
Stevens, C. M., 268, 270
Stewart, A. L., 27, 28
Stolnitz, George J., 220, 253, 330, 352

Struening, E., 207
Sullivan, D. F., 30, 43, 56
Summers, Robert, 253
Sundaresan, T. K., 50, 56
Survey of Household Expenditure, South Africa, 321, 327, 328
Sutter, Gerald E., 89, 109
Switzer, B. R., 124

Tanahashi, T., 50, 56
Tanner, J. M., 104, 109
Taubman, P., 184, 185, 204, 206
Taylor, C. E., 358
Tefft, B. M., 28
The Provincial Auditor's Report, South Africa, 328
Theodore, Christ N., 89, 109
Thrall, R. M., 27
Tilden, Robert, 242
Tomes, Nigel, 106, 109
Trengrove-Jones, S., 312, 328
Truett, J., 207

Uemura, K., 50, 56
United Nations, 252, 253
United Nations, Economic Commission for Africa, 285, 309
Upton, Charles, 97, 106, 109
U.S. Bureau of the Census, 281, 309

van de Walle, E., 269
Vaupel, J., 183, 207
Venediktov, D. D., 330, 352

Wachter, M., 205, 207
Ware, Jr., J. E., 27, 28
Watson, W., 358
Weiss, W., 204, 207
Welch, Finis, 79, 103, 107, 109, 142, 143, 178, 180
Wells, L. G., 319, 328
Wen, C. P., 307, 309
Williams, A. D., 143, 147, 171, 181

Williams, K. N., 27, 30, 31, 56
Williams, R. L., 168, 181
Willis, R. J., 178, 181
Willis, Robert, 79, 109
Wilson, Ronald W., 101, 108
World Bank, 252, 258, 268, 270, 279, 280, 281, 282, 285, 303, 306, 309, 351, 318

World Health Organization, 27, 74, 253, 258, 259, 266, 270, 271, 272, 278, 282, 283, 295, 305, 306, 309, 351
Wray, J. D., 205, 206, 207
Wynder, E. L., 204, 207

Zimmer, B. G., 78, 109

Research Annuals in
ECONOMICS

Consulting Editor for Economics

Paul Uselding
Chairman, Department of Economics
University of Illinois

Advances in Applied Micro-Economics
Series Editor: V. Kerry Smith,
University of North Carolina

Advances in Econometrics
Series Editors: R. L. Basmann,
Texas A & M University
and George F. Rhodes, Jr.,
Colorado State University

Advances in the Economics of Energy and Resources
Series Editor: John R. Moroney,
Tulane University

Advances in Health Economics and Health Services Research
(Volume 1 published as Research in Health Economics)
Series Editor: Richard M. Scheffler, *George Washington University.* Associate Series Editor: Louis F. Rossiter, *National Center for Health Services Research*

Applications of Management Science
Series Editor: Randall L. Schultz, *University of Texas at Dallas*

Research in Corporate Social Performance and Policy
Series Editor: Lee E. Preston,
University of Maryland

Research in Domestic and International Agribusiness Management
Series Editor: Ray A. Goldberg,
Harvard University

Research in Economic Anthropology
Series Editor: George Dalton, *Northwestern University*

Research in Economic History
Series Editor: Paul Uselding,
University of Illinois

Research in Experimental Economics
Series Editor: Vernon L. Smith,
University of Arizona

Research in Finance
Series Editor: Haim Levy,
The Hebrew University

Research in Human Capital and Development
Series Editor: Ismail Sirageldin,
The Johns Hopkins University

Research in International Business and Finance
Series Editor: Robert G. Hawkins,
New York University

Research in Labor Economics
Series Editor: Ronald G. Ehrenberg, *Cornell University*

Research in Law and Economics
Series Editor: Richard O. Zerbe, Jr.,
University of Washington

Research in Marketing
Series Editor: Jagdish N. Sheth, *University of Illinois*

Research in Organizational Behavior
Series Editors: Barry M. Staw, *University of California at Berkeley*
and L. L. Cummings, *University of Wisconsin—Madison*

Research in Philosophy and Technology
Series Editor: Paul T. Durbin,
University of Delaware

Research in Political Economy
Series Editor: Paul Zarembka, *State University of New York—Buffalo*

Research in Population Economics
Series Editor: Julian L. Simon,
University of Illinois

Research in Public Policy Analysis and Management
Series Editor: John P. Crecine,
Carnegie-Mellon University

Research in Real Estate
Series Editor: C. F. Sirmans,
University of Georgia

Research in Urban Economics
Series Editor: J. Vernon Henderson, *Brown University*

Please inquire for detailed brochure on each series.

 JAI PRESS INC.

Research in
Human Capital and Development

Edited by **Ismail Sirageldin**
Departments of Population Dynamics and Political Economy
The Johns Hopkins University

Research in Human Capital and Development brings together theoretical and empirical developments in the field of human capital formation that are relevant to developmental issues including education, manpower training, fertility behavior, health, and the important triangle of equity, efficiency and development.

Although the general focus of the Series is economics, contributions of an interdisciplinary nature are encouraged. Most of these areas were touched on in the first volume (1979). The plan for future volumes is to focus on each of these topics individually. For example, Volume 2 (1981) examines equity and distribution issues in the context of development and human capital formation. Volume 3 (1982) is devoted to issues of health, human capital and development. Future volumes include Volume 4, on migration theory and development, and Volume 5, on education in the context of human capital and development.

Volume 1, 1979, 258 pp.
ISBN 0-89232-019-2

REVIEW: "The idea of developing RHCD into a forum for important empirical and theoretical research is a brilliant one and will definitely be fruitful in the years to come... Without a doubt the book is an extremely valuable contribution to research on human capital theory."
— *The Paristan Development Review*

CONTENTS: Introduction, *Ismail Sirageldin.*
Part 1. HEALTH AND FERTILITY. Relevance of Human Capital Theory to Fertility Research: Comparative Findings for Bangladesh and Pakistan, *Ali Khan, The Johns Hopkins University.* **Health and Economic Development: A Theoretical and Empirical Review,** *Robin Barlow, University of Michigan.* **Health, Nutrition, and Mortality in Bangladesh,** *W. Henry Mosley, Cholera Research Laboratory, Dacca.* **Discussion,** *Ismail Sirageldin, The Johns Hopkins University.*
Part 2. EDUCATION AND MANPOWER. College Quality and Earnings, *James N. Morgan and Greg J. Duncan, University of Michigan.* **Some Theoretical Issues in Manpower and Educational Training,** *Mohiuddin Alamgir, Bangladesh Institute of Developmental Studies.* **Barriers to Educational Development in Underdeveloped Countries: With Special Reference to Venezuela,** *Kristin Tornes, University of Bergen.* **Manpower Planning and the Choice of Technology,** *S.C. Kelley, Center for Human Resource Research, Ohio State.* **The Growth of Professional Occupations in U.S. Manufacturing: 1900-1973,** *Carmel Ullman Chiswick, University of Illinois, Chicago Circle.* **Summary and Discussion,** *Alan Sorkin, University of Maryland.*
Part 3. DISTRIBUTION AND EQUITY. Equity Social Striving, and Rural Fertility, *Ismail Sirageldin and John Kantner, The Johns Hopkins University.* **Index.**

Volume 2, Equity, Human Capital and Development
1981,   244 pp.
ISBN 0-89232-098-2

Edited by **Ali Khan** and **Ismail Sirageldin,** *The Johns Hopkins University*
CONTENTS: Introduction and Summary, *Ali Khan and Ismail Sirageldin, The Johns Hopkins University.*
Part 1. MEASURES AND PARADOXES. Measures of Poverty and their Policy Implications, *Koichi Hamada and Noriyuki Takayama, University of Tokyo.* **Paradoxes of Work and Consumption in Late 20th Century America,** *Nathan Keyfitz, Harvard University.*
Part 2. THEORETICAL ISSUES. A Model of Economic Growth with Investment in Human Capital, *Ronald Findlay and Carlos A. Rodriguez, Columbia University.* **The Influence of Nonhuman Wealth on the Accumulation of Human Capital,** *John Graham, University of Illinois.* **The Changing Role of Breastfeeding in Economic Development: A Theoretical Exposition,** *William P. Butz, The Rand Corporation.*
Part 3. CASE STUDIES. The Effects of Income Maintenance on School Performance and Educational Attainment in the U.S.A., *Charles D. Mallar and Rebecca A. Maynard, Mathematica Policy Research.* **An Analysis of Education, Employment and Income Distribution Using an Economic Demographic Model of the Phillippines,** *G.B. Rodgers, International Labour Organization.* **The Welfare Implications of Relative Price Distortions and Inflation: An Analysis of the Recent Argentine Experience,** *Ke-young Chu and Andrew Feltenstein, International Monetary Fund.* **Index.**

Supplement 1 - Manpower Planning in the Oil Countries, Edited by **Naiem A. Sherbiny,** *The World Bank*

1981, 276 pp.
ISBN 0-89232-129-6

"This book constitutes an attempt at *pioneering on several fronts* simultaneously. To begin with it tackles a new complex of problems, which may be characterized as "planning for development in a situation of *manpower scarcity.*"... The novelty of the subject means, to begin with, that, for the countries concerned, the *goals* of development have to be reformulated."

Jan Tingergen. Chapter 1.

CONTENTS: Series Editor's Introduction, *Ismail Sirageldin, Johns Hopkins University.* **Editor's Introduction,** *Naiem A. Sherbiny, The World Bank.*
Part 1. OVERVIEW AND METHODOLOGY. The Issues, *Jan Tinbergen. The Netherlands.* **Structural Changes in Output and Employment in the Arab Countries,** *Maurice Girgis, Ball State University and Kuwait Institute for Scientific Research.* **The Modeling and Methodology of Manpower Planning in the Arab Countries,** *M. Ismail Serageldin, The World Bank.* **Vocational and Technical Education and Development Needs in the Arab World,** *Atif Kubursi, McMaster University.*
Part 2. CASE STUDIES. An Econometric/Input-Output Approach for Projecting Sectoral Manpower Requirements: The Case of Kuwait, *M. Shokri Marzouk, York University and Kuwait Institute for Scientific Research.* **A Macroeconomic Simulation Model of High Level Manpower Requirements in Iraq,** *George T. Abed and Atif Kubursi, IMF, McMaster University.* **Sectoral Employment Projections with Minimum Data Base: The Case of Saudi Arabia,** *Naiem A. Sherbiny, The World Bank.*
Part 3. POLICY ISSUES. Labor Adaptation in the Oil Exporting Countries, *M. Ismail Serageldin, The World Bank.* **Labor and Capital Flows in the Arab World: A Policy Perspective,** *Naiem A. Sherbiny, The World Bank.* **Index.**

University libraries depend upon their faculty for recommendations before purchasing. Please encourage your library to subscribe to this series.

INSTITUTIONAL STANDING ORDERS will be granted a 10% discount and be filled automatically upon publication. Please indicate initial volume of standing order.

INDIVIDUAL ORDERS must be prepaid by personal check or credit card. Please include $2.00 per volume for postage and handling on all domestic orders; $3.00 for foreign.

JAI PRESS INC., 36 Sherwood Place, P.O. Box 1678
Greenwich, Connecticut 06836
Telephone: 203-661-7602   Cable Address: JAIPUBL

Research in Labor Economics

Edited by **Ronald G. Ehrenberg**
School of Industrial and Labor Relations, Cornell University

Volume 1, 1977, 376 pp.
ISBN 0-89232-017-6

CONTENTS: Preface, *Ronald G. Ehrenberg, Cornell University.* **Human Capital: A Survey of Empirical Research,** *Sherwin Rosen, University of Rochester.* **The Incentive Effects of the U.S. Unemployment Insurance Tax,** *Frank Brechling, Northwestern University.* **A Life Cycle Approach to Migration: Analysis of the Perspicatious Peregrinator,** *Solomon W. Polacheck and Francis W. Horvath, University of North Carolina.* **Manpower Requirements and Substitution Analysis of Labor Skills: A Synthesis,** *Richard B. Freeman, Harvard University.* **Models of Labor Market Turnover: A Theoretical and Empirical Survey,** *Donald O. Parsons, Ohio State University.* **Work Effort, On-The-Job Screening and Alternative Methods of Remuneration,** *John H. Pencavel, Stanford University.* **A Simulation Model of the Demographic Composition of Employment, Unemployment and Labor Participation,** *Ralph E. Smith, The Urban Institute.* **Extensions of a Structural Model of the Demographic Labor Market,** *Richard S. Toikka, William J. Scanlon and Charles C. Holt, The Urban Institute.* **The Institutionalist Analysis of Wage Inflation: A Critical Appraisal,** *John Burton, Kingston Polytechnical and John Addison, Aberdeen University.*

Volume 2, 1978, 381 pp.
ISBN 0-89232-097-4

CONTENTS: Introduction, *Ronald Ehrenberg, Cornell University.* **The United Mine Workers and the Demand for Coal: An Econometric Analysis of Union Behavior,** *Henry S. Farber, Massachusetts Institute of Technology.* **Queuing for Union Jobs and the Social Returns to Schooling,** *John Bishop, University of Wisconsin.* **Labor Supply Under Uncertainty,** *Kenneth Burdett, University of Wisconsin and Dale T. Mortensen, Northwestern University.* **Governmentally Imposed Standards: Some Normative Aspects,** *Russell F. Settle, University of Delaware and Burton Weisbrod, University of Wisconsin.* **Cyclical Earnings Changes of Low Wage Workers,** *Wayne Vroman, The Urban Institute.* **Earnings, Transfers, and Poverty Reduction,** *Peter T. Gottschalk, Bowdin College.* **The Influence of Fertility on Labor Supply of Married Women: Simultaneous Equation Estimates,** *T. Paul Schultz, Yale University.* **The Labor Market Adjustments of Trade Displaced Workers: The Evidence from the Trade Adjustment Assistance Program,** *George R. Neumann, University of Chicago.*

Supplement 1 - Evaluating Manpower Training Programs (Revisions of Papers Originally Presented at the Conference on Evaluating Manpower Training Programs, Princeton University, May 1976) Edited by **Farrell Bloch,** *Princeton University*

1979, 375 pp.
ISBN 0-89232-046-X

CONTENTS: Introduction, *Farrell E. Bloch, Princeton University.* **A Decision Theoretic Approach to the Evaluation of Training Programs,** *Frank P. Stafford, University of Michigan.* **A Sensitivity Analysis to Determine Sample Sizes for Performing Impact Evaluation of the CETA Programs,** *Hugh M. Pitcher, U.S. Department of Labor.* **Choice of Sample Size in Evaluating Manpower Programs: Comments on Pitcher and Stafford,** *John Conlisk, University of California, San Diego.* **Estimating the Effect of Training Programs on Earnings with Longitudinal Data,** *Orley Ashenfelter, Princeton University.* **Earnings and Employment Dynamics of Manpower Trainees: An Exploratory Econometric Analysis,** *Thomas F. Cooley, University of California, Santa Barbara, Timothy W. McGuire, Carnegie-Mellon University and Edward C. Prescott, Carnegie Mellon University.* **An Evaluation of Two Evaluations,** *Ronald G. Ehrenberg, Cornell University.* **The Economic Benefits from Four Government Training Programs,** *Nicholas M. Kiefer, Princeton University.* **Estimates of the Benefits of Training for Four Manpower Training Programs,** *Gordon P. Goodfellow, Department of Health, Education, and Welfare.* **Comments on Kiefer and Goodfellow,** *Daniel S. Hamermesh, Michigan State University.* **The Labor Market Displacement Effect in the Analysis of the Net Impact of Manpower Training Programs,** *George E. Johnson, The University of Michigan.* **Comment on "The Labor Market Displacement Effect in the Analysis of the Net Impact of Manpower Training Programs" by George Johnson,** *R.E. Hall, Massachusetts Institute of Technology.* **Potential use of Markov Process Models to Determine Program Impact,** *Hyman B. Kaitz, Westat, Inc.* **Markov Processes and Public Service Employment: A Comment,** *Michael L. Wachter, University of Pennsylvania.* **Theoretical Issues in the Estimation of Production Functions in Manpower Programs,** *Burt S. Barnow, U.S. Department of Labor.* **Information Issues in Department of Labor Program Evaluation,** *Ernst W. Stromsdorfer, Indiana University and ABT Associates, Inc.* **Information Issues in Department of Labor Program Evaluation: Discussion,** *Michael E. Borus, Michigan State University, Robert S. Gay, Brooklyn College.*

Volume 3,   1980,   389 pp.
ISBN 0-89232-157-1

CONTENTS: Introduction, *Ronald G. Ehrenberg*. **Overview of the Tucson Conference Paper,** *Orley Ashenfelter, Princeton University and Ronald L. Oaxaca, University of Arizona.* **Expectations, Realizations and the Aging of Young Men,** *Zvi Griliches, Harvard University.* **Discussion,** *Daniel H. Saks, Michigan State University.* **Discussion,** *Burt S. Barnow, U.S. Department of Labor.* **Unionism, Labor Turnover and Wages of Young Men,** *Henry S. Farber, Massachusetts Institute of Technology.* **Discussion,** *Duane E. Leigh, University of Virginia.* **Minimum Wage Legislation and the Educational Outcomes of Youths,** *Ronald G. Ehrenberg and Alan J. Marcus, Cornell University.* **Discussion,** *Clifford E. Reid, Grinnell College.* **Predicated vs Potential Work Experience in an Earnings Function for Young Women,** *Nancy Garvey and Cordelia Reimers, Princeton University.* **The Impact of Maternal Characteristics and Significant Life Events on the Work Orientation of Adolescent Women,** *Ann Statham Macke and Frank L. Mott, Ohio State University.* **Discussion,** *Wendy C. Wolf, University of Arizona.* **Discussion,** *Daniel S. Hamermesh, Michigan State University.* **Income Prospects for Job Mobility of Younger Men,** *George J. Borjas, University of California-Santa Barbara and Sherwin Rosen, University of Chicago.* **Discussion,** *Stephen Nickell, London School of Economics.* **Discussion,** *Walter Y. Oi, University of Rochester.* **The Structure of Wages and Turnover in the Federal Government,** *George J. Borjas, University of California-Santa Barbara.* **The Demographic Structure of Unemployment Rates and Labor Market Transition Probabilities,** *Ronald G. Ehrenberg, Cornell University.* **The Design of Social Experiments: A Critique of the Conlisk-Watts Assignment Model and its Application to the Seattle and Denver Income Maintenance Experiments,** *Michael C. Keeley and Philip K. Robins, SRI International.* **The Use of Time and Technology by Households in the United States,** *Frank Stafford and Greg J. Duncan, University of Michigan.*

Volume 4,   1981,   488 pp.
ISBN 0-89232-220-9

CONTENTS: Introduction, *Ronald Ehrenberg, Cornell University.* **A Survey of Labor Supply Models: Theoretical Analysis and First-Generation Empirical Results,** *Mark Killingsworth, Rutgers University.* **New Methods For Estimating Labor Supply Functions: A Survey,** *James J. Heckman, University of Chicago and Thomas E. McCurdy, Stanford University.* **The Effect of Tax and Transfer Programs on Labor Supply: The Evidence From The Income Maintenance Experiments,** *Robert A. Moffitt, Rutgers University and Kenneth C. Kehrer, Mathematica Policy Research.* **Youth Labor Supply During the Summer: Evidence for Youths From Low-Income Households,** *George Farkas, Ernst W. Stromsdorfer, Abt Associates and Randall J. Olsen, Yale University.* **The Effect of Unions on Wages in Hospitals,** *Glen G. Cain and Catherine G. McLaughlin, University of Wisconsin, Brian E. Becker, SUNY, Buffalo and Albert E. Schwenk, U.S. Bureau of Labor Statistics.* **The Effects of Collective Bargaining on Economic and Behavioral Job Outcomes,** *Thomas A. Kochan, Massachusetts Institute of Technology and David E. Helfman, Pennsylvania State Education Association.* **Trade Adjustment Assistance for Workers: Results of a Survey of Recipients Under the Trade Act of 1974,** *Walter Corson and Walter Nicholson, Amherst College.* **The Influence of Effective Human Capital on the Wage Equation,** *Randall Keith Filar, Brandeis University.*

**JAI PRESS INC., 36 Sherwood Place, P.O. Box 1678
Greenwich, Connecticut 06836**
Telephone: 203-661-7602   Cable Address: **JAIPUBL**

Advances in
Industrial and Labor Relations

Edited by **David B. Lipsky**
New York State School of Industrial and Labor Relations, Cornell University

This series will publish major, original research on all subjects within the field of industrial relations, including union behavior, structure, and government; collective bargaining, in both the private and public sectors; labor law and public policies affecting the employment relationships; the economics of collective bargaining; and international and comparative labor movements. Reflecting the multi-disciplinary nature of industrial relations, its contributors will include economists, sociologists, and other social scientists as well as lawyers and specialists in labor relations. Although there are now several journals that publish research on industrial relations, the space limitations of these journals preclude their publishing longer—and possibly more reflective—studies. Many industrial relations scholars have sought a forum for the publication of research that is too long for a journal article but not enough for a book or monograph.

Volume 1, In preparation, Spring 1983
ISBN 0-89232-250-0

CONTENTS: Organizations and Expectations: Organizational Determinants of Union Membership Demands, *Samuel Bacharach and Stephen M. Mitchell, Cornell University.* **Perceptions of Academic Bargaining Behavior,** *Robert Birnbaum, Columbia University.* **The Unionization Process: A Review of the Literature,** *Richard Block and Steven L. Premack, Michigan State University.* **Unionization in Secondary Labor Markets: The Historical Case of Building Services Employees,** *Peter Doeringer, Boston University.* **Public Sector Impasses Procedures: A Six State Study,** *Paul Gerhart and John Drotning, Case Western Reserve University.* **The Effects of Civil Service Systems and Unionism on Pay Outcomes in the Public Sector,** *David Lewin, Columbia University.* **The Relationship Between Seniority, Ability, and the Promotion of Union and Nonunion Workers,** *Craig A. Olson and Chris J. Berger, State University of New York at Buffalo.* **Towards a Theory of the Union's Role in an Enterprise,** *Donna Sockell, Columbia University.* **Union Organizing in Manufacturing: 1973-76,** *Ronald Seeber, Cornell University.*

University libraries depend upon their faculty for recommendations before purchasing. Please encourage your library to subscribe to this series.

INSTITUTIONAL STANDING ORDERS will be granted a 10% discount and be filled automatically upon publication. Please indicate initial volume of standing order.

INDIVIDUAL ORDERS must be prepaid by personal check or credit card. Please include $2.00 per volume for postage and handling on all domestic orders; $3.00 for foreign.

JAI PRESS INC., 36 Sherwood Place, P.O. Box 1678
Greenwich, Connecticut 06836
Telephone: 203-661-7602   Cable Address: JAIPUBL